Paving the Future: Sustainable Road Design and Urban Mobility in Smart Cities

Paving the Future: Sustainable Road Design and Urban Mobility in Smart Cities

Guest Editors

Maria Luisa Tumminello
Elżbieta Macioszek
Anna Granà
Tullio Giuffrè

MDPI

Basel • Beijing • Wuhan • Barcelona • Belgrade • Novi Sad • Cluj • Manchester

Guest Editors
Maria Luisa Tumminello
Department of Engineering
University of Palermo
Palermo
Italy

Elżbieta Macioszek
Department of Transport
Systems, Traffic Engineering
and Logistics
Silesian University
of Technology
Katowice
Poland

Anna Granà
Department of Engineering
University of Palermo
Palermo
Italy

Tullio Giuffrè
Faculty of Engineering
and Architecture
University of Enna Kore
Enna
Italy

Editorial Office
MDPI AG
Grosspeteranlage 5
4052 Basel, Switzerland

This is a reprint of the Special Issue, published open access by the journal *Smart Cities* (ISSN 2624-6511), freely accessible at: https://www.mdpi.com/journal/smartcities/special_issues/F1G3D78C8Q.

For citation purposes, cite each article independently as indicated on the article page online and as indicated below:

Lastname, A.A.; Lastname, B.B. Article Title. *Journal Name* **Year**, *Volume Number*, Page Range.

ISBN 978-3-7258-6864-3 (Hbk)
ISBN 978-3-7258-6865-0 (PDF)
https://doi.org/10.3390/books978-3-7258-6865-0

Contents

About the Editors

Maria Luisa Tumminello

Maria Luisa Tumminello graduated with honors in Civil Engineering in 2014 from the University of Palermo, Italy. In February 2018, she obtained her PhD in Civil, Environmental, and Material Engineering from the same university. In February 2022, she became a researcher in the Scientific Disciplinary Sector CEAR-03/A (Roads, Railways, and Airports) at the Department of Engineering, University of Palermo, Italy. Her main research interests include analyzing the operational and safety performance of road infrastructures, integrating connected and autonomous driving technologies through traffic microsimulation models and surrogate safety measures. She also focuses on the design and assessment of road intersection solutions and efficient traffic management strategies. Additionally, her research activities encompass the sustainability and environmental impact of road infrastructure, as well as traffic calming measures in the context of smart roads. She is a co-author of several scientific publications in journals and conference proceedings, and she has presented her research findings at international conferences. She serves on the organizing committee of several international conferences and is a member of the scientific committee for the RoadLCA 2026 international congress. Moreover, she is a Guest Editor for international journals and a PhD co-advisor for a student enrolled in the PhD program in AIM HIGHEST, XXXVII cycle, at the University of Palermo.

Elżbieta Macioszek

Elżbieta Macioszek graduated with a degree in transport in 2002. Currently, she is working as a University Professor in the Department of Transport Systems, Traffic Engineering and Logistics, Faculty of Transport and Aviation Engineering, Silesian University of Technology, Poland. She is a member of the scientific board for civil engineering, geodesy, and transportation in the Faculty of Transport and Aviation Engineering, Silesian University of Technology. She has received five awards from the Rector of the Silesian University of Technology for scientific achievements and four awards from the Rector of the Silesian University of Technology for organizational achievements. Her professional experience includes more than ten national and international contract jobs in the field of transportation. Her scientific interests combine the problems of road traffic engineering (including traffic analysis and prognosis, transport systems modeling, and optimization of transport networks) with the development of commuting behaviors within cities. She is an expert in performing assessments of project applications in Poland. Moreover, she is the author and co-author of more than 230 papers and chapters in books and a member of numerous organizational, program, technical, and scientific committees of scientific conferences organized in various countries around the world, as well as a member of the Editorial Board and reviewer in journals connected with transport, traffic, and civil engineering. She has completed several scientific internships, e.g., in the Department of Civil Engineering, Faculty of Science and Technology, Tokyo University of Science, Japan, and in the Department of Engineering at the University of Palermo, and she has also taken part in and represented the Silesian University of Technology in many European Programs.

Anna Granà

Anna Granà graduated with honors in Civil Engineering in April 1997, receiving the "Marida Correnti Prize" for her master's thesis on road safety at the University of Palermo, Italy. She completed a higher training course in Public Management at CERISDI in Palermo in 1999 and earned her PhD in "Road Infrastructure Engineering" from the University of Palermo in February 2002. She was appointed Assistant Professor in December 2002, became an Associate Professor in November 2011, and was promoted to Full Professor in the Scientific Sector of Road, Railways, and Airports on May 4, 2020, at the University of Palermo. Anna Granà has actively participated in numerous national and international congresses, presenting her research. Her scholarly work includes many articles in journals and proceedings, and she serves as a Guest Editor for international journals. Her research focuses on the functional design of road geometry, particularly for motorways and intersections, as well as road safety analysis, crash modeling, microscopic traffic simulation, and connected automated vehicles. Additionally, she studies roundabout capacity evaluation, surrogate safety measures, and the sustainability of road transport infrastructures. She also collaborates on various national research projects and is involved in initiatives supported by the Sicilian Region and PNRR MUR. Additionally, she has been a member of the International Scientific Advisory Board for the Priority Research Area on Smart Cities and Future Mobility at the Silesian University of Technology in Poland since January 2021. She served as the coordinator of the bachelor's and master's degree programs in Civil Engineering for the three-year period from 2022 to 2025 at the University of Palermo.

Tullio Giuffrè

Tullio Giuffrè obtained his PhD in the field of Geodesy and Geomatics from the PhD School of Politecnico di Milano (Italy) in 2006. Since May 2014, he has been an Associate Professor of Road, Railways, and Airports at Università Kore di Enna. At present, he teaches a course on "Road, Railway, and Airport Design and Construction" in the Civil Engineering master's degree and Architecture master's degree programs. Since 2010, he has been the Manager of the road materials testing laboratory at Università Kore di Enna. He has more than ten years of academic and consulting activities within infrastructure sectors, including traffic analysis, road safety analysis, road and airport design, infrastructure pavement management, and intelligent transport systems. He has achieved great skills and expertise in urban infrastructure planning, both from a design and operations point of view. He has been a member of the Italian Scientific Association for Infrastructures and Transport since 2015; the Italian Professional Engineer Register since 2002; the Board of Road and Airport Research Interuniversity Italian Association since 2009; the Board of PhDs in Design of Mobility System since 2009; the Board of PhDs in Aeronautical Technologies and Infrastructures since 2009; and the Italian Society for Transportation Policy since 2014. As a consultant, he has carried out several infrastructure designs, traffic analyses and studies, and infrastructure environmental evaluations. He is the author of more than 50 papers in international and national journals on issues related to the planning and management of transport infrastructures and their relationship with land use and environment.

Preface

The Reprint of this Special Issue, entitled "Paving the Future: Sustainable Road Design and Urban Mobility in Smart Cities," collects 10 research contributions published in the *Smart Cities* journal, in the section "Smart Urban Mobility, Transport, and Logistics." It explores innovative solutions in road engineering that incorporate advanced technologies for the design of smart roads, efficient traffic management strategies, cooperative driving, and autonomous vehicles.

Some research presented in this Reprint focuses on studying the interactions between different road users, proposing inclusive infrastructure solutions such as traffic calming measures, integrated warning systems for pedestrians, or strategies for balancing the presence of cyclists and autonomous vehicles, aiming to develop safer and more sustainable, livable urban road spaces. Strategies for developing efficient urban mobility are also explored in this Reprint by investigating the offer of public transport and proposing interventions to encourage its use.

Furthermore, to underpin the design of human-centered road infrastructures and smart mobility systems, topics on the accuracy of traffic modeling are addressed by exploring the integration of some behavioral parameters inside these models. Another research work based on what-if scenario simulation through virtual twins proposes comparative road design assessment methodologies incorporating smart vehicle technologies. Aspects of road safety and maintenance management are also addressed using computer vision techniques.

In line with the sustainable development goals of smart cities, the Reprint also includes themes on road resilience and the adaptive ability of autonomous vehicles in the face of complex road contexts.

This Reprint offers an overview of methodological approaches, replicable models, lessons, and tools for early-stage evaluation of design solutions that integrate smart road and urban mobility technologies based on global case studies, useful for supporting the decision-maker process on sustainable, inclusive, resilient, and safe road infrastructures and mobility systems.

Maria Luisa Tumminello, Elżbieta Macioszek, Anna Granà, and Tullio Giuffrè
Guest Editors

smart cities

Article

Turning Features Detection from Aerial Images: Model Development and Application on Florida's Public Roadways

Richard Boadu Antwi [1,*], Michael Kimollo [2], Samuel Yaw Takyi [1], Eren Erman Ozguven [1], Thobias Sando [2], Ren Moses [1] and Maxim A. Dulebenets [1]

1 Florida State University College of Engineering, Florida State University, 2525 Pottsdamer Street, Tallahassee, FL 32310, USA
2 School of Engineering, University of North Florida, Jacksonville, FL 32224, USA
* Correspondence: rantwi@fsu.edu

Abstract: Advancements in computer vision are rapidly revolutionizing the way traffic agencies gather roadway geometry data, leading to significant savings in both time and money. Utilizing aerial and satellite imagery for data collection proves to be more cost-effective, more accurate, and safer compared to traditional field observations, considering factors such as equipment cost, crew safety, and data collection efficiency. Consequently, there is a pressing need to develop more efficient methodologies for promptly, safely, and economically acquiring roadway geometry data. While image processing has previously been regarded as a time-consuming and error-prone approach for capturing these data, recent developments in computing power and image recognition techniques have opened up new avenues for accurately detecting and mapping various roadway features from a wide range of imagery data sources. This research introduces a novel approach combining image processing with a YOLO-based methodology to detect turning lane pavement markings from high-resolution aerial images, specifically focusing on Florida's public roadways. Upon comparison with ground truth data from Leon County, Florida, the developed model achieved an average accuracy of 87% at a 25% confidence threshold for detected features. Implementation of the model in Leon County identified approximately 3026 left turn, 1210 right turn, and 200 center lane features automatically. This methodology holds paramount significance for transportation agencies in facilitating tasks such as identifying deteriorated markings, comparing turning lane positions with other roadway features like crosswalks, and analyzing intersection-related accidents. The extracted roadway geometry data can also be seamlessly integrated with crash and traffic data, providing crucial insights for policymakers and road users.

Keywords: turning lanes; deep learning; roadway characteristic index (RCI); pavement markings; machine learning (ML); roadway geometry features

1. Introduction

Advancement of computer vision technology is rapidly empowering traffic agencies to streamline the collection of roadway geometry data, providing savings in both time and resources. Traditionally, image processing has been viewed as a time-consuming and error-prone method for capturing roadway data. However, with the evolution of computational capabilities and image recognition techniques, new possibilities have emerged for accurately detecting and mapping various roadway attributes from diverse imagery sources. A study conducted in 2015 [1] demonstrated that utilizing aerial and satellite imagery for gathering geometry data offered significant advantages over traditional field observations in terms of equipment costs, data accuracy, crew safety, data collection expenses, and time required for collection.

Since the introduction of the Highway Safety Manual (HSM) in 2010, many state departments of transportation (DOTs) have placed a strong emphasis on enhancing safety by aligning with the manual's requirements for geometric data [2]. However, the manual collection of data across extensive roadway networks has presented a considerable challenge for numerous state and local transportation agencies. These agencies have a critical role in maintaining up-to-date roadway data, which is essential for effective planning, maintenance, design, and rehabilitation efforts [3]. Roadway characteristics index (RCI) data encompasses a comprehensive inventory of all elements that constitute a roadway, including highway performance monitoring system (HPMS) data, roadway geometry, traffic signals, lane counts, traffic monitoring locations, turning restrictions, intersections, interchanges, rest areas with or without amenities, high occupancy vehicle (HOV) lanes, pavement markings, signage, pavement condition, driveways, and bridges. DOTs utilize various methodologies for RCI data collection, ranging from field inventory and satellite imagery to mobile and airborne light detection and ranging (LiDAR), integrated geographic information system (GIS)/global positioning system (GPS) mapping systems, static terrestrial laser scanning, and photo/video logging [2].

These methods fall into two main categories: ground observation/field surveying and aerial imagery/photogrammetry. The process of gathering RCI data involves considerable expenses, with each approach requiring specific equipment and time allocation for data collection, thereby impacting overall costs. Ground surveying entails conducting measurements along the roadside using total stations, while field inventory involves traversing the roadway to document current conditions and inputting this information into the inventory database. On the other hand, aerial imagery or photogrammetry entails extracting RCI data from georeferenced images captured by airborne systems such as satellites, aircraft, or drones. Each approach has its own advantages and disadvantages in terms of data acquisition costs, accuracy, quality, constraints on collection, storage requirements, labor intensity, acquisition and processing time, and crew safety. Additionally, pavement markings typically have a lifespan of 0.5 to 3 years [4]. Because of their short lifespan, frequent inspection and maintenance are necessary. However, road inspectors generally adopt a periodic manual inspection approach to assess pavement marking conditions, which is both time-consuming and risky.

Despite the prevalent use of direct field observations by highway agencies and DOTs for gathering roadway data [3], these methods pose numerous challenges, including difficulty, time consumption, risk, and limitations, especially in adverse weather conditions. Therefore, there is a critical need to explore alternative and more efficient approaches for collecting roadway inventory data. Researchers have explored novel and emerging technologies, such as image processing and computer vision methods [2], for this purpose. Over the past decade, satellite and aerial imagery have been extensively utilized for acquiring earth-related information [2]. High-resolution images obtained from aircraft or satellites can quickly provide RCI data upon processing [5–7]. In a previous study, artificial intelligence (AI) was employed to extract school zones from high-resolution aerial images [7]. The increasing availability of such images, coupled with advancements in data extraction techniques, highlights the importance of optimizing their use for efficiently extracting roadway inventory data. While there may be challenges in extracting obscure or small objects, recent advancements in machine learning research have lessened this limitation. Well-trained models can significantly enhance output accuracy. Furthermore, this method entails minimal to no field costs, making it economically advantageous.

RCI data encompass critical features like center, left, or right turning lanes, which are significant to be known to both state DOTs and local transportation agencies in the context of traffic operations and safety. These pavement markings, primarily situated near intersections, indicate permissible turns from each lane, effectively guiding and managing traffic flow. They offer continuous assistance in vehicle positioning, lane navigation, and maintaining roadway alignment, as highlighted in a USDOT study [8]. Many DOTs have expressed keen interest in automated methods for detecting and evaluating these markings [9]. Although turning lane information plays a pivotal role in optimizing roadway efficiency and minimizing accidents, particularly at intersections, a comprehensive geospatial inventory of turning lanes is currently lacking, spanning both local- and state-operated roadways in Florida. Therefore, it is crucial to develop innovative, efficient, and rapid methods for gathering such data. This is of paramount importance for the Florida DOT, serving various purposes such as identifying outdated or obscured markings, juxtaposing turning lane positions with other geometric features like crosswalks and school zones, and analyzing accidents occurring in these zones and at intersections.

To the best of the authors' knowledge, there has been no prior research exploring the utilization of computer vision techniques to extract and compile an inventory of turning lane markings using high-resolution aerial images. As such, this study seeks to address this gap by developing automated tools for detecting these crucial roadway features through deep learning-based object detection models. Specifically, the focus will be on creating a You Only Look Once (YOLO)-based artificial intelligence model and framework to identify and extract turning lane markings, including left, right, and center lanes, from high-resolution aerial images. The primary objective is to devise an image processing and object detection methodology tailored to identify and extract these lane features across the State of Florida using high-resolution aerial imagery. The goals are to:

(a) develop a multiclass object detection model,
(b) evaluate the performance of the model with ground truth data, and
(c) apply the model to aerial images to detect the left, right, and center lane features on Florida's public roadways.

This initiative holds significant importance for the Florida Department of Transportation (FDOT) and other transportation agencies for various reasons, including:

i. infrastructure management, such as identifying aging or obscured markings.
ii. comparing turning lane locations with other geometric features like crosswalks and school zones and analyzing accidents occurring within these zones and at intersections.
iii. automatically extracting road geometry data that can be seamlessly integrated with crash and traffic data, offering valuable insights to policymakers and roadway users.

2. Literature Review

Research on using AI techniques to extract RCI data—particularly those pertaining to pavement markings—has significantly increased in the last few years. By using detection models, AI approaches are essential in obtaining information about roadways from such data. For example, a study [7] used high-resolution aerial pictures and YOLO to locate school zones. Although a lot of research has focused on employing LiDAR technology to capture RCI data, this method has drawbacks, including expensive equipment, complicated data processing, and long processing periods. For instance, a study [3] used precision navigation and powerful computers to use LiDAR to compile inventories of various roadway elements. Furthermore, in other studies, mobile terrestrial laser scanning (MTLS) has been utilized to gather data from highway inventories [5]. Furthermore, computer vision, sensor technology, and AI techniques have been combined in recent autonomous driving research to recognize pavement markings for vehicle navigation [10–14].

The majority of previous research has been on conventional approaches, which are linked to longer data gathering times, traffic flow disruptions, and possible dangers to crew members. Conversely, there has not been much focus on developing technologies like computer vision, object detection, and deep learning in the domain of roadway geometry feature extraction using high-resolution aerial images. With the help of these cutting-edge technologies, roadway features may be extracted from high-resolution aerial photos, doing away with the necessity for labor-intensive fieldwork and drastically cutting down on data gathering time. In addition, the data collected using these cutting-edge techniques are readily available, protect crew safety by averting traffic-related risks, and quickly capture roadway characteristics over wide areas. Standardized image collection is not the only criterion to consider when choosing the best object detection model [15].

2.1. Utilizing Computer Vision and Deep Learning for Roadway Geometry Feature Extraction

Experts in transportation are beginning to recognize convolutional neural networks (CNNs), a new instance of computer vision and deep learning methods [16]. Notable progress has been made in the extraction of roadway geometry data by researchers using CNN and region-based convolutional neural networks (R-CNN). These methods have proven to be effective in quickly identifying, detecting, and mapping roadway geometry features over large regions with no need for human interaction. For example, in ref. [4], photogrammetric data from Google Maps was utilized to automatically identify pavement marking issues on roadways using R-CNN to evaluate pavement conditions. With an average accuracy of 30%, the developed model in this study was trained to detect flaws in pavement markings, such as incorrect alignment, ghost marking, missing or fading edges, corners, segments, and cracks.

In one study, Google Street View (GSV) photos taken along interstate segments were used to recognize and categorize traffic signs using computer vision algorithms [17]. In order to classify road signs according to their pattern and color information, the researchers used a machine learning technique that paired histograms of oriented gradients (HOG) data with a linear support vector machine (SVM) classifier. A classification accuracy of 94.63% was attained by them. However, neither local nor non-interstate roads were used to test the methodology. The research by [18] used ground-level photos and a technique known as ROI-wise reverse reweighting network to recognize road markers. Using ROI-wise reverse reweighting and multi-layer pooling operation based on faster RCNN, the study achieved a 69.95% detection accuracy for road markers in images by highlighting multi-layer features.

In ref. [12], inverse perspective mapped photos were used to estimate roadway geometry and identify lane markings on roadways using CNN. Researchers used CNN in [19] to detect, measure, and extract the geometric properties of hidden cracks in roads using ground-penetrating radar pictures. CNN was used by the authors in [20] to extract roadway features from remotely sensed photos. These methods have been used to detect images for a variety of purposes. For example, driverless cars use them to instantly identify objects and road infrastructure. Using the Caltech-101 and Caltech pedestrian datasets, a novel hybrid local multiple system (LM-CNN-SVM) incorporating CNN and SVMs was tested for item and pedestrian detection [21]. Several CNNs were tasked with learning the properties of local areas and objects after the full image was divided into local regions. Significant characteristics were chosen using principal component analysis, and in order to improve the classifier system's ability to generalize, these features were then fed into several SVMs using empirical and structural risk reduction strategies rather than only using CNN. Moreover, an original CNN architecture and a pre-trained AlexNet were both used, and the SVM output was mixed. The approach showed an accuracy range that was between 89.80% and 92.80%.

CNNs based on deep learning have demonstrated impressive capabilities in object detection recently [22–25]. These algorithms have been greatly enhanced by their progression from region-based CNN (R-CNN) to fast R-CNN [26] and ultimately faster R-CNN [27]. The CNN method basically involves computing CNN features after extracting region proposals as possible object placements. By adding another region proposal network (RPN) to the object detection network, faster R-CNN improves performance even further. In a study [28], the vehicle detection and tracking skills of machine learning algorithms and deep learning techniques—faster R-CNN and ACF—were thoroughly compared. The results show that quicker R-CNN performs better in this area than ACF. Therefore, choosing the algorithm to perform the analysis requires significant thought. The output of vehicle identification and tracking algorithms may include parameters like speed, volume, or vehicle trajectories, depending on the analysis's goals. Vehicle detection algorithms differ from point tracking techniques in that they can also classify automobiles.

2.2. Obtaining Roadway Geometry Data through LiDAR and Aerial Imagery Techniques

Road markings can be extracted from photos by taking advantage of their high reflectivity, which is a characteristic that they usually display. In ref. [29], researchers developed an algorithm that uses pixel extraction to recognize and locate road markers in pictures taken by cameras installed on moving cars. Initially, a median local threshold (MLT) image filtering technique was used by the system to extract marking pixels. After that, a recognition algorithm was used to identify markings according to their dimensions and forms. For the identified markers, the average true positive rate was 84%. Researchers in [30] extracted highway markers from aerial photographs using a CNN-based semantic segmentation technique. The paper presented a novel method for processing high-resolution images for lane marking extraction that makes use of a discrete wavelet transform (DWT) and a fully convolutional neural network (FCNN). In contrast to earlier techniques, this strategy—which used a symmetric FCNN—focused on small-sized lane markers and investigated the best places for DWT insertion sites inside the architecture. The accuracy of lane detection could be enhanced by the model's ability to collect high-frequency information with the integration of DWT into the FCNN architecture. Aerial LaneNet architecture consisted of a sequence of pooling, wavelet transform, and convolutional layers. The model's effectiveness was assessed using an aerial lane detection dataset that was made available to a general audience. The study found that washed-out lane markers and shadowed areas presented constraints.

Researchers used the scan line method in [31], which allowed them to recover road markings from LiDAR point clouds. The team efficiently structured the data by timestamp-sequentially ordering the LiDAR point clouds and grouping them into scan lines based on the scanner angle. They then isolated seed roadway points, which formed the foundation for determining complete roadway points, by taking advantage of the height difference between trajectory data and the road surface. Using these points, a line was drawn that encircled the scan line's seed points as well as every other point. After that, points that fell within a given range of this line were kept and classified as asphalt points or road markings according to how intense they were. In order to smooth out intensity values and improve data quality by lowering noise, a dynamic window median filter was used. Finally, the research team identified and extracted roadway markers using edge detection algorithms and limitations. Roadway data were gathered by researchers in [3] using two-meter resolution aerial images and six-inch accuracy LiDAR data saved in ASCII comma-delimited text files. Using ArcView, they added height information to LiDAR point shapefiles and then converted the data points into Triangular Irregular Networks (TIN). The analytical process was started by combining the aerial photos and LiDAR boundaries. To find obstructions in the observer's line of sight, ArcView's object obstruction tool was utilized. Tracing the line of sight—visible terrain represented by green line segments, obscured terrain by red line segments—was used to calculate the stopping sight distance. With ArcView's identification and contour capabilities, side slope and contours were determined. The grade was also determined by the researchers by calculating the elevation difference between a segment's two ends. Conversely, ref. [32] built a deep learning model for the automated recognition and extraction of road markings using a 3D profile of pavement data obtained by laser scanning. They were able to detect with 90.8% accuracy. As part of the project, a step-shaped operator was designed to identify possible locations for pavement marking edges. The road markings were subsequently extracted using CNN by combining the geometric properties and convolution information of these locations.

In ref. [33] pavement markings were extracted from camera images and pavement conditions were assessed using computer vision techniques. Preprocessing was applied to videos that were taken of cars from the frontal perspective for the research. The preprocessing involved the extraction of image sequences, the Gaussian blur median filter for filtering and smoothing the images, and the inverse perspective transform for correcting the photos. A hybrid detector that relied on color and gradient features was used to find pavement markings. After that, regions with marks were identified using image segmentation and then categorized into edge, dividing, barrier, and continuity lines according to their characteristics. Building on a prior study [34] that used Yolov3 to extract turning lane features from aerial images, this research will use additional image processing techniques along with a more advanced version of yolo to extract turning lane features, including left, right, and center, from extremely high-resolution aerial images in Leon County, Florida.

To the authors' knowledge, there has not been any research carried out on how to create such a large inventory of geocoded left, right, and center turning lane markings on a state- or countywide scale using an integrated image processing and machine learning techniques on very high-resolution aerial images. Thus, by creating automated methods for identifying these turning lane markings from very high-resolution images on a large scale (e.g., left, right, and center), our research attempts to close this gap. To achieve this, the study will specifically concentrate on developing an artificial intelligence model based on the You Only Look Once (YOLO) methodology.

3. Study Area

The case study region for turning lanes and configurations detection chosen is Leon County (Figure 1). There are 292,198 people residing in Leon County, which is home to the City of Tallahassee, the capital city of Florida, and it is 668 square miles in total size [35]. It shares boundaries with Georgia's Grady and Thomas counties as well as other Florida counties like Gadsden, Wakulla, Liberty, and Jefferson. The county was taken into consideration because of the variety of its roadway infrastructure development. The models' performance was verified using ground truth data from Leon County, Florida.

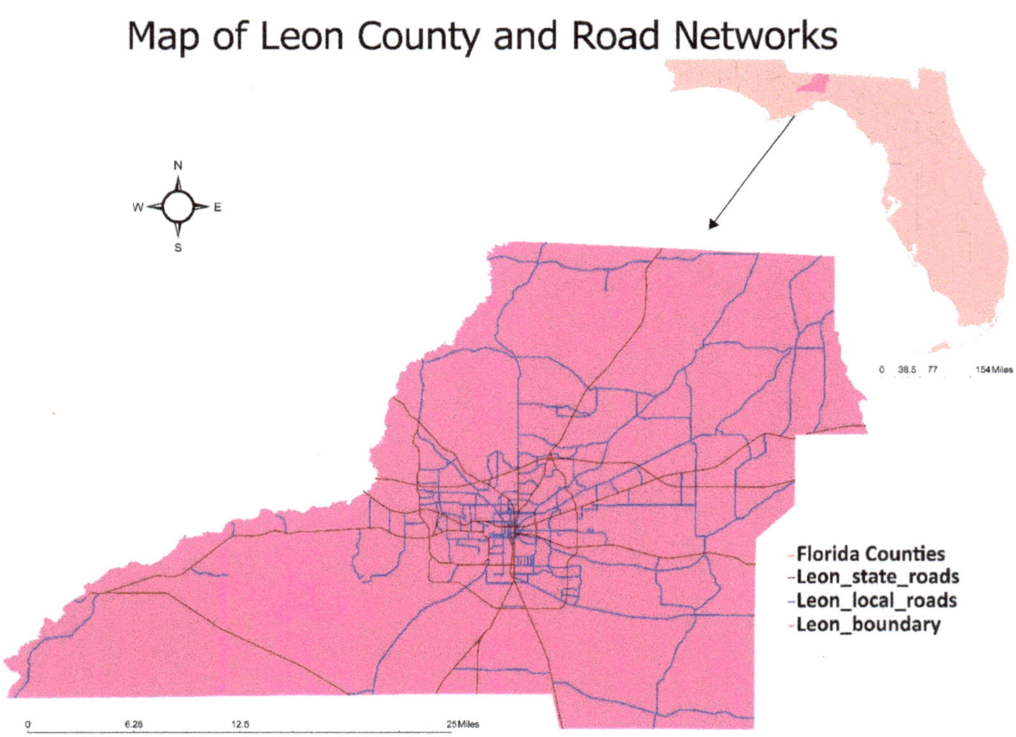

Figure 1. Map of Leon County, Florida with the roadway network.

4. Materials and Methodology

The selection of the methodology for gathering roadway inventory data are contingent upon various factors, including data collection time (e.g., data collection, reduction, and processing) and cost (e.g., data collection, and reduction), as well as accuracy, safety, and data storage demands. In this study, our objective was to develop a deep learning object detection model tailored to identify turning lane markings from high-resolution aerial images within Leon County, Florida.

4.1. Data Description

The proposed algorithm for detecting turning lanes relies on high-resolution aerial images and recent advancements in computer vision and object detection techniques. The goal is to automate the process of identifying pavement markings for turning lanes, offering a novel solution for transportation agencies. The aerial images used in this study were sourced from the Florida Aerial Photo Look-Up System (APLUS), an aerial imagery database managed by the Surveying and Mapping Office of the Florida Department of Transportation (FDOT). Specifically, images from Leon (2018) and Miami-Dade (2017) Counties, Florida, were utilized. These images have a resolution ranging from 1.5 ft down to 0.25 ft, ensuring sufficient detail for accurate detection of turning lanes. The model's resolution requirement allows for the utilization of any imagery falling within or exceeding this resolution threshold. The majority of the images were in 0.5 ft/pixel resolution, with a size of 5000 $\times$ 5000 and a 3-band (RGB) image format, although precise resolutions varied by county. Additionally, state and local roadway shapefiles were obtained from the FDOT's GIS database, providing further data for the detection algorithm. The images are given in MrSID format, which enables GIS projection onto maps, enhancing spatial analysis capabilities. Also, state, and local roadway shapefiles were acquired from the FDOT's GIS database, complementing the image data, and enriching the dataset for further analysis.

This study focuses on identifying turning lanes present on roadways managed by counties or cities, as well as those situated on state highway system routes, excluding interstate highways. State highway system routes are termed ON system roadways or state roadways, while county or city-managed roads are referred to as OFF system roadways or local roadways, according to their FDOT classification. Our proposed approach involves merging all centerlines from the shapefile of county and city-managed roadways. While the FDOT's GIS data can provide various geometric data points crucial for mobility and safety assessments, they lack information on the locations of turning configurations on the local roadways. Therefore, the main objective of this project is to utilize an advanced object identification model to compile an inventory of turning lane markers, specifically left only, right only, and center lanes, on both state and local roadways in Leon County, Florida.

4.2. Pre-Processing

Due to the volume of data and the complexity of object recognition, preprocessing is essential. This method involves selecting and discarding images that do not intersect with a roadway centerline and mask out pixels outside a buffer zone. In this approach, the number of photos was decreased, and the image masking model excluded objects that were 100 feet away from state and local roadways (Figure 2). Prior to masking the images, the roadway shapefile is buffered to create overlapping polygons, serving as reference for cropping intersecting regions of aerial images. During masking, pixels outside the reference layer's boundary are removed. The resulting cropped images, with fewer pixels, are then mosaiced into a single raster file, facilitating easier handling and analysis.

First, all aerial images from the selected counties are imported into a mosaic dataset using ArcGIS Pro. Geocoded photos are then managed and visualized using mosaic databases, allowing for the intersection with additional vector data to select specific image tiles based on location. Following this, a subset of photos is extracted, comprising images that intersect with roadway centerlines. An automated image masking tool is subsequently developed using ArcGIS Pro's ModelBuilder interface. This tool systematically processes the photo folder, applying a mask based on a 100-foot buffer around roadway centerlines, and saves the resulting masked images as JPG files for the object detection process. Exported images are further processed for model training and detection.

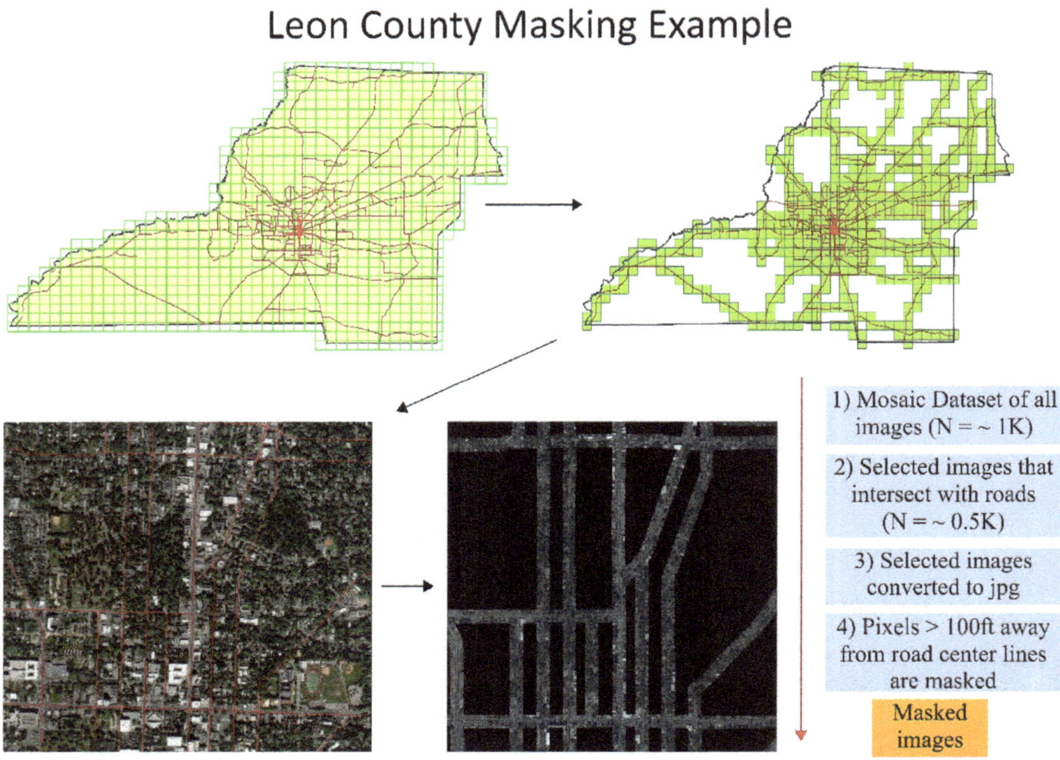

Figure 2. Preprocessing approach, and automated image masking model.

4.3. Data Preparation for Model Training and Evaluation

High-quality image data are crucial for training deep learning models to identify various turning lane features. The process of image preparation, depicted in Figure 3, outlines the steps from turning lane identification to the extraction of training data or images. Significant time and effort were dedicated to generating training data for model creation, as a substantial portion of the model's performance hinges on the quantity, quality, and diversity of the training data used. Similarly, additional time and effort were invested in the model training process, involving testing different parameters, and selecting the optimal ones to train the model. This is depicted by the larger box sizes for "Manually label turning lane features" and "training data" in Figure 3.

For the initial investigation, a 4-class object detection model was developed to identify the turning configurations. The training data with varying classes were prepared for the model. The training data prepared for the turning lane model comprised 4 classes: "left_only", "right_only", "center", and "none", denoted as classes 1, 2, 3, and 0, respectively. These labels were represented by rectangular bounding boxes encompassing the turning lane markings (Table 1). The training data initially prepared for the turning lane model were manually labelled 4687 features on the aerial images of Miami–Dade county using the Deep-Learning Toolbox in ArcGIS Pro. The model had 1418 features (30.3%) for the "left_only" class, 1328 features (28.3%) for the "right_only" class, 662 features (14.1%) for the "center" class, and 1279 features (27.3%) for the "none" class. The "none" class labels for the model were made up of all other visible markings or features except "left_only", "right_only", and "center". Some features which were very common, for instance, left only and right only, were more prevalent in the training data for the turning lane model, while less common features like "center" had lower representation. To address this imbalance and enhance dataset diversity, some of these less common features were duplicated. However, to prevent potential issues like model overfitting and bias resulting from duplication, data augmentation techniques such as rotation were employed. Rotation involved randomly rotating the training data features at various angles to generate additional training data, thereby increasing dataset diversity. This approach enabled the model to recognize objects in different orientations and positions, which was particularly advantageous in this study, given that features like left only and right only lanes can appear in multiple travel directions at intersections. A 90 degrees rotation was applied to the training data. Finally, 11,036 exported image chips containing 16,564 features were used to train the model. It is important to mention that a single image chip may contain multiple features.

Each class within the training data was distinctively labeled to ensure clarity. These classes distinctly categorized the turning lane features and non-features identified in the input image by the detector, thereby enhancing the model's detection accuracy. The output data were then sorted into categories such as left only, right only, center, and none, utilizing a class value field. The metadata generated for these labels, such as the image chip size, object class, and bounding box dimensions, were stored in an xml file. A function was then created to further convert the metadata into yolov5 format to obtain the transformed bounding box center coordinates, the box width, and height, using the following equations:

$$c_x = \frac{1}{2}\left(\frac{x_{min} + x_{max}}{w}\right) \tag{1}$$

$$c_y = \frac{1}{2}\left(\frac{y_{min} + y_{max}}{h}\right) \tag{2}$$

$$b_w = \frac{x_{max} - x_{min}}{w} \tag{3}$$

$$b_h = \frac{y_{max} - y_{min}}{h} \tag{4}$$

where c_x is the x coordinate of the bounding box center, c_y is the y coordinate of the bounding box center, b_w is width the bounding box, b_h is height the bounding box, w is width the image chip, h is height the image chip, x_{max} is maximum x value of the bounding box, x_{min} is minimum x value of the bounding box, y_{max} is maximum y value of the bounding box, and y_{min} is minimum y value of the bounding box. The normalization of coordinates was performed using the image dimensions to maintain the bounding box positions relative to the input image.

It is worth noting that object detection models perform optimally when trained on clear and distinctive features. Given that left only and right only turning markings often exhibit a lateral inversion of one another, rendering them less distinguishable, the model's performance in detecting them is expected to be lower. To ensure uniqueness, the labels for training the center lane class featured both left arrows facing each other from different travel directions. This approach was similarly applied to label all other features to maintain uniqueness in the training features. The input mosaic data comprised high-resolution aerial images covering Miami–Dade County, Florida, with a tile size of 5000 × 5000 square feet.

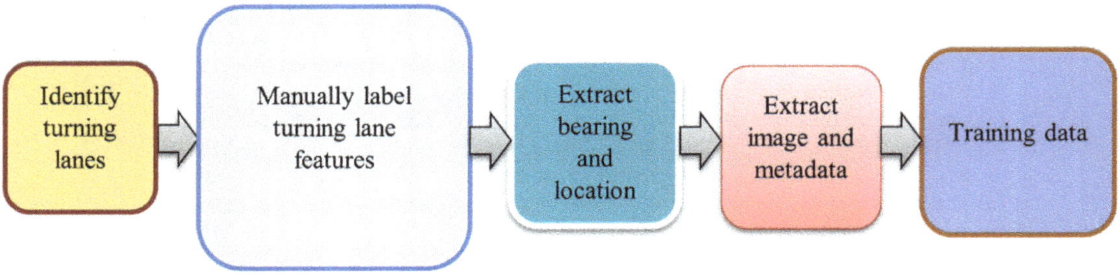

Figure 3. Model training data preparation framework.

Table 1. Training data description.

ID	Class Name	Description	Example/Picture	Training Example
1	left_only	Left only		
2	right_only	Right only		
3	center	Center lane (left turn possible in both direction or two way)		
0	none	None: Yield, left through, right through, bicycle, merge, u-turn, stop, parking, speed limits, etc.	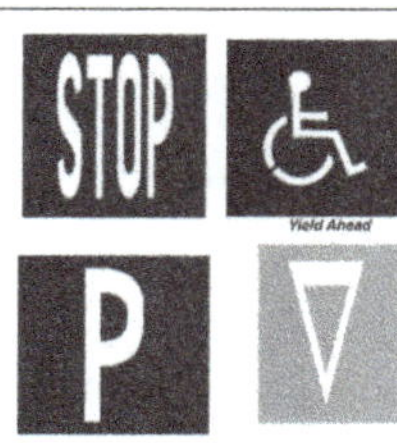	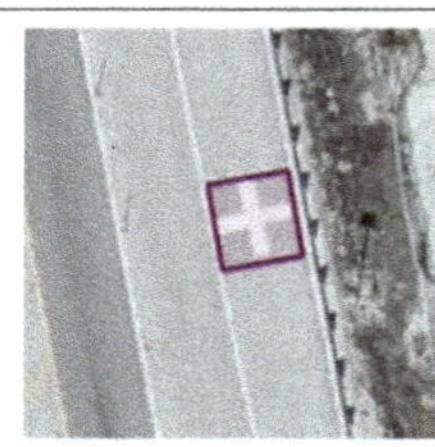

4.4. YOLOv5—Turning Lane Detection Model

You Only Look Once (YOLO) is predominantly utilized for real-time object detection. Compared to other models such as R-CNN or faster R-CNN, YOLO's standout advantage lies in its speed. Specifically, YOLO surpasses R-CNN and faster R-CNN architectures by a factor of 1000 and 100, respectively [36]. This notable speed advantage stems from the fact that while other models first classify potential regions and then identify items based on the classification probability of those regions, YOLO can predict based on the entire image context and perform the entire image analysis with just one network evaluation. Initially introduced in 2016 [37], YOLO has seen subsequent versions, including YOLOv2 [38] and YOLOv3, which introduced improvements in multi-scale predictions. YOLOv4 [39] and YOLOv5 [40] were subsequently released in 2020 and 2022, respectively.

YOLOv5 architecture is made up of the backbone, the neck (PANet), and the head (output). Unlike other detection models, YOLOv5 uses cross-stage partial network (CSP)-Darknet-53 as its backbone, PANET as its neck, and a yolo layer as its head. CSP-Darknet is a better version of Darknet-53, which was the backbone of YOLOv3 with higher accuracy and processing speed [40]. The architecture of the backbone and 6 × 6 convolution 2d structure improves model's efficiency [41], making the feature extractor perform better and faster than other object detection models including the previous versions [42]. Compared to the previous versions, it has a higher speed, higher precision, and a smaller volume. YOLOv5's backbone enhances both the accuracy and speed of the model, performing twice as fast as ResNet152 [36]. Utilizing multiple pooling and convolution, YOLOv5 forms image features at different levels using CSP and spatial pyramid pooling (SPP) to extract different-sized features from input images [43]. The architecture has a significant advantage in the area of computational complexity: on the one hand, the BottleneckCSP architecture reduces complexity and improves speed in the inference process; on the other hand, SPP allows for the extraction of features from images in different parts of the input image, creating three-scale feature maps which improve detection quality. Moreover, the neural network neck is a series of layers that are zones between the input and output layers, which is made up of feature pyramid network (FPN) and path aggregation network (PAN) structures, the task of which is to integrate and synthesize image features in various proportions before transferring it to the prediction. FPN enables semantic features of high level to be passed to the lower feature maps, and PAN enables localization features of high accuracy to be passed to the higher feature maps, finally moving into the detection phase where targets of different sizes in feature maps are generated in the output head [43]. Figure 4 shows the network architecture of YOLOv5, illustrating the three-scale detection process. This process involves applying a 1 × 1 detection kernel on feature maps situated at three distinct areas and sizes within the network.

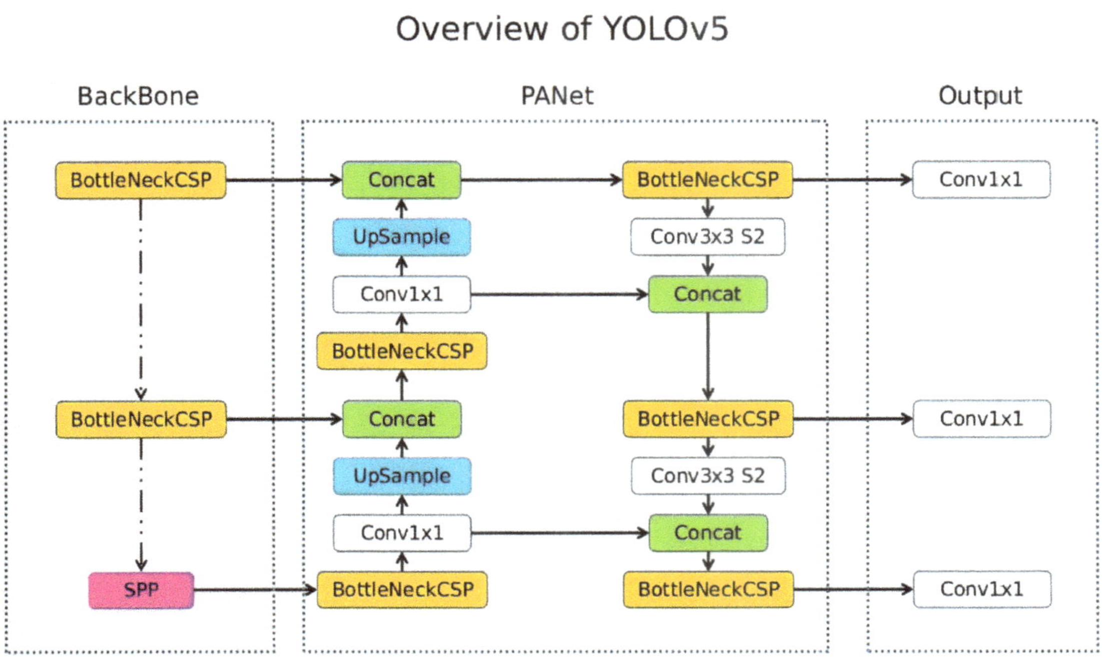

Figure 4. YOLOv5 network architecture, adapted from [40].

4.5. Turning Lane Detector

The object detection model's adjustable parameters and hyperparameters encompass various factors such as the learning rate, input image dimensions, number of epochs, batch size, anchor box dimensions, and ratios, as well as the percentages of training and testing data. Visualization of the machine learning model's evaluation metrics was presented through graphs depicting precision–recall, validation and training loss, F1 confidence curve, precision confidence, and confusion matrix. Validation loss and mean average precision were computed on the validation dataset, which constituted 10% of the input dataset. Key parameters influencing object detection performance include the batch size, learning rate, and training epoch. The learning rate dictates the rate at which the model acquires new insights from training data, striking a balance between precision and convergence speed. Optimal learning rates of 3.0×10^{-4} were used to train the model. Batch size refers to the number of training samples processed per iteration, with larger sizes facilitating parallel processing for faster training, albeit requiring more memory. In contrast, smaller batch sizes enhance model performance on new data by increasing randomness during training. Given the multi-class nature and high data complexity of the developed model, a batch size of 64 was adopted to enhance performance. Also, the anchor box represents the size, shape, and location of the object being detected. Four sizes of initial detection anchor boxes were used to train the model. The epoch number denotes the iterations the model undergoes during training, representing the number of times the training dataset is processed through the neural network. The model was trained using 100 epochs. In this study, 80% of the input data were utilized for model training, selected randomly to ensure representativeness. A 20% (10% validation and 10% test) split of the input dataset was used for validation during training and testing after training to measure the model's performance. This test set derived from the input dataset was used to compute the evaluation metrics such as accuracy, precision, recall, and F1 score after model training. In other words, part of 2207 image chips was used to tune the model's hyperparameters during the training process and part of the image chips was used to assess the model's accuracy after the training process.

The determination of the training and test data split primarily hinged on the size of the training dataset. For datasets exceeding 10,000 samples, a validation size of 20–30% was deemed sufficient to provide randomly sampled data for evaluating the model's performance. A default 50% overlap between the label and detection bounding boxes was considered a valid prediction criterion. Subsequently, recall and precision were computed to assess the true prediction rate among the original labels and all other predictions, respectively. During the model training, a binary cross-entropy loss was used for each label to calculate the classification loss, rather than using mean square error. Therefore, logistic regression was used to predict both object confidence and class predictions. This approach reduces the complexity of computations involved and improves the model's performance [36]. The calculation methods for the loss function for Yolov5 is the summation of the class loss, object confidence loss, and the location or box loss, which are formulated as follows [40]:

$$Loss = \lambda_1 L_{cls} + \lambda_2 L_{obj} + \lambda_3 L_{loc} \quad (5)$$

$$L_{cls} = \sum_{i=0}^{s^2} \ell_i^{obj} \sum_{j=0}^{B} \left[(p_i(c) - \hat{p}_i(c))^2 \right] \quad (6)$$

$$L_{obj} = \sum_{i=0}^{s^2} \sum_{j=0}^{B} \ell_i^{obj} \left[(c_i - \hat{c}_i)^2 \right] + \lambda_{noobj} \sum_{i=0}^{s^2} \sum_{j=0}^{B} \ell_i^{noobj} \left[(c_i - \hat{c}_i)^2 \right] \quad (7)$$

$$L_{loc} = 1 - IoU + \frac{\rho^2(b, b^{gt})}{c^2} + \alpha v \quad (8)$$

$$IoU = \frac{|b \cap b^{gt}|}{|b \cup b^{gt}|} \quad (9)$$

$$\alpha = \frac{v}{(1 - IoU) + v} \quad (10)$$

$$v = \frac{4}{\pi^2}\left(arctan\frac{\omega^{gt}}{h^{gt}} - arctan\frac{\omega}{h}\right)^2 \quad (11)$$

where $Loss$ is the overall loss, L_{cls} is the classification loss, L_{obj} is the object confidence loss, L_{loc} is the location or box loss, λ_1, λ_2, λ_3 are the weights for the classification loss, object confidence loss, and location or box loss respectively. s is the size of the grid (if the input image is divided into $s \times s$-grid cells). B is the number of bounding boxes predicted per grid cell. ℓ_i^{obj} is the probability predictor for the object category, c is the diagonal distance of the smallest closure area that can contain both the predicted box and the real box, c_i is the confidence score, $\hat{c}_i$ is the intersection of prediction boundary box and basic facts, λ_{noobj} is the weights for the classification error, ℓ_i^{noobj} is the probability predictor of the classification error, IoU is the intersection over union, ρ^2 is the Euclidean distance between the centroids of the prediction box and the real box, b, b^{gt} are the center point of the predicted box and the real box, α is the positive trade-off parameter, v measures the consistency of aspect ratio, and $\frac{\omega^{gt}}{h^{gt}}$ and $\frac{\omega}{h}$ are the aspect ratios of the target and predictions.

After model training, the performance was evaluated. The confusion matrix describes the true positive rate as well as the misclassified rate. The diagonal values show very high true positive predictions for each class in the confusion matrix (Figure 5a). A background class can be observed on the confusion matrix. This class represents areas where no object is detected. Therefore, this accounts for instances where no object is detected in a particular region. It can also be observed that the model has extremely low misclassified features with center features showing a 100% correct classification and 0 misclassification. The graph of the train-validation loss for the model is shown in Figure 5b. The validation loss examined how well the model fits the validation set derived from the input data, whereas the training loss evaluated how well the model fits training data. A high loss indicates that the product of the model has errors, while a low loss value shows there are fewer errors present in the model. From the graph, it can be noted that the model had a very low training and validation loss difference with the training loss is slightly higher than validation loss (Figure 5b). This shows that the model is very accurate and performs well on new and existing data; however, the model performed better on the new dataset. The model has a very good precision–recall curve as shown in Figure 5c. This describes how sensitive the model is in detecting true positives and how well it predicts the positive values. A good classifier has both high precision and high recall across the graph. The developed model has a precision of 0.969 and a recall value of 0.970. F1 value is a metric calculated from the harmonic mean between the precision and accuracy values. Higher F1 values indicate a better performance. When the F1 value is compared with the confidence thresholds, the optimum threshold can be identified. From the F1 confidence curve (Figure 5d), the model performs very well between 0.90 and 0.98 F1 score when the confidence threshold is set between 0.05–0.8. The optimum threshold of 0.622 returns an F1 value of 0.97 for all classes. Observing the precision–confidence curve, the precision gently increases with higher confidence thresholds. Therefore, the model is very precise even at lower confidence of 0.2 reaching an ideal precision at 0.98 (Figure 5e). The average accuracy of the developed model was 0.987. Therefore, we can conclude that the developed detector performs quite well.

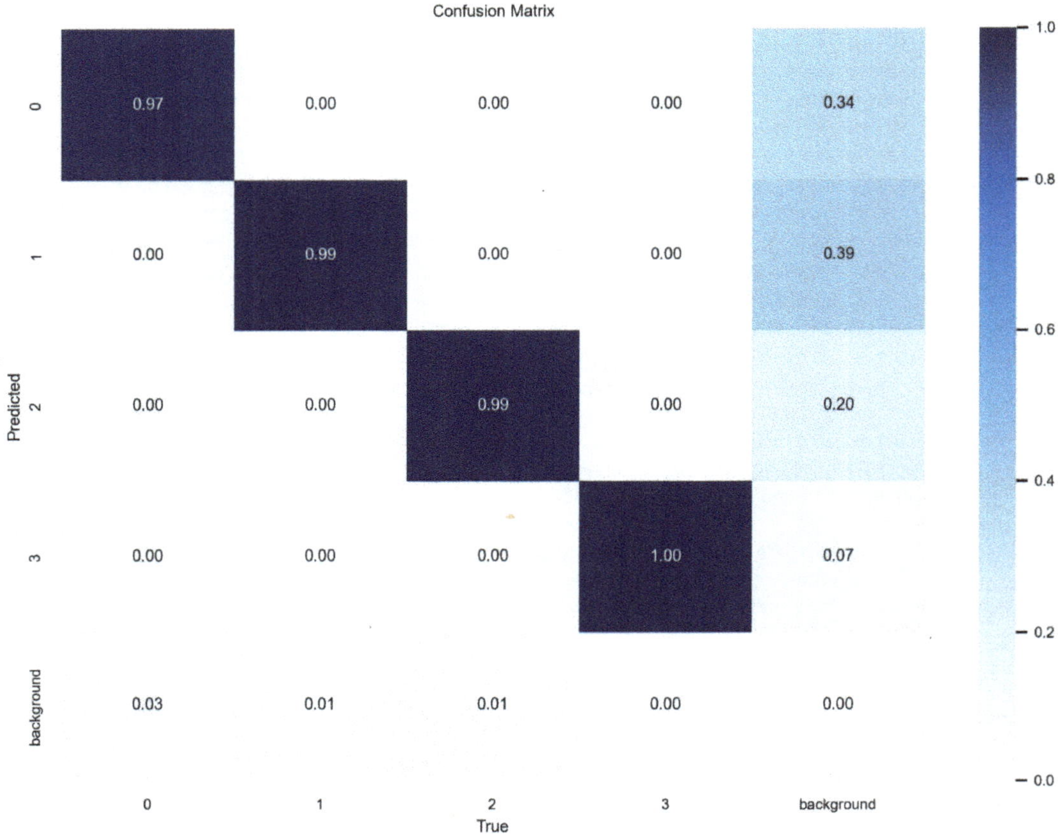

(a)

Figure 5. *Cont.*

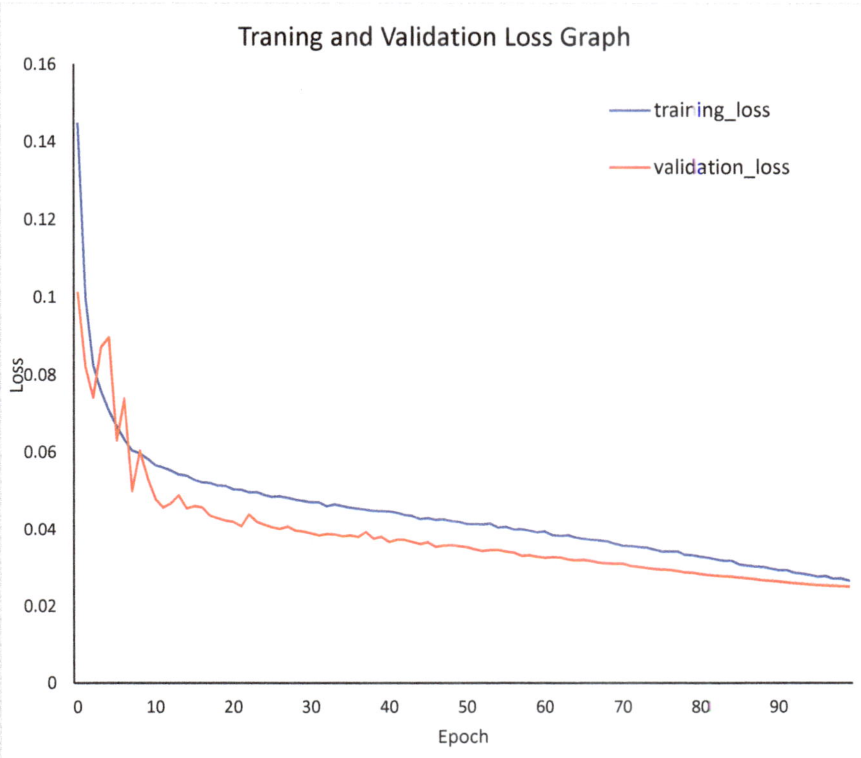

(b)

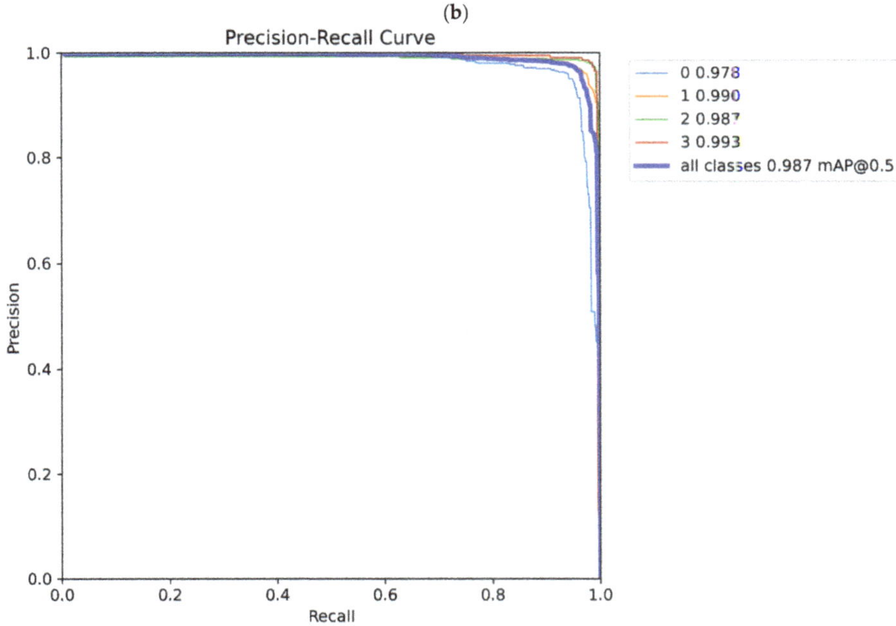

(c)

Figure 5. *Cont.*

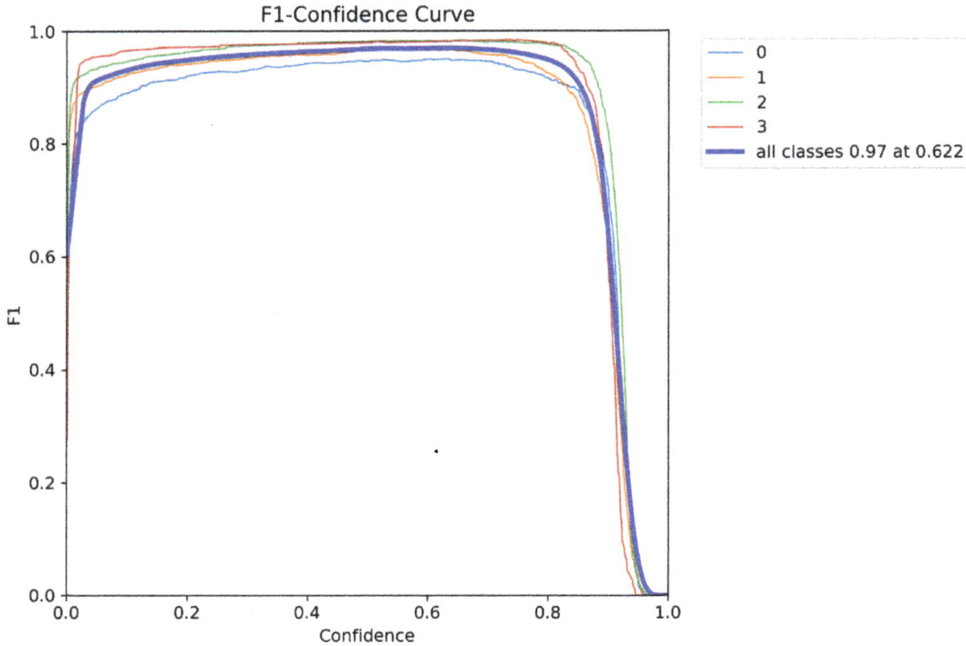

(d)

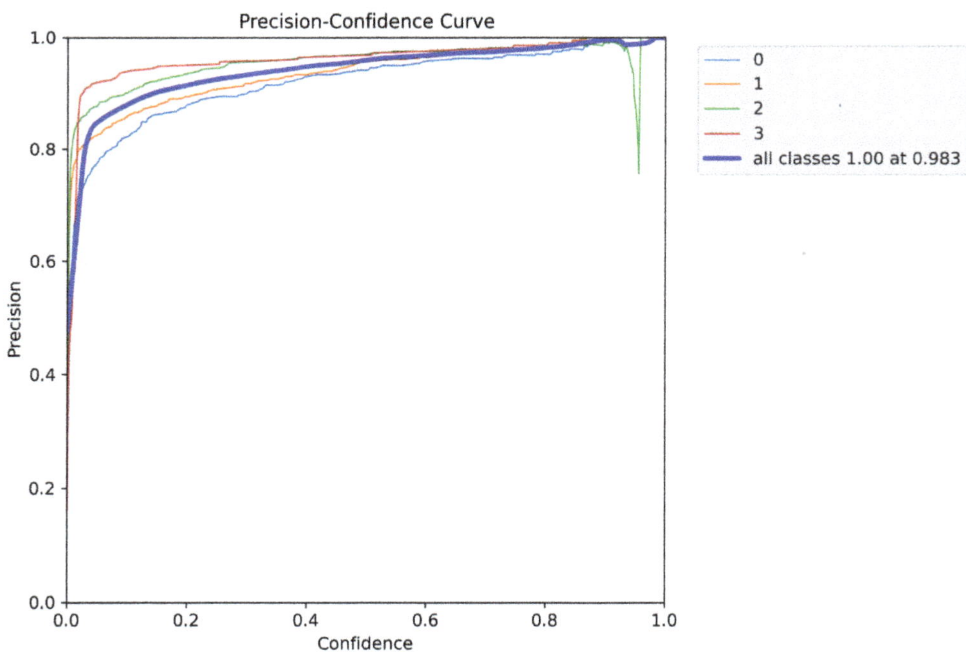

(e)

Figure 5. Developed YOLOv3 turning lane model (**a**) confusion matrix, (**b**) training and validation loss graph (**c**) precision–recall curve, (**d**) F1 confidence curve, and (**e**) precision–confidence curve.

4.6. Mapping Turning Lanes

The turning lane detector was initially tested on individual photos. Figure 6b demonstrates how the detector correctly outlines turning lanes with bounding boxes using the detection's confidence score. A threshold of 0.1 was employed to capture all features detected with very low confidence levels. It is important to note that lower detection thresholds generally increase false positives, which results in lower precision and increases computational workload as well as detection time, and irrelevant features or noise. On the other hand, lower detection thresholds increase the model's sensitivity to detecting faint or partially visible features while increasing recall. With higher recall, the model is more likely to identify and detect all instances of the target object class and reduce the chances of missing any objects. More than 10% overlap between two bounding boxes was avoided to minimize duplicate detections. The detector was trained on 512 × 512 sub-images and a resolution of 0.5 feet per pixel. It should be noted that utilizing huge photos with object detection techniques is impractical, since the cost of computing grows rapidly. Therefore, another image processing step was utilized to split detection images into tiles of 512 × 512 or less to carry out tile detections on the input aerial images. The output detection labels with coordinates were converted into shapefiles and visualized in ArcGIS for further processing and analysis.

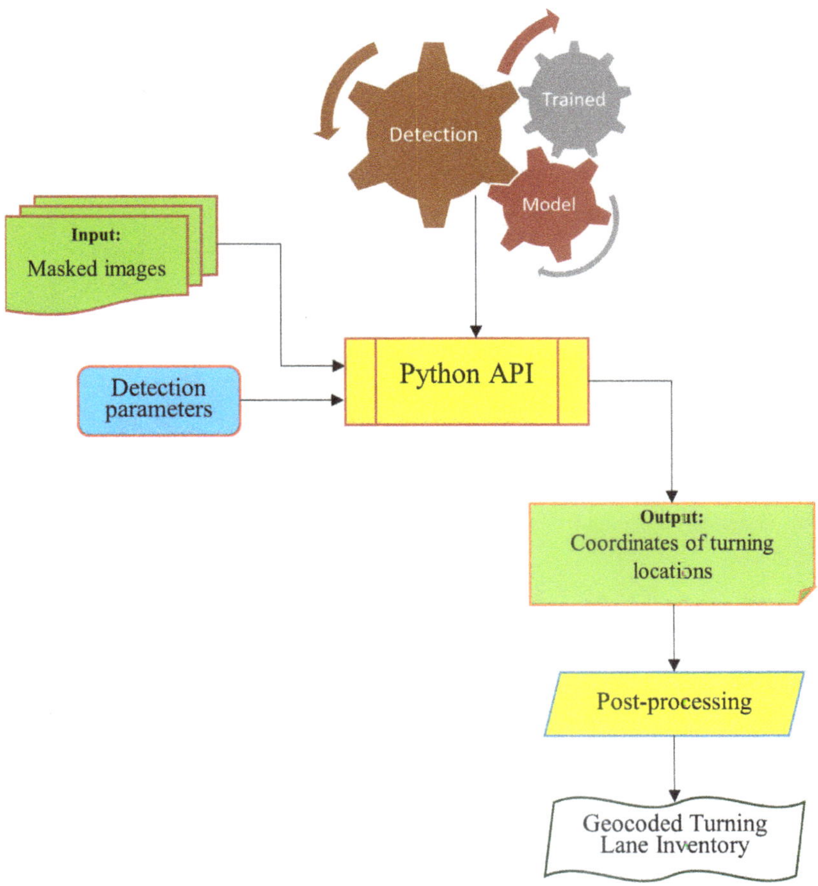

(a)

Figure 6. *Cont.*

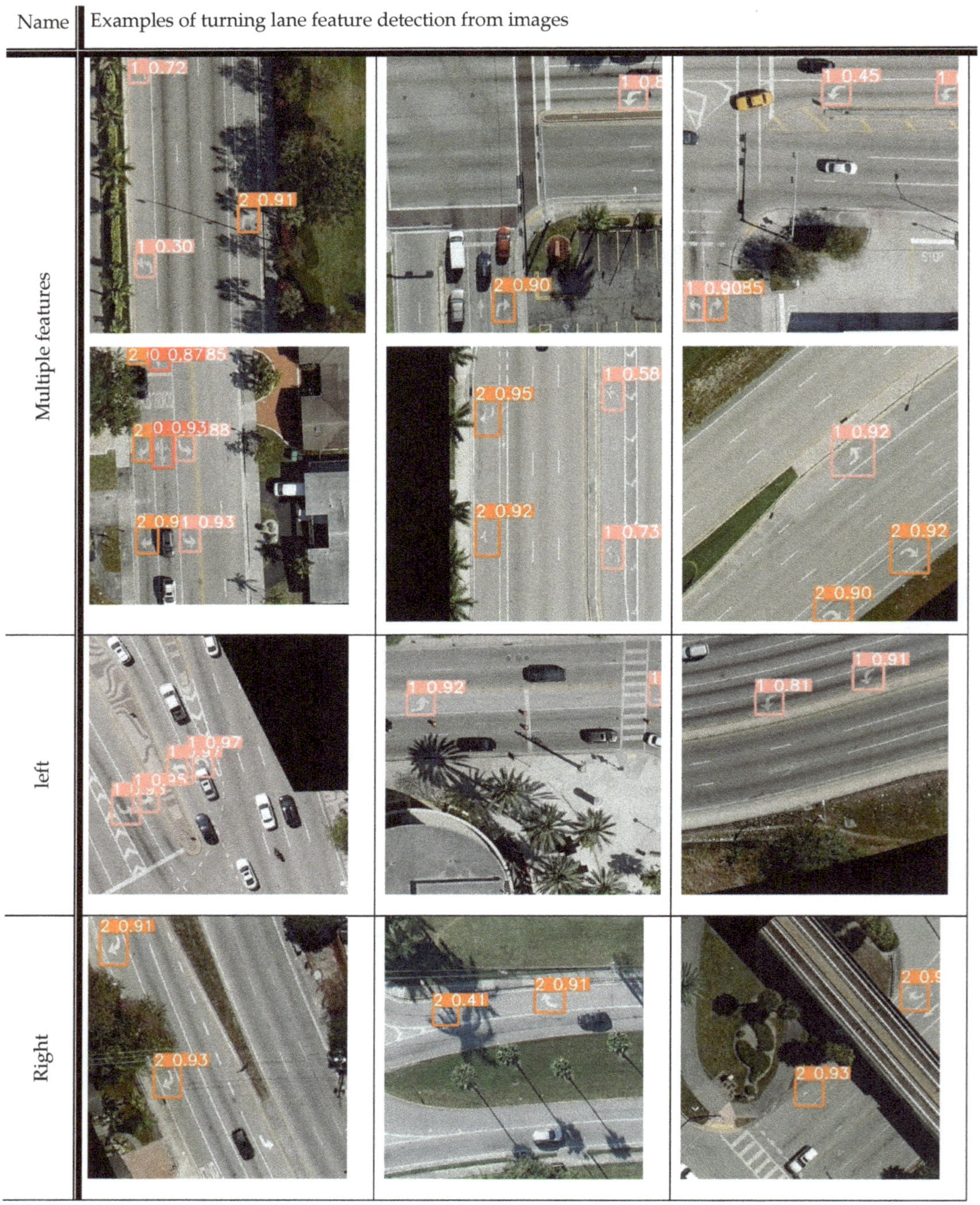

Figure 6. *Cont.*

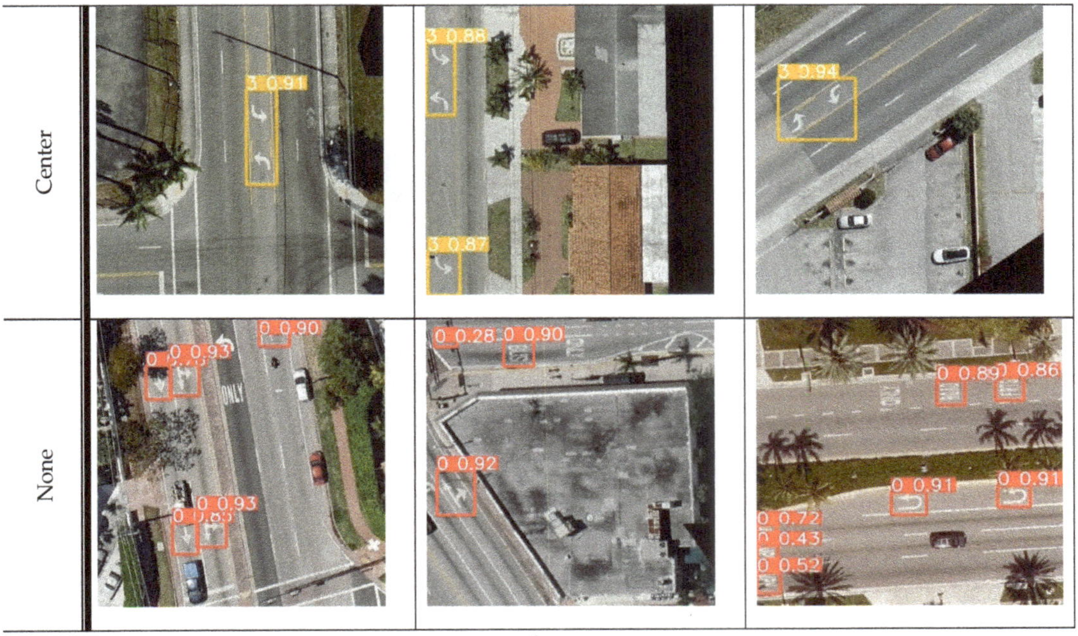

(b)

Figure 6. (a) Turning lane detection framework, and (b) turning lane detection polygons and confidence scores on images (left_only—pink, right_only—orange, center—yellow, none—red).

The detection and mapping procedure was carried out at the county level, since the detector performed very well on single photos. Figure 6a provides a summary of this procedure. The photographs in the folder with a label of "masked images" were first picked out and iterated through the detector. An output file of all the identified turning lanes in that county was created once all photos had been sent to the detector. Confidence scores were included in the output file. This file was used to map turning lanes. Note that the model can detect turning lane markings from images with a resolution ranging from 1.5 ft down to 0.25 ft or higher. However, the model has not been tried on any images with resolution lower than the ones provided by the Florida APLUS system. From the observations, the model made some false detections in a few instances. These were outlined and discussed in the results. Figure 6b shows some examples of the detected features.

4.7. Post-Processing

Finally, after applying the model to the obtained aerial images in Leon County, the total number of observed left, right, and center detections in Leon County was found to be 4795 using the model. Redundant detections caused by the overlapping distance on image were removed at the post-processing stage. The turning configurations for state and local roadways can also be classified into groups depending on analytical objectives. The filter's non-maximum suppression was applied, which kept the detection that overlapped and had the highest confidence level. Therefore, all the detected turning lane markings with over 10% overlap with lower confidence levels were removed. The detected features were converted from polygon shapefiles to point shapefiles for analysis.

5. Results and Discussion

The model's performance was assessed during the model training using precision, recall, and F1 scores. Afterwards, Leon County was utilized as the county where ground truth (GT) data were collected, and the developed model's performances were assessed with respect to completeness, correctness, quality, and F1 score. The GT dataset of turning lane features on Leon County roadways were used as a proof of concept. After detection of left, right and center features, a total of 4795 detections were observed, where 3101 were classified as left, 1478 were classified as right, and 216 were classified as center lane features. After visual inspection, a total of 5515 visible turning lane markings for left, right, and center lane features were collected as a GT dataset using the masked photographs as the background. Note that each turning lane may exhibit a range of 1–6 features, such as left only or right only markings. Consequently, a single lane may contain multiple consecutive turning features aligned in a straight line, and each lane may consist of several features that collectively define that lane. However, to perform a robust analysis, the assessment was carried out by directly comparing individual turning features on the aerial images. Figure 7 shows the GT dataset and detected turning lane markings in Leon County. Although some of the turning lane markings, especially the left only marking, were missed by the model, the overall performance of the model was reliable. This is mainly because of various reasons such as occlusions, faded markings, shadows, poor image resolution, and variety of the pavement marking design.

As noted, the suggested model identified turning lane markers with a minimum confidence score of 10%. For the purposes of this case study, the model identified turning lane markers (M) on the local roads in Leon County were retrieved. On the GT, a similar location-based selection methodology was used. The suggested model's performance was assessed by examining the points that were discovered within the polygons and vice versa at various confidence levels of 90%, 75%, 50%, 25%, and 10%, respectively (Table 2). A confusion matrix was used to visualize the model's performance against the GT dataset of Leon County (Figure 8a).

This study's major goal was to assess the accuracy and performance of the proposed model's predictions and contrast them with a ground truth dataset. Separate evaluation analysis was performed using each "left_only", "right_only", and "center" detections of the developed model. The accuracy and performance on turning lane model were assessed after testing the model's correctness (precision) and completeness (recall) using a complete ground truth dataset and measuring the F1 score. The F1 score calculates the harmonic mean using the precision and recall values. It is the appropriate evaluation metric when dealing with imbalanced datasets. It is crucial in object detection tasks where missing actual objects is more detrimental than incorrectly classifying background regions as objects.

Finally, the suggested model's performance was assessed using the criteria of completeness (recall), correctness (precision), quality (intersection over union), and F1. The model's performance was also visualized using a circus plot [44] in Figure 8b. These criteria were initially utilized in [45,46] for the goal of highway extraction, and they are now often used for performance evaluation of the related models [47,48]. The following selection criteria are necessary to determine the performance evaluation metrics:

i. GT: Number of turning lane polygons (ground truth),
ii. M: Number of model-detected turning lane points
iii. False Negative (FN): Number of GT turning lane polygon without M turning lane point
iv. False Positive (FP): Number of M turning lane points not found within GT turning lane polygon
v. True Positive (TP): Number of M turning lane points within GT turning lane polygon

Performance evaluation metrics:
Completeness = $\frac{GT-FN}{GT} * 100\%$, true detection rate among GT turning lane (recall)
Correctness = $\frac{M-FP}{M} * 100\%$, True detection rate among M turning lane (precision)
Quality = $\frac{GT-FN}{GT+FP} * 100\%$, True detection among M turning lane plus the undetected GT turning lane (Intersection over Union: IoU)

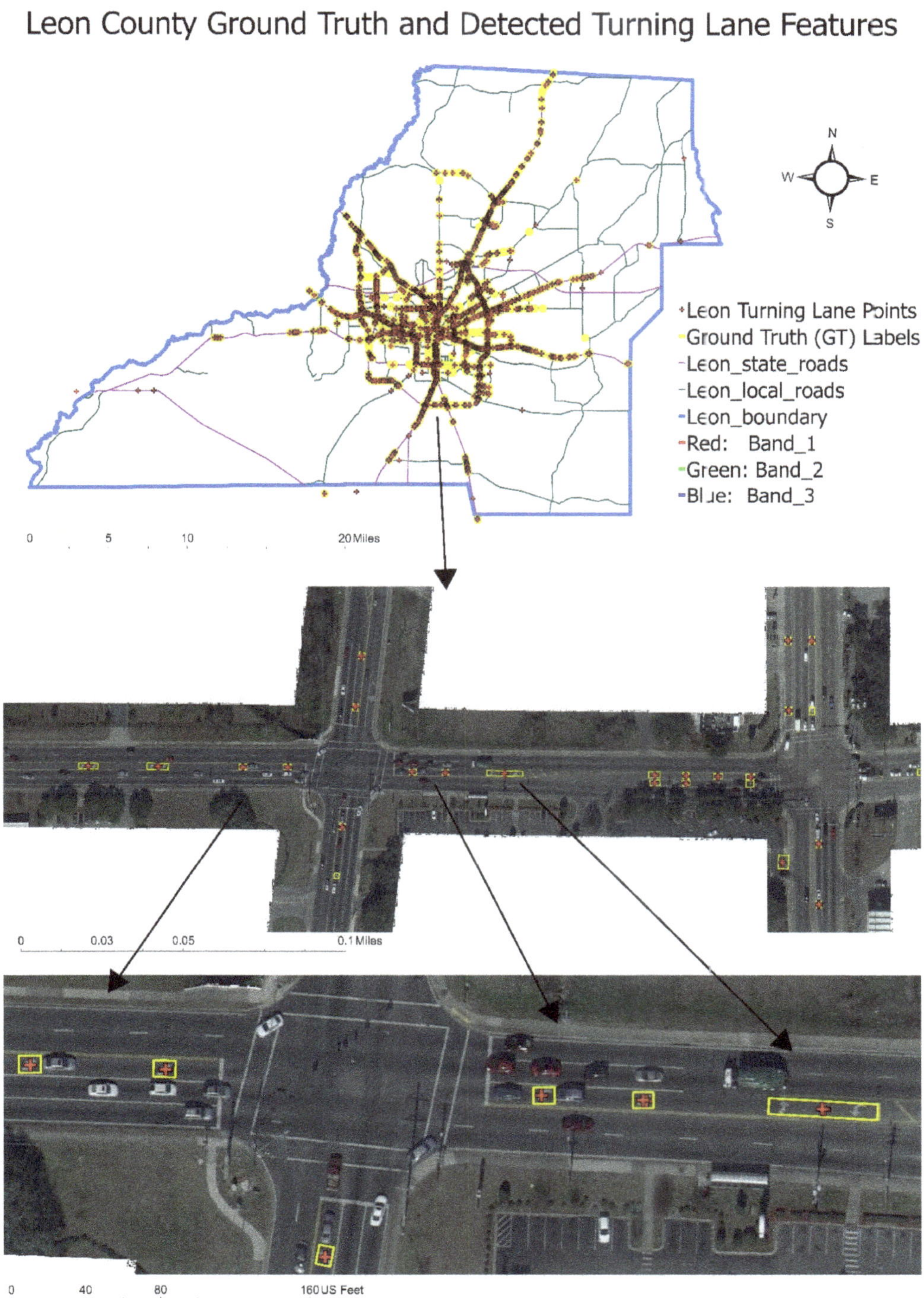

Figure 7. Manually labeled ground truth turning lane markings (GT) and detected turning lane markings in Leon County, Florida.

Based on the findings, we observe that this automated turning lane detection and mapping model can on average detect and map 57% of the turning lanes with 99% precision at 75% confidence level and an F1 score of 71%. At a lower confidence level of 25%, it can detect ~80% of the turning lanes with 96% precision and an F1 score of 87%. At a 75% confidence level, the average quality of the model's detection is 58% whereas, the average quality of the model is also 77% at 25% confidence level.

Higher accuracy was achieved at low confidence levels, since there is a higher recall and more room is given to increase the number of detections. That is, from the observations, detected turning lane markings that had occlusions from vehicle or trees, shadows, and faded markings generally had lower confidence levels. Therefore, reducing the confidence level threshold adds these detected features to the total number of detections. The new detections allowed into the pool for valuation relatively includes more true positives, fewer or zero false positives, and fewer or zero false negatives. With the increase in the number of true positives as confidence decreases, the accuracy of the model increases, since the accuracy is described based on the relationship between the number of true positives and the total number of detections. It can also be observed that the poor distinctiveness of the detection features affected detection performance. As stated earlier, the observed difference between the left only and right only turning markings, which is just a lateral inversion of the other marking, made it less unique and therefore resulted in lower detection confidence levels. When the left turn is flipped horizontally, it becomes a right turn and vice versa. On the other hand, the center lane, which was trained using a relatively distinct shape, recorded better detection results than the left only and right only lanes.

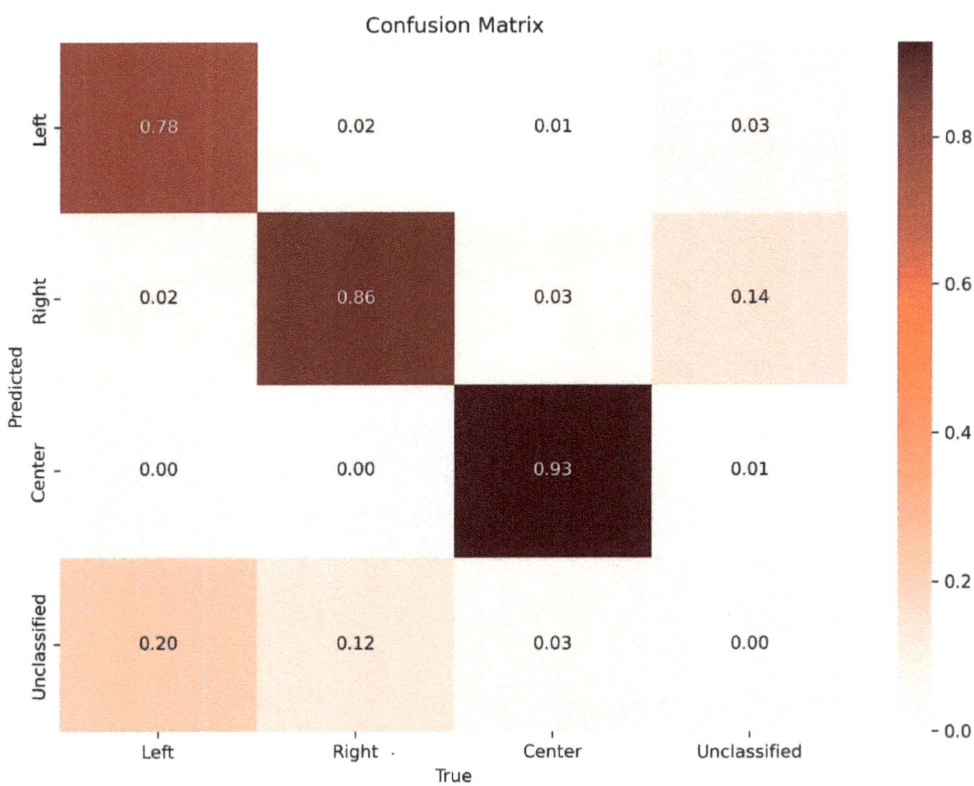

(a)

Figure 8. *Cont.*

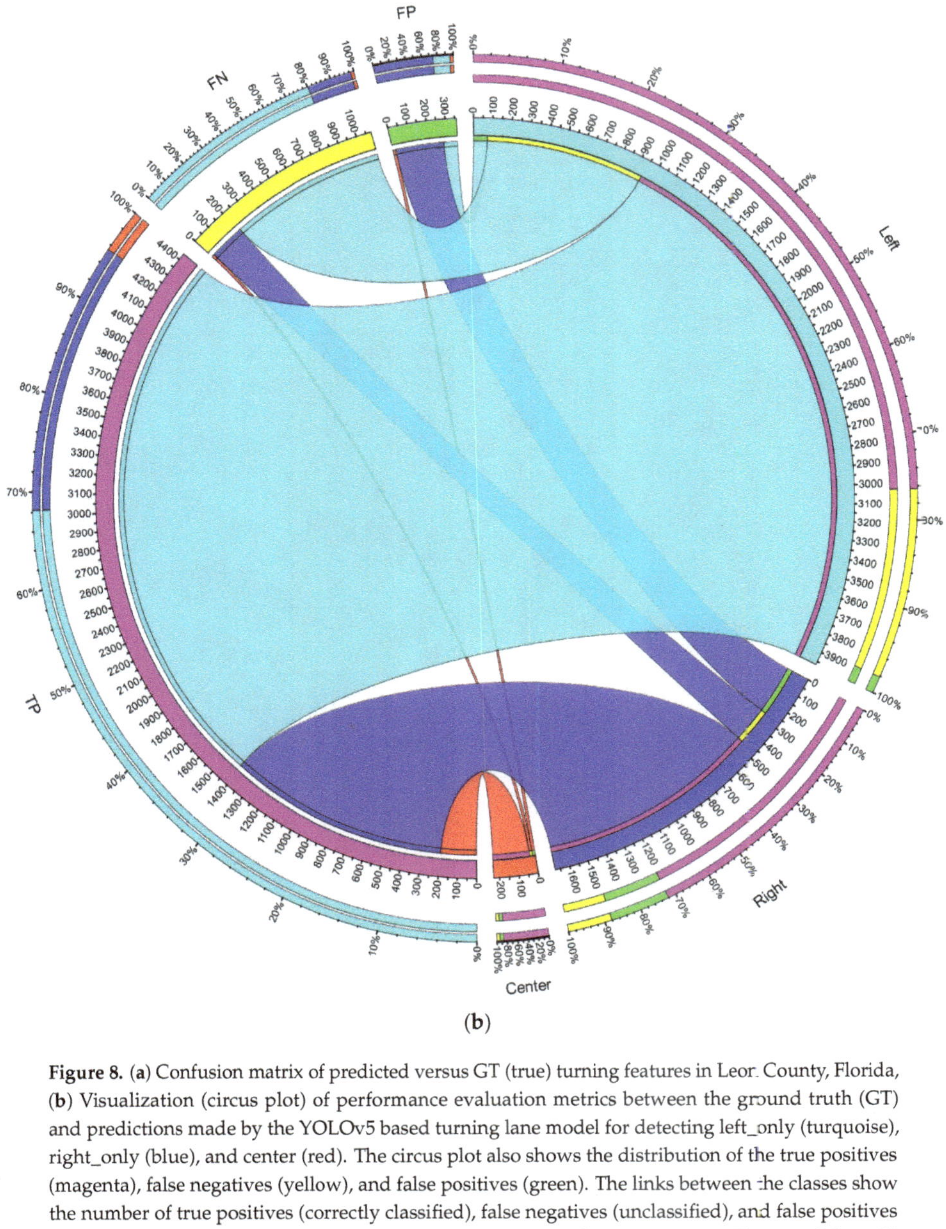

(**b**)

Figure 8. (**a**) Confusion matrix of predicted versus GT (true) turning features in Leon County, Florida, (**b**) Visualization (circus plot) of performance evaluation metrics between the ground truth (GT) and predictions made by the YOLOv5 based turning lane model for detecting left_only (turquoise), right_only (blue), and center (red). The circus plot also shows the distribution of the true positives (magenta), false negatives (yellow), and false positives (green). The links between the classes show the number of true positives (correctly classified), false negatives (unclassified), and false positives (misclassified) in each class; the thickness of the links describe their percentages. The size of the radii of the inner segments depicts the total value of the fields in ascending order. The outer concentric bars depict the percentages of the values in descending order. From the plot, about 72% of right only detections are true positives while about 18% are false negatives. Also, over 76% of left_only are true positives.

The detected turning lanes were classified under different confidence levels. The final list is shown in Table 2. The extracted roadway geometry data can be integrated with crash and traffic data, especially at intersections, to advise policy makers and roadway users.

That is, they can be used for a variety of purposes such as identifying those markings that are old and invisible, comparing the turning lane locations with other geometric features like crosswalks and school zones, and analyzing the crashes occurring around the turning lanes at intersections

Table 2. Model performance evaluations.

Leon County Ground Truth Comparison Analysis								
Turning Lane Model								
	left_only: GT = 3891							
Confidence (%)	M	TP	FP	FN	Completeness (%)	Correctness (%)	Quality (%)	F1 score (%)
90	251	251	0	3640	6.45	100.00	6.45	12.12
75	1392	1383	9	2508	35.54	99.35	35.46	52.36
50	2146	2125	21	1766	54.61	99.02	54.32	70.40
25	2683	2648	35	1243	68.05	98.70	67.45	80.56
10	3101	3026	75	865	77.77	97.58	76.30	86.56
	right_only: GT = 1409							
Confidence (%)	M	TP	FP	FN	Completeness (%)	Correctness (%)	Quality (%)	F1 score (%)
90	197	196	1	1213	13.91	99.49	13.90	24.41
75	887	871	16	538	61.82	98.20	61.12	75.87
50	1108	1066	42	343	75.66	96.21	73.47	84.70
25	1262	1153	109	256	81.83	91.36	75.96	86.33
10	1478	1210	268	199	85.88	81.87	72.15	83.82
	center: GT = 215							
Confidence (%)	M	TP	FP	FN	Completeness (%)	Correctness (%)	Quality (%)	F1 score (%)
90	13	13	0	202	6.05	100.00	6.05	11.40
75	162	160	2	55	74.42	98.77	73.73	84.88
50	191	188	3	27	87.44	98.43	86.24	92.61
25	200	195	5	20	90.70	97.50	88.64	93.98
10	216	200	16	15	93.02	92.59	86.58	92.81

6. Conclusions and Future Work

This study investigates the utilization of computer vision tools for roadway geometry extraction focusing on Florida turning lanes as a proof of concept. This is a creative approach that uses computer vision technology to possibly replace labor- and error-intensive traditional manual inventory. The created system can extract recognizable turning marks from images using high quality images. By removing the requirement for a human inventory procedure and improving highway geometry data quality by removing mistakes from manual data entry, the findings will assist stakeholders in saving money. The benefits of such roadway data extraction from imagery for transportation agencies are numerous and include identifying markings that are outdated and invisible, comparing the locations of turning lanes with other geometric features like crosswalks, school zones, and analyzing crashes that take place close to these locations.

However, the study also identifies notable limitations and offers recommendations for future research. Challenges arise from aerial images of roadways obstructed by tree canopies, limiting the identification of turning lane markings. Moving forward, the developed model can be integrated into roadway geometry inventory datasets, such as those used in the HSM and the Model Inventory of Roadway Elements (MIRE), to identify and rectify outdated or missing lane markings. Future research endeavors will also focus on refining and expanding the capabilities of the model to detect and extract additional roadway geometric features. Additionally, there are plans to integrate the extracted left only, right only, and center lanes with crash data, traffic data, and demographic information for a more comprehensive analysis.

Author Contributions: Study conception and design: R.B.A., M.K., E.E.O., R.M., S.Y.T., M.A.D. and T.S.; Data collection: R.B.A., M.K. and E.E.O.; Analysis and interpretation of results; Manuscript preparation: R.B.A. and E.E.O. All authors have read and agreed to the published version of the manuscript.

Funding: This research was funded by the State of Florida Department of Transportation (DOT) grant BED30-977-02 and The APC was waived for Authors.

Data Availability Statement: Data from this study are unavailable due to privacy or ethical restrictions. FDOT plans to publish the data on their website and make it available to the public through their online sources including the project final report and the associated deliverables.

Acknowledgments: The contents of this paper and discussion represent the authors' opinions and do not reflect the official views of the Florida Department of Transportation.

Conflicts of Interest: The authors declare no conflict of interest.

References

1. Jalayer, M.; Hu, S.; Zhou, H.; Turochy, R.E. Evaluation of Geo-Tagged Photo and Video Logging Methods to Collect Geospatial Highway Inventory Data. *Pap. Appl. Geogr.* **2015**, *1*, 50–58. [CrossRef]
2. Jalayer, M.; Zhou, H.; Gong, J.; Hu, S.; Grinter, M. A comprehensive assessment of highway inventory data collection methods. *J. Transp. Res. Forum* **2014**, *53*, 73–92. [CrossRef]
3. Shamayleh, H.; Khattak, A. Utilization of LiDAR technology for highway inventory. In Proceedings of the 2003 Mid-Continent Transportation Research Symposium, Ames, Iowa, 21–22 August 2003.
4. Alzraiee, H.; Leal Ruiz, A.; Sprotte, R. Detecting of pavement marking defects using faster R-CNN. *J. Perform. Constr. Facil.* **2021**, *35*, 04021035. [CrossRef]
5. Gong, J.; Zhou, H.; Gordon, C.; Jalayer, M. Mobile terrestrial laser scanning for highway inventory data collection. In *Computing in Civil Engineering*; ASCE: Reston, VA, USA, 2012; pp. 545–552.
6. Zhou, H.; Jalayer, M.; Gong, J.; Hu, S.; Grinter, M. *Investigation of Methods and Approaches for Collecting and Recording Highway Inventory Data*; FHWA-ICT-13-022; Illinois Center for Transportation: Rantoul, IL, USA, 2013.
7. Antwi, R.B.; Takyi, S.; Karaer, A.; Ozguven, E.E.; Moses, R.; Dulebenets, M.A.; Sando, T. Detecting School Zones on Florida's Public Roadways Using Aerial Images and Artificial Intelligence (AI2). *Transp. Res. Rec.* **2023**, *2678*, 622–636. [CrossRef]
8. Carlson, P.J.; Park, E.S.; Andersen, C.K. Benefits of pavement markings: A renewed perspective based on recent and ongoing research. *Transp. Res. Rec.* **2009**, *2107*, 59–68. [CrossRef]
9. Cho, Y.; Kabassi, K.; Pyeon, J.H.; Choi, K.; Wang, C.; Norton, T. Effectiveness study of methods for removing temporary pavement markings in roadway construction zones. *J. Constr. Eng. Manag.* **2013**, *139*, 257–266. [CrossRef]
10. Cheng, W.; Luo, H.; Yang, W.; Yu, L.; Li, W. Structure-aware network for lane marker extraction with dynamic vision sensor. *arXiv* **2020**, arXiv:2008.06204.
11. Lee, S.; Kim, J.; Shin Yoon, J.; Shin, S.; Bailo, O.; Kim, N.; So Kweon, I. Vpgnet: Vanishing point guided network for lane and road marking detection and recognition. In Proceedings of the IEEE International Conference on Computer Vision, Venice, Italy, 29 October 2017; pp. 1947–1955.
12. Li, J.; Mei, X.; Prokhorov, D.; Tao, D. Deep neural network for structural prediction and lane detection in traffic scene. *IEEE Trans. Neural Netw. Learn. Syst.* **2016**, *28*, 690–703. [CrossRef] [PubMed]
13. He, B.; Ai, R.; Yan, Y.; Lang, X. Accurate and robust lane detection based on dual-view convolutional neutral network. In Proceedings of the 2016 IEEE Intelligent Vehicles Symposium (IV), Gothenburg, Sweden, 19–22 June 2016; IEEE: Piscataway, NJ, USA, 2016; pp. 1041–1046.
14. Huval, B.; Wang, T.; Tandon, S.; Kiske, J.; Song, W.; Pazhayampallil, J.; Ng, A.Y. An empirical evaluation of deep learning on highway driving. *arXiv* **2015**, arXiv:1504.01716.
15. Campbell, A.; Both, A.; Sun, Q.C. Detecting and mapping traffic signs from Google Street View images using deep learning and GIS. *Comput. Environ. Urban Syst.* **2019**, *77*, 101350. [CrossRef]
16. Aghdam, H.H.; Heravi, E.J.; Puig, D. A practical approach for detection and classification of traffic signs using convolutional neural networks. *Robot. Auton. Syst.* **2016**, *84*, 97–112. [CrossRef]
17. Balali, V.; Ashouri Rad, A.; Golparvar-Fard, M. Detection, classification, and mapping of US traffic signs using google street view images for roadway inventory management. *Vis. Eng.* **2015**, *3*, 15. [CrossRef]

18. Zhang, X.; Yuan, Y.; Wang, Q. ROI-wise Reverse Reweighting Network for Road Marking Detection. In Proceedings of the BMVC Newcastle upon Tyne, UK, 3–6 September 2018; p. 219.
19. Tong, Z.; Gao, J.; Zhang, H. Recognition, location, measurement, and 3D reconstruction of concealed cracks using convolutional neural networks. *Constr. Build. Mater.* **2017**, *146*, 775–787. [CrossRef]
20. Panboonyuen, T.; Jitkajornwanich, K.; Lawawirojwong, S.; Srestasathiern, P.; Vateekul, P. Road segmentation of remotely-sensed images using deep convolutional neural networks with landscape metrics and conditional random fields. *Remote Sens.* **2017**, *9* 680. [CrossRef]
21. Uçar, A.; Demir, Y.; Güzeliş, C. Object recognition and detection with deep learning for autonomous driving applications *Simulation* **2017**, *93*, 759–769. [CrossRef]
22. Tang, T.; Zhou, S.; Deng, Z.; Zou, H.; Lei, L. Vehicle detection in aerial images based on region convolutional neural networks and hard negative example mining. *Sensors* **2017**, *17*, 336. [CrossRef] [PubMed]
23. Vattapparamban, E.; Güvenç, I.; Yurekli, A.I.; Akkaya, K.; Uluağaç, S. Drones for smart cities: Issues in cybersecurity, privacy, and public safety. In Proceedings of the 2016 International Wireless Communications and Mobile Computing Conference (IWCMC) Paphos, Cyprus, 5–9 September 2016; IEEE: Piscataway, NJ, USA, 2016; pp. 216–221.
24. Xie, X.; Yang, W.; Cao, G.; Yang, J.; Zhao, Z.; Chen, S.; Shi, G. Real-time vehicle detection from UAV imagery. In Proceedings of the 2018 IEEE Fourth International Conference on Multimedia Big Data (BigMM), Xi'an, China, 13–16 September 2018; IEEE Piscataway, NJ, USA, 2018; pp. 1–5.
25. Xu, Y.; Yu, G.; Wang, Y.; Wu, X.; Ma, Y. Car detection from low-altitude UAV imagery with the faster R-CNN. *J. Adv. Transp.* **2017** *2017*, 2823617. [CrossRef]
26. Girshick, R. Fast r-cnn. In Proceedings of the IEEE International Conference on Computer Vision, Santiago, Chile, 7–13 December 2015; pp. 1440–1448.
27. Ren, S.; He, K.; Girshick, R.; Sun, J. Faster r-cnn: Towards real-time object detection with region proposal networks. *Adv. Neural Inf. Process. Syst.* **2015**, *28*, 1137–1149. [CrossRef] [PubMed]
28. Kim, E.J.; Park, H.C.; Ham, S.W.; Kho, S.Y.; Kim, D.K. Extracting vehicle trajectories using unmanned aerial vehicles in congested traffic conditions. *J. Adv. Transp.* **2019**, *2019*, 9060797. [CrossRef]
29. Foucher, P.; Sebsadji, Y.; Tarel, J.P.; Charbonnier, P.; Nicolle, P. Detection and recognition of urban road markings using images. In Proceedings of the 2011 14th International IEEE Conference on Intelligent Transportation Systems (ITSC), Washington, DC, USA 5–7 October 2011; IEEE: Piscataway, NJ, USA, 2011; pp. 1747–1752.
30. Azimi, S.M.; Fischer, P.; Körner, M.; Reinartz, P. Aerial LaneNet: Lane-marking semantic segmentation in aerial imagery using wavelet-enhanced cost-sensitive symmetric fully convolutional neural networks. *IEEE Trans. Geosci. Remote Sens.* **2018**, *57* 2920–2938. [CrossRef]
31. Yan, L.; Liu, H.; Tan, J.; Li, Z.; Xie, H.; Chen, C. Scan line based road marking extraction from mobile LiDAR point clouds. *Sensors* **2016**, *16*, 903. [CrossRef] [PubMed]
32. Zhang, D.; Xu, X.; Lin, H.; Gui, R.; Cao, M.; He, L. Automatic road-marking detection and measurement from laser-scanning 3D profile data. *Autom. Constr.* **2019**, *108*, 102957. [CrossRef]
33. Xu, S.; Wang, J.; Wu, P.; Shou, W.; Wang, X.; Chen, M. Vision-based pavement marking detection and condition assessment—A case study. *Appl. Sci.* **2021**, *11*, 3152. [CrossRef]
34. Antwi, R.B.; Takyi, S.; Kimollo, M.; Karaer, A.; Ozguven, E.E.; Moses, R.; Dulebenets, M.A.; Sando, T. Computer Vision-Based Model for Detecting Turning Lane Features on Florida's Public Roadways from Aerial Images. *Transp. Plan. Technol.* **2024** [CrossRef]
35. United States Census Bureau (US Census). Population Estimates. 2020. Available online: https://www.census.gov/quickfacts/leoncountyflorida (accessed on 20 June 2023).
36. Redmon, J.; Farhadi, A. Yolov3: An incremental improvement. *arXiv* **2018**, arXiv:1804.02767.
37. Redmon, J.; Divvala, S.; Girshick, R.; Farhadi, A. You only look once: Unified, real-time object detection. In Proceedings of the IEEE Conference on Computer Vision And Pattern Recognition, Las Vegas, NV, USA, 27–30 June 2016; pp. 779–788.
38. Redmon, J.; Farhadi, A. YOLO9000: Better, faster, stronger. In Proceedings of the IEEE Conference on Computer Vision and Pattern Recognition, Honolulu, HI, USA, 21–26 July 2017; pp. 7263–7271.
39. Bochkovskiy, A.; Wang, C.Y.; Liao, H.Y.M. Yolov4: Optimal speed and accuracy of object detection. *arXiv* **2020**, arXiv:2004.10934
40. Jocher, G. YOLOv5 by Ultralytics. Released Date. 2020, pp. 5–29. Available online: https://github.com/ultralytics/yolov5 (accessed on 30 October 2023).
41. Wang, C.Y.; Liao, H.Y.M.; Wu, Y.H.; Chen, P.Y.; Hsieh, J.W.; Yeh, I.H. CSPNet: A new backbone that can enhance learning capability of CNN. In Proceedings of the IEEE/CVF Conference on Computer Vision and Pattern Recognition Workshops, Seattle WA, USA, 14–19 June 2020; pp. 390–391.
42. Liu, K.; Tang, H.; He, S.; Yu, Q.; Xiong, Y.; Wang, N. Performance validation of YOLO variants for object detection. In Proceedings of the 2021 International Conference on Bioinformatics and Intelligent Computing, Harbin, China, 22–24 January 2021; pp. 239–243.
43. Horvat, M.; Gledec, G. A comparative study of YOLOv5 models performance for image localization and classification. In Proceedings of the 33rd Central European Conference on Information and Intelligent Systems (CECIIS), Dubrovnik, Croatia 21–23 September 2022; p. 349.

4. Krzywinski, M.; Schein, J.; Birol, I.; Connors, J.; Gascoyne, R.; Horsman, D.; Marra, M.A. Circos: An information aesthetic for comparative genomics. *Genome Res.* **2009**, *19*, 1639–1645. [CrossRef] [PubMed]
5. Wiedemann, C.; Heipke, C.; Mayer, H.; Jamet, O. Empirical evaluation of automatically extracted road axes. *Empir. Eval. Tech. Comput. Vis.* **1998**, *12*, 172–187.
6. Wiedemann, C.; Ebner, H. Automatic completion and evaluation of road networks. *Int. Arch. Photogramm. Remote Sens.* **2000**, *33*, 979–986.
7. Sun, K.; Zhang, J.; Zhang, Y. Roads and intersections extraction from high-resolution remote sensing imagery based on tensor voting under big data environment. *Wirel. Commun. Mob. Comput.* **2019**, *2019*, 6513418. [CrossRef]
8. Dai, J.; Wang, Y.; Li, W.; Zuo, Y. Automatic method for extraction of complex road intersection points from high-resolution remote sensing images based on fuzzy inference. *IEEE Access* **2020**, *8*, 39212–39224. [CrossRef]

smart cities

MDPI

Article

Video-Based Analysis of a Smart Lighting Warning System for Pedestrian Safety at Crosswalks

Margherita Pazzini *, Leonardo Cameli, Valeria Vignali, Andrea Simone and Claudio Lantieri

Department of Civil, Environmental and Material (DICAM) Engineering, University of Bologna, 40136 Bologna, Italy; leonardo.cameli2@unibo.it (L.C.); valeria.vignali@unibo.it (V.V.); andrea.simone@unibo.it (A.S.); claudio.lantieri2@unibo.it (C.L.)

* Correspondence: margherita.pazzini2@unibo.it

Highlights:

What are the main findings?

- The presence of an integrated lighting warning system near crosswalks significantly increases yielding compliance and allows drivers to respect the stopping distance before a pedestrian crossing.
- Five-month continuous monitoring of different warning lighting systems proved that the combination of lighting with in-curb fixed LED strips, orange flashing beacons, and dedicated LED lighting reduces drivers' speed compared to standard street lighting systems.

What are the implications of the main findings?

- Smart lighting systems detecting pedestrians at night, both in urban and suburban areas, are a valid solution to increase safety near pedestrian crossings. These smart systems improve drivers' vision and significantly reduce the risk of accidents between vehicles and pedestrians.
- In-curb fixed LED strips and flashing beacons increase driver attention towards pedestrians, thus making drivers stop vehicles at crosswalks, yielding to pedestrians.

Abstract: This study analyses five months of continuous monitoring of different lighting warning systems at a pedestrian crosswalk through video surveillance cameras during nighttime. Three different light signalling systems were installed near a pedestrian crossing to improve the visibility and safety of vulnerable road users: in-curb LED strips, orange flashing beacons, and asymmetric enhanced LED lighting. Seven different lighting configurations of the three systems were studied and compared with standard street lighting. The speed of vehicles for each pedestrian–driver interaction was also evaluated. This was then compared to the speed that vehicles should maintain in order to stop in time and allow pedestrians to cross the road safely. In all of the conditions studied, speeds were lower than those maintained in the five-month presence of standard street lighting (42.96 km/h). The results show that in conditions with dedicated flashing LED lighting, in-curb LED strips, and orange flashing beacons, most drivers (72%) drove at a speed that allowed the vehicle to stop safely compared to standard street lighting (10%). In addition, with this lighting configuration, the majority of vehicles (85%) stopped at pedestrian crossings, while in standard street lighting conditions only 26% of the users stopped to give way to pedestrians.

Keywords: pedestrian crosswalk; pedestrian safety; nighttime road safety; driver–pedestrian interaction; LED road lighting; road safety measures; lighting warning system

1. Introduction

In the European Union, pedestrians are involved in one out of five fatal road accidents. In 2019, 72% of fatal crashes occurred in urban areas and 20% involved pedestrians, reaching 4628 deaths [1]. This figure is considerably high when considering other vulnerable road users, such as motorcyclists, amounting to 16%, cyclists, to 9%, and mopeds, to 3%. In Italy, the number of road victims in 2021 amounted to 471 pedestrians, with a 15% increase compared to 2020 [2]. The number of fatal road accidents increased further in 2022, while remaining below pre-COVID-19 levels [3]. Data show that since 2009 almost 90% of the increase in pedestrian casualties occurred at night [4,5], probably due to high vehicle speeds. High speed is one of the main causes of road accidents involving pedestrians, especially at night in the dark. In addition, late braking, left turns on two-lane roads, pedestrian crossings, lighting poles reducing road visibility, and unprotected pedestrian areas should be mentioned. Pedestrian safety research has long recognised that darkness is a major risk factor in road crashes involving pedestrians [6–8]. Night collisions are usually more severe than day crashes since lighting conditions have an important impact on pedestrian safety. Literature and research in the field provide clear evidence of this. Schneider (2020) [9] showed that the death rate of pedestrians in dark conditions has increased steadily since the late 1970s. Since the early 1980s, the death rate has risen from about 63% to 76% in 2018 [5]. The highest risk of death for pedestrians is between 3 a.m. and 6 a.m. [10]. Night crashes at intersections have an 83% risk of being fatal in the absence of street lighting, while falling to 54% with street lighting [11]. Studies focused on how darkness affects drivers' vision, determining driving speed and the ability to detect an obstacle, thus estimating the correct stopping distance [12,13]. Sullivan and Flannagan (2001) [14] identified high speed at night as a high-risk factor for pedestrians, compounded by excessive alcohol use, both by pedestrians and motorists.

Several studies also showed that the retroreflection of road signs, correct lighting, and bio-movement significantly increase the distance from which a driver can detect a pedestrian [15]. Other studies proved that age or any visual impairment can negatively affect the ability of drivers to distinguish pedestrians at night [16]. The increase in fatal accidents at night is also closely linked to the type of infrastructure. Unsigned pedestrian crossings increase fatalities by 80.8%; 40–45 mph roads are often responsible for 54.6% of serious accidents, five-lane roads for 40.7%, urban streets for 99,7%, and high-flow roads for 81.1% [17]. Extending the curb could be the first infrastructure solution to the problem. Bella and Silvestri (2015) [18] found that a curb extension leads drivers to maintain the most appropriate speed when approaching a pedestrian crossing. More than 80% of drivers perceived curb extensions as effective and showed a greater willingness to yield to pedestrians, since the pedestrian crossing was better visible.

Speed, age, and type of infrastructure may affect drivers' vision, but it is definitely better in light conditions. The first big difference in terms of visibility on the road is between day and night [19,20]. Alogaili and Mannering (2022) [21] showed that nighttime crashes have more severe consequences for pedestrians over time. Technologies that seek to reproduce conditions of daytime visibility even at night (improved lighting, infrared pedestrian detection, etc.) may bring significant safety benefits, even if the increase in vehicle speed plays an important role in causing accidents. High speed reduces the overall ability of a driver to see and perceive [22] because they tend to focus on the centre of the road rather than the sides, from which pedestrians usually come [23]. At night, drivers do not moderate their speed to compensate for reduced visibility, while pedestrians overestimate the ability of drivers to see them [24]. In addition, the extensive use of mobile phones and other distracting devices can cause further safety problems for both drivers and pedestrians [25,26]. Proper street lighting lowers the risk of fatal crashes for pedestrians [27,28], while mitigating road crashes under different driving conditions [29]. Focusing on pedestrian crossings, several countermeasures could be taken to increase visibility so that drivers can stop in time and yield to pedestrians.

Hakkert et al. (2002) [30] added in-curb flashing lights near a pedestrian crossing. The average speed of vehicles approaching the area decreased by 2 to 5 km/h, while the number of drivers giving priority to pedestrians increased accordingly. This resulted in a significant reduction in vehicle–pedestrian crashes, as well as in the number of pedestrians crossing outside the marked pedestrian zone. Hussain et al. (2023) [31] showed that both LED light and advanced variable message signal (VMS) were useful in increasing visibility by up to 98.4% and reducing vehicle–pedestrian conflicts. Both devices were effective in motivating drivers to reduce vehicle speed in advance. Vignali et al. (2019) [32] studied the inclusion of flashing beacons near unmarked pedestrian crossings. They showed that flashing beacons increased fixations of "Yield here to pedestrian" vertical signs and improved the distance at which drivers could detect the presence of a pedestrian crossing the road. Hoye and Laureshyn (2019) [33] added the "SeeMe" light with automatic pedestrian detection to pedestrian crossing signs in residential areas. This lighting system on two-lane roads improved driver performance in areas with moderate motor vehicle volumes and speed limits of up to 50 km per hour. Shurbutt et al. (2009) [34] installed a rectangular side-mounted stutter-flash LED beacon system near a pedestrian crossing to increase the safe braking performance of vehicles toward pedestrians. Results showed a considerable increase in safety distances for a two-beacon system. Patella et al. (2020) [35] added white LED strips to evaluate drivers' speed at night. Results showed that the average travel speed was reduced by about 7.8 km/h. Studies reported that rectangular rapid flash beacons (RRFBs) installed at crosswalks improved drivers' braking performance and yielding rates at midblock crosswalks [36,37]. Goswamy et al. (2023) [38] demonstrated how rectangular rapid flashing beacons (RRFBs) reduce the severity of night crashes. Fitzpatrick and Park (2021) [28] showed that there were no substantial differences between daytime and nighttime driver-yielding performance with pedestrian hybrid beacons (PHBs), while rectangular rapid flashing beacons (RRFBs) were more effective at night.

Different light signalling systems have been extensively studied in the literature. However, few studies have analysed which configuration is the most effective for increasing the safety of weak road users at night near a pedestrian crossing. This study aims to assess the impact of integrated light warning systems on driver behaviour to identify the most effective configuration and increase the safety of vulnerable road users. The systems included a dedicated lighting vertical sign on both sides, side-mounted flashing beacons and in-curb LED strips for both warning and lighting pedestrian crossings. Seven different lighting configurations were studied and compared to current standard street lighting. Cameras installed near the pedestrian crossing monitored different lighting configurations continuously for five months. Compared to previous studies [39], continuous analysis of the lighting systems allowed users to be monitored without being aware of the experiment. The results obtained are evidence of the reliability of the integrated light warning system.

2. Materials and Methods

2.1. Pedestrian Crossing Lighting System

The pedestrian crossing chosen for the experimental system is located in Via Andrea Costa, in the Porto-Saragozza district of Bologna, Italy. This two-way road conveys the vehicles in the direction of the city avenues, and allows the flow both towards Bologna city centre and in the opposite direction, towards the nearest suburbs. In its current dual-strip configuration, the pedestrian crossing serves both pedestrians coming from and those heading to bus stops through the pedestrian safety island, as well as users swapping between the two existing pavements (Figure 1).

For the analysis of pedestrian crossings, a lighting system was introduced (Figure 2) consisting of the following:

- Movement sensor for the detection of pedestrians near the curb-side area.
- In-curb LED strips: Flexible in-curb LED strips of the Flexi-led line type, and adaptable to the desired shape, installed at both ends of the curbs near the pedestrian crossing. Two alternative activations for the LED system were studied: LED strips always on with constant light emission as the default state (fixed); LED strips lit with flashing light emission as the default state (flashing).
- Orange beacons: Made with a backlit LED panel, they were placed on both sides of the pedestrian crossing. Very-high-refraction film was applied above the panel,

showing the symbol of the pedestrian crossing [40] and two pairs of highly bright amber flashing LEDs were specularly installed on it.

- Asymmetric enhanced LED lightning: A metal pole was installed on both sides of the pedestrian crossing, six metres high from the paved road. A die-cast aluminium lighting body was mounted on it with a white LED system and asymmetric optics focused on the zebra crossing.

Figure 1. Urban context of the experimental pedestrian crosswalk (source: Google Maps).

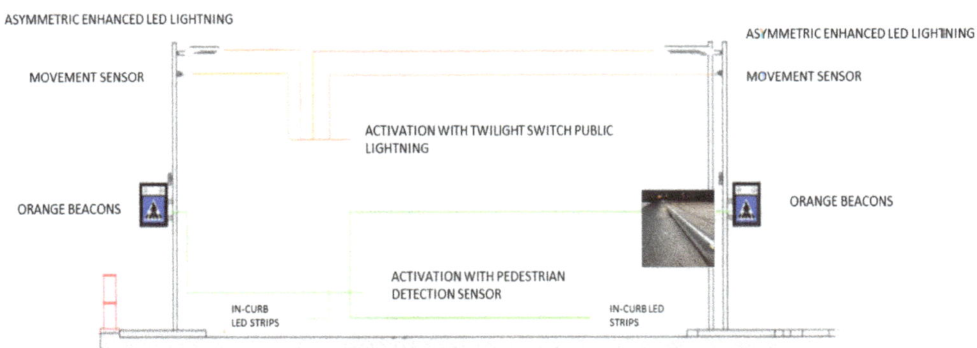

Figure 2. Section of the pedestrian crossing analysed with different lighting warning systems.

An autonomous mechanism was used as a monitoring system. The components used to operate the system were placed in a box installed on a pole of the lighting system connected to the new crossing technology. The video surveillance system consisted of two metal-optic cameras with a viewing angle of about 70 degrees, which allowed for recording colour video even in low-light conditions (Figure 3).

Through a network-switching device, surveillance cameras of the pedestrian crossing were connected to a DVR recorder with a memory capacity of about 140 GB, corresponding to seven days of video recording. The cameras and video recorder were active all night, according to the times imposed by public lighting. Both data and power were supplied and transmitted via a LAN cable.

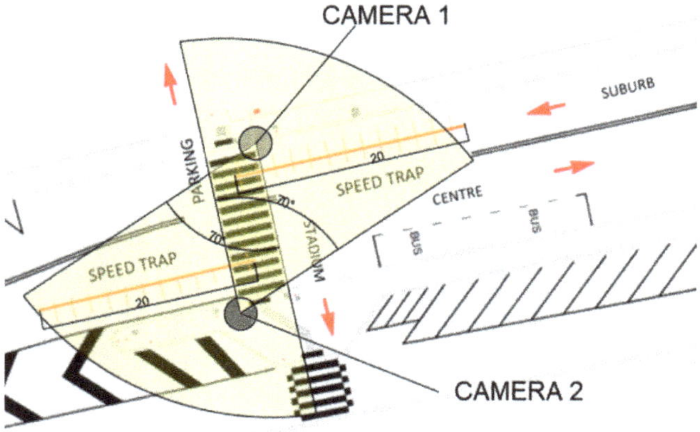

Figure 3. Crossing camera position and operating radius. The red arrows indicate the two road directions (suburb and centre), the parking area and the stadium.

The video surveillance system was placed on both sides of the pedestrian crossing to analyse the behaviour of all drivers in the seven different experimental lighting configurations of the crossing. Figure 4 shows the output of the monitoring system recorded during the night (from 7 p.m. to 6 a.m., considering seasonal fluctuations of 2 h).

Figure 4. Extracted frame showing vehicle–pedestrian interaction.

2.2. *The Integrated Lighting Warning System and the Monitoring System*

Figure 5 shows four phases of the study to evaluate different lighting warning systems. Starting from the standard street lighting configuration, phase 1 involved the evaluation of 7 different lighting configurations and the subdivision of smart lighting accordingly. In phase 2, the different systems were monitored continuously for five months. Phase 3 involved the analysis of video data collected to study vehicle speeds and braking distances. Finally, in phase 4 the 7 different lighting configurations were statistically compared to assess which was best suited to ensure pedestrian safety at night.

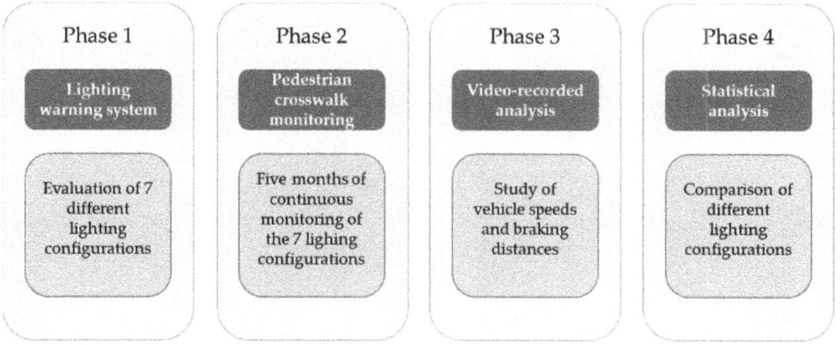

Figure 5. Steps of the study to evaluate different lighting warning systems.

Table 1 shows the seven different light configurations considered during the monitoring phase to assess the behaviour of drivers near the crossing.

Table 1. On–off status of the pedestrian crossing lighting systems in the seven conditions.

Configuration	Standard Road Lightning	Orange Beacons	Enhanced LED Lightning	In-Curb LED Strips—Fixed	In-Curb LED Strips—Flashing
0	On	Off	Off	Off	Off
1	On	On	Off	Off	Off
2	On	On	On	Off	Off
3	On	On	Off	On	Off
4	On	On	Off	Off	On
5	On	On	On	On	Off
6	On	On	On	Off	On

Equipped with a single standard streetlamp without any experimental lighting element, configuration 0 was considered the control configuration. The lighting was always active regardless of the presence of pedestrians and was the basic configuration for all subsequent experimental configurations. In configuration 1, orange beacons were added to the standard street lighting. They provided dedicated lighting with LED projectors operating only via pedestrian sensors. In configuration 2, enhanced LED lighting was added to the orange beacon system, providing an orange flashing illumination of the vertical pedestrian crossing signal. Configurations 3 and 4 reflected configuration 1 with the addition of in-curb LED strips. In configuration 3 the LED strips were fixed with constant emission (fixed). In configuration 4, the LED strips flashed (flashing). Configurations 5 and 6 resembled configuration 2 with the addition of fixed and flashing LED strips.

The first step was to define which type of pedestrian–vehicle interaction to analyse. Based on the monitoring carried out, pedestrians can be involved in different scenarios defined by the behaviour of the vehicle approaching the pedestrian crossing:

- Slowdown: once the pedestrian is detected, the vehicle slows down to allow crossing without stopping completely.
- Stop: the vehicle stopped completely to allow crossing.
- No priority: the vehicle did not stop or slow down, so the pedestrian was forced not to cross the road.

Vehicle slowing down and stopping scenarios were taken into consideration in order to obtain the following indicators:

- Speed of vehicles approaching the pedestrian crossing.
- Number of crossings, diversified by categories of users (pedestrians, cyclists).
- Percentage of correct interactions, in relation to crossing settings.

Pedestrian–vehicle interactions were monitored for five months, without considering the separate direction of vehicle flows. In order to correctly position some of the infrastructure for speed analysis, a part of the lane length was identified for each direction of travel. The time taken by vehicles to travel that section of road was measured to obtain the instantaneous speed. Considering the field of view of the camera and the results of the survey, two points on the road platform, at a distance of 20 m from the zebra crossing line, were identified for measurements. Once the reference road section was identified, the video metadata were analysed. The sampling rate used was 15.00 frames per second, equivalent to a time interval of 0.0667 s between each frame (Figure 6).

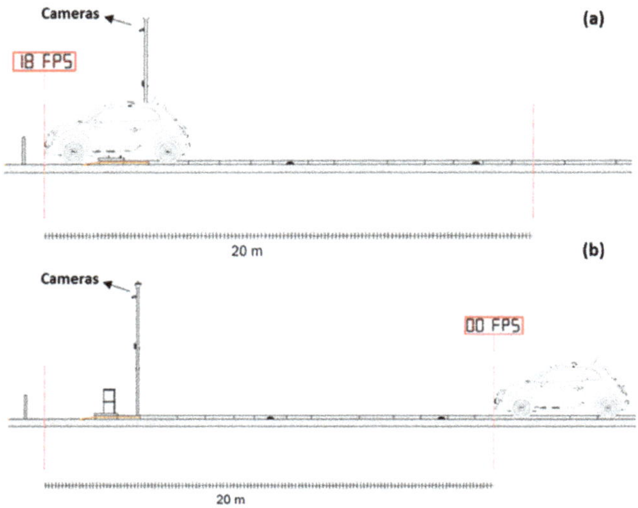

Figure 6. Detection technique for speed estimation: (**a**) entrance zone of the reference section; (**b**) exit zone of the reference section with the number of frames per second detected.

Once the time interval for each frame was identified, measurements of all pedestrian crossings with pedestrian–vehicle interaction were taken for each vehicle. The number of frames needed to cover the reference section of 20 m in the direction of travel of the vehicle was recorded. The instant speed formula was then applied, provided by the ratio between the length of the reference stroke (equal to 20 m) and the number of detected frames (fps), with the time interval between each frame equal to 0.0667 s.

The curve of the speed of approach to the pedestrian crossing was calculated using the formula of the Italian Ministerial Decree DM 2001 (MIT, 2001) [41] (1), which depends on the space covered in time τ (D_1), the braking distance (D_2), the vehicle speed at the start of braking and the final speed (V_0), final speed of the vehicle, where $V_1 = 0$ in case of stopping, the resistance forces (aerodynamics (R_a), coefficient of surface adhesion (f_l), track slope (i), gravitational acceleration (g), and vehicle mass (m)), and the total reaction time (τ) (perception, reflection, reaction, and actuation).

$$D_A = D_1 + D_2 = \frac{V_0}{3.6} \times \tau - \frac{1}{3.6^2} \int_{V_0}^{V_1} \frac{V}{g \times \left[f_l(V) \pm \frac{i}{100} \right] + \frac{R_a(V)}{m} + r_0(V)} dV \qquad [m] \qquad (1)$$

Five months of video was analysed by five independent researchers who examined the experimental crossing for all seven configurations (considering the standard configuration, 0). Every light configuration was present for 20 days. The raters were not blind to the conditions they were examining. They watched the videos and detected the speed of the vehicles approaching the reference point of 20 m. They did not participate in the choice of lighting configurations and in the final comparison of the data obtained. When the driver–pedestrian interaction met the conditions to be considered valid (i.e., the correct approach to the crossing perpendicular to the sidewalk), the type of interaction and speed were studied with the method described above.

3. Results

The analysis of the seven configurations defined the number of vehicle–driver interactions and their approach speeds. The speed of vehicles recorded at the experimental pedestrian crossing, according to the different configurations, is analysed in Table 2.

Table 2. Descriptive statistics of the speed data for the seven configurations.

Configuration	N of Interactions	Minimum Speed (km/h)	Maximum Speed (km/h)	Average Speed (km/h)		Standard Deviation	Variance
	Statistics	Statistics	Statistics	Statistics	SD	Statistics	Statistics
0	42	17.41	59.97	42.96	1.53	9.94	98.86
1	27	14.20	56.81	28.24	2.07	10.74	115.33
2	50	14.39	59.97	35.69	1.61	11.39	129.80
3	59	8.24	59.97	34.08	1.78	13.70	187.70
4	136	7.29	56.81	31.41	1.19	13.93	194.02
5	117	6.50	59.00	20.37	1.04	11.22	125.99
6	90	12.71	60.00	40.32	1.27	12.08	146.03

For each configuration, the number of interactions changed accordingly. This depended on the period of the year in which the interactions were recorded and the number of crossings that insisted on the configuration considered in the period analysed. Table 2 shows that the average speeds of configurations 1, with orange beacons (28.24 km/h), and 5, with flashing led strips (20.37 km/h), allow for a safe pedestrian crossing. A decrease in speed is evident for all configurations compared to condition 0 (standard road lighting), which has no innovative crossing elements, except for configuration 6, where the average speed is 40.32 km/h. Changes in speed data were reported for each configuration. Standard deviations show low variability in speed data between different configurations, without abnormal values. This confirms that all users, in the absence of the new lighting systems described above, were proceeding at speeds very close to the average of 42 km/h.

The speed recorded on the road under analysis was always below the 50 km/h limit. The aim of this study was to verify whether the actual speed maintained by drivers was adequate to allow safe passage for vulnerable users. The speed calculated for each pedestrian–vehicle interaction was compared to the speed vehicles should maintain to yield to pedestrians. A safe stop manoeuvre would start 20 m before the central line of the pedestrian crossing at a speed of 7 m/s (25 km/h). This condition would give a driver time to perceive the pedestrian crossing, assess the emergency manoeuvre to be performed, and safely stop within the available space. Percentages of events involving vehicles with speeds between 25 km/h and over 45 km/h analysed for each configuration are given below (Figure 7).

As shown in Figure 7, under all conditions, a higher percentage of vehicles maintain a speed of 25 km/h or even lower than the standard condition without innovative equipment. The results also show that conditions 1 (orange beacons) and 5 (enhanced LED lighting, orange beacons, and in-curb fixed LED) recorded more vehicles with speeds within the acceptable range to detect a pedestrian and to brake accordingly. From the calculation given by Formula (1), three speed classes were defined within which a driver is able to perform complete manoeuvres, partial manoeuvres, or no manoeuvres in order to stop the vehicle and let a pedestrian cross. With a speed of up to 25 km/h, a driver can perceive a user at a pedestrian crossing, assess any emergency manoeuvres to be made, and stop the vehicle safely within the available space. For speeds between 25 km/h and 35 km/h, a driver can only perceive the pedestrian crossing, assess the emergency manoeuvre to be carried out, and activate the emergency braking without completing the transitional braking phase. Finally, for speeds above 35 km/h, the driver of a vehicle is not able to perceive a user crossing the road and implement the emergency manoeuvre, since the required journey time is less than the psychophysical reaction time set at 2.10 s [18]. Figure 7 shows that, for the 25–35 km/h range, configuration 1 has the highest event rate of 34% compared to other light configurations. Although configuration 2 has a percentage of events equal to condition 6, it shows peak values in the range 35–45 km/h. In the next stage, a series of statistical tests were carried out to validate the analysed samples. As shown in Table 2, the highest speeds are in configurations 0 and 6, where the highest percentage of events occur in the range > 45 km/h (Figure 7).

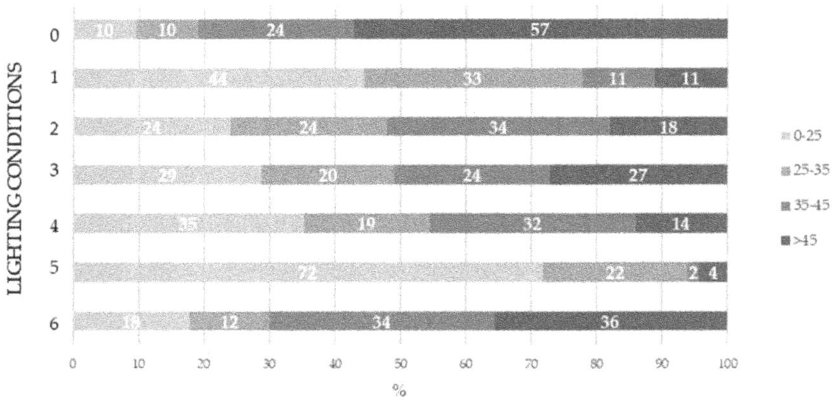

Figure 7. Percentage of events in the different speed ranges for each condition.

A one-way ANOVA was run with the seven configurations as the independent variable. This test statistically confirmed that there were significant differences among the seven configurations' average speeds. Table 3 compares all of the different configurations to find statistically significant differences that allow for the effectiveness of new experimental lighting systems in relation to speed to be validated. It has been suggested that adding a feature to the system will reduce speed. Table 3 summarises the comparison between the seven conditions, reporting the ANOVA result.

Table 3. Single-way ANOVA among the different configurations.

ANOVA	Sum of Squares	df	Mean Square	F	Sig.
Between groups	28,467.52	6	4744.58	31.22	0.000
Within groups	78,104.55	514	151.95		
Total	106,572.07	520			

The results of the ANOVA showed a significant difference between the speed of the seven configurations ($F(6, 514) = 31.22$, $p < 0.001$). To further investigate the correlation among the different configurations in terms of speed, a Scheffe post hoc analysis (Table 4) was carried out. Configuration 0 showed significant differences with almost all of the others. From configurations 0 to 5, adding new light devices, there were significant differences. By increasing the light signalling devices, the average speed decreased significantly compared to other configurations. Configuration 6, with flashing in-curb LED strips, showed an unusual trend as it did not have substantial changes compared to configuration 0. In configuration 5 (n = 117, M = 20.37, SD = 1.04), there is a significant difference in average speed compared to all of the configurations (not for configuration 1).

Table 4. Scheffe post hoc analysis.

(I) Condition		Mean Difference (I–J)	Std. Error	Sig.	95% Confidence Interval Lower Bound	Upper Bound
0	1	14.72 *	3.04	0.000	5.44	24.01
	2	7.26	2.58	0.106	−0.61	15.15
	3	8.88 *	2.49	0.008	1.23	16.48
	4	11.54 *	2.18	0.000	4.90	18.19
	5	22.58 *	2.22	0.000	15.82	29.36
	6	2.63	2.30	1.000	−4.39	9.67
1	0	−14.72 *	3.04	0.000	−24.01	−5.44
	2	−7.45	2.94	0.244	−16.44	1.53
	3	−5.84	2.86	0.881	−14.53	2.90
	4	−3.17	2.59	1.000	−11.11	4.75
	5	7.86	2.63	0.062	−0.17	15.90
	6	−12.08 *	2.70	0.000	−20.34	−3.83
2	0	−7.26	2.58	0.106	−15.15	0.61
	1	7.45	2.94	0.244	−1.53	16.44
	3	1.61	2.37	1.000	−5.62	8.85
	4	4.28	2.04	0.764	−1.94	10.50
	5	15.32 *	2.08	0.000	8.96	21.68
	6	−4.63	2.17	0.706	−11.27	2.01
3	0	−8.88 *	2.48	0.008	−16.48	−1.28
	1	5.84	2.86	0.881	−2.90	14.59
	2	−1.61	2.37	1.000	−8.85	5.62
	4	2.66	1.92	1.000	−3.20	8.53
	5	13.71 *	1.97	0.000	7.69	19.72
	6	−6.25	2.06	0.055	−12.55	0.059
4	0	−11.55 *	2.18	0.000	−18.19	−4.90
	1	3.18	2.59	1.000	−4.75	11.11
	2	−4.28	2.04	0.764	−10.50	1.95
	3	−2.66	1.92	1.000	−8.53	3.20
	5	11.04 *	1.55	0.000	6.29	15.79
	6	−8.91 *	1.67	0.000	−14.02	−3.79
5	0	−22.59 *	2.22	0.000	−29.36	−15.82
	1	−7.87	2.63	0.062	−15.90	0.17
	2	−15.32 *	2.08	0.000	−21.68	−8.96
	3	−13.71 *	1.97	0.000	−19.72	−7.70
	4	−11.04 *	1.55	0.000	−15.79	−6.29
	6	−19.95 *	1.72	0.000	−25.23	−14.68
6	0	−2.64	2.30	1.000	−9.67	4.39
	1	12.09 *	2.70	0.000	3.83	20.34
	2	4.63	2.17	0.706	−2.01	11.27
	3	6.25	2.06	0.055	−0.058	12.55
	4	8.91 *	1.68	0.000	3.79	14.02
	5	19.95 *	1.723	0.000	14.67	25.23

* The mean difference is significant for a p-value less than 0.05.

Significant changes were found among the configurations, especially between configuration 0 and the others, with a significant difference when compared to configuration 5 ($p < 0.001$), according to the ANOVA with the Scheffé post hoc analysis. Configuration 6 is distinct from a number of the other configurations in an important manner. The addition of the dedicated enhanced LED lighting and the fixed-in-curb LED device shows that the difference between configurations 0 and 5 was significant in terms of speed. This suggests that the use of all of the smart instrumentation has real effectiveness in decreasing speed.

The addition of dedicated enhanced LED lighting and a fixed in-curb LED device showed significant speed differences between conditions 0 and 5. This means that the use of smart devices had a real effectiveness in reducing speed. The use of orange beacons (configuration 1) or enhanced LED lighting, orange beacons, and a fixed in-curb LED device (configuration 5) proved to be the best for reducing speed. In any case, they did not show a significant difference in the average values. For this reason, results were interfaced with two other factors: the number of stops made or not made by drivers at the pedestrian crossing. Figure 8 shows the different types of interaction as "Stop" and "No-Stop" manoeuvres of drivers.

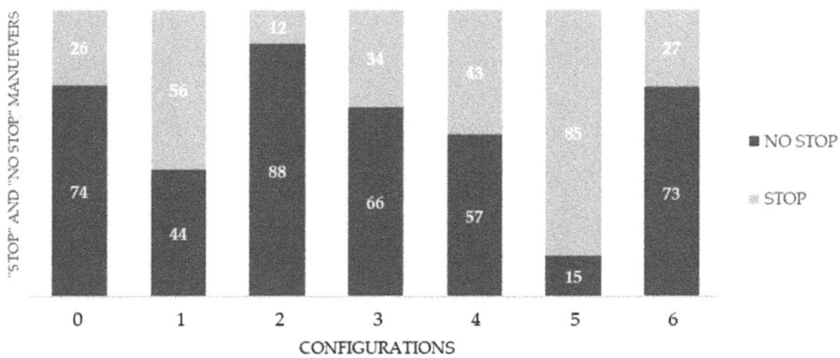

Figure 8. Percentage of vehicles' manoeuvres approaching the pedestrian crossing.

The analysis was carried out by observing the videos when drivers yielded to pedestrians crossing the road. The number of stops did not increase with the addition of innovative signalling elements near the pedestrian crossing. However, for condition 5, 85% of vehicle–pedestrian interactions allowed the pedestrian to cross, while in condition 1 only 56% were able to do so. The statistically significant speed difference previously recorded by the one-way ANOVA was due to the greater number of "stops" recorded in the interactions analysed.

4. Discussion

This study presents the analysis of data collected over five months from monitoring a pedestrian crossing. The "natural" interaction between pedestrians and vehicles was evaluated to assess which lighting configuration was the best for vulnerable road users' safety. Since neither pedestrians nor drivers were aware of participating in this study, their behaviour was not influenced by the characteristics of demand. Seven lighting configurations were analysed, including condition 0 with only standard street lighting. Configurations differed only for the activation of a different device or the specific feature of a device, such as the continuity of light emission for in-curb LED lighting.

Based on the established parameters, analysis of the collected data shows that, during the 5-month observation period, the total number of recorded crossings amounted to 519. The distribution of these crossings was not homogeneous, because both pedestrian flow and hours of darkness affected the number of observations. The average speed and standard deviation were calculated for each configuration. Speeds were always lower than that of configuration 0 (only standard road lighting and all innovative devices switched off) and the speed limit imposed on the road. The slowest speeds were reached in configuration 1 with orange beacons (n = 27, Vmean = 28.24, SD = 2.07) and configuration 5, with a fixed in-curb LED system (n = 117, Vmean = 20.37 SD = 1.04). An average speed of 40.32 km/h was recorded for configuration 6 only, similar to configuration 0. With all devices on and an in-curb flashing LED system, drivers felt safe and able to detect any passing user. For this reason, they did not slow down and maintained speeds similar to those of configuration 0. This result can also be seen in Figure 8, which shows that the percentage of vehicles stopping at the pedestrian crossing is 27%, as well as for configurations 0 and 6. The presence of lighting warning systems reduced vehicles' speed in configurations 1, 2, 3, 4, and 5, analysed compared to the standard road configuration, 0 [42,43]. Despite the different number of samples analysed for each configuration, the data show a comparable variance.

After this, it was assessed whether a driver was able to stop the vehicle at a distance of 20 m from the pedestrian crossing in relation to the driving speed maintained in the presence of a pedestrian crossing the road. Considering the braking distance, at a speed below 25 km/h a driver should be able to stop the vehicle to let a pedestrian cross the road. The results show the trend at average speeds. For each condition, there is a percentage of vehicles travelling at speeds lower than 25 km/h. This speed was higher in configurations 0 and 6.

Analysing the different lighting combinations, the use of orange beacons (configuration 1) or enhanced LED lighting, orange beacons, and fixed in-curb LED strips (configuration 5) are those able to guarantee a greater degree of road safety in terms of the speed of approach. Again, configurations 5 (72%) and 1 (44%) show a greater number of vehicles able to stop in time at the pedestrian crossing.

A one-way ANOVA test was used to validate the effectiveness of the best lighting configurations in reducing vehicle speed. The ANOVA test with a Scheffe post hoc analysis showed a significant difference among the seven configurations. This confirmed that the road lightning system influenced the speed of the drivers. This study confirmed that configuration 5 showed a significant difference with all other configurations except for configuration 1. The comparisons between these two configurations were investigated by analysing vehicle–pedestrian interactions. The percentages of "stop" and "non-stop" vehicles were evaluated. This analysis concluded that configuration 5 with 85% stops allowed more crossings than configuration 1 with only 56%.

The results obtained are in line with those of Carrese et al. [44], which show that an integrated LED lighting system generates a significant speed-reducing effect on vehicles, improving pedestrian safety at pedestrian crossings. Pena-Garcia et al. [45] also found that well-lit streets (i.e., where lighting is uniform) with improved levels of illumination make people feel safer. By inserting fixed in-curb LED strips, which are activated only when a pedestrian crosses the road, the crossing is even more visible and therefore safer.

5. Conclusions

In all cities, more attention is being paid to pedestrian safety and strengthening measures have been taken to achieve Vision Zero targets for road traffic fatalities or serious injuries [44]. In Italy, the number of pedestrians killed in road accidents was 17% (534 victims) of total road accidents in 2019 [45]. The enhanced intensity of smart street lighting increases the visibility of pedestrians at night, especially near pedestrian crossings. This type of intervention was associated with a significant reduction in night pedestrian accidents. This research analysed the impact on pedestrians of an experimental smart lighting warning system placed near a pedestrian crossing to increase road safety. The pedestrian crossing was equipped with different signalling systems controlled by pedestrian detection sensors or flashing devices: in-curb LED strips, orange beacons, and enhanced LED lighting. Two video surveillance cameras were installed on the two poles of the enhanced LED lighting system, placed at the two ends of the pedestrian crossing. With a viewing angle of about 70 degrees and a high-resolution 1920 × 1080 colour format, these cameras detected the pedestrian crossing and the part of the road platform that approaches it even in poor-visibility conditions. As streetlamps are operated by a twilight switch adjusting the voltage only at night, a 12/24 volt battery with a charger was installed in the orange beacon system with pedestrian presence detectors even during the daytime. Vehicles approaching the pedestrian crossing could see it thanks to the light beam projected by the enhanced LED lighting system and the light of the backlit panels with the symbol of the crossing. In addition to these devices, vehicles were also able to detect pedestrians entering the crossing by flashing LED orange beacons and in-curb LED strips. This articulated system appropriately warned drivers approaching a pedestrian crossing of pedestrians already engaged or about to cross it, so that they could safely adjust their speed by means of appropriate devices. For the evaluation of the best-integrated lighting system, each device was evaluated in relation to the approaching driver's behaviour. The smart lighting warning system, in combination with both orange flashing beacons and in-curb flashing LED lighting, resulted in being particularly effective in reducing the speed at night [28,38]. This research confirmed the validity of previous studies using mobile eye tracking [39,46–48]. Again, the use of enhanced LED lights, orange beacons, and in-curb LED strips (configuration 5) showed a higher percentage of permitted crossings (85%) [49]. A video (S1) of the present configuration (configuration 5) of the pedestrian crossing lighting warning system was added to the Supplementary Materials. This research also showed how lighting conditions affect vehicle performance regardless of driving or environmental conditions. The calculation of the stopping distance did not consider other factors that may affect a driver, such as pedestrian clothing, fatigue, and age of the driver. Future analyses should predict how these elements affect the stopping distance. Finally, to improve road safety, there is a need for in-depth studies on aspects closely related to pedestrian crossings, such as location, traffic volume, vehicle speed, number of pedestrians, width of the pavement, and distracting elements on the side of the road.

Supplementary Materials: The following supporting information can be downloaded at: https://www.mdpi.com/article/10.3390/smartcities7050114/s1, Video S1: The pedestrian crossing lighting warning system.

Author Contributions: Conceptualization, C.L.; methodology, C.L.; software, M.P. and L.C.; validation, V.V. and A.S.; formal analysis, L.C. and M.P.; investigation, C.L. and M.P.; data curation, M.P.; writing—original draft preparation, M.P. and L.C.; writing—review and editing, C.L. and V.V.; supervision, C.L.; funding acquisition, A.S. All authors have read and agreed to the published version of the manuscript.

Funding: This research was funded by Spoke 7 of the MOST—Sustainable Mobility National Research Center, and received funding from the European Union NextGenerationEU (PIANO NAZIONALE DI RIPRESA E RESILIENZA (PNRR)—MISSIONE 4 COMPONENTE 2, INVESTIMENTO 1.4—D.D. 1033 17/06/2022, CN00000023). This manuscript reflects only the authors' views and opinions. Neither the European Union nor the European Commission can be considered responsible for them.

Data Availability Statement: The data presented in this study is available upon request from the corresponding author.

Acknowledgments: The University of Bologna would like to thank the Municipality of Bologna for the proposal to investigate an integrated lighting warning system and the assistance in collecting data.

Conflicts of Interest: The authors declare no conflicts of interest.

References

1. European Comission. EU Road Safety Thematic Report–Pedestrians. Available online: https://road-safety.transport.ec.europa.eu/system/files/2021-07/road_safety_thematic_report_pedestrians_tc_final.pdf (accessed on 18 July 2024).
2. Istututo Nazionale di Statistica (ISTAT) (National Statistical Institute). Road Accidents in Italy, in 2021. Available online: https://www.istat.it/it/archivio/273324 (accessed on 5 June 2024).
3. International Transport Forum, Road Safety Annual Report 2022. Available online: https://www.itf-oecd.org/sites/default/files/docs/irtad-road-safety-annual-report-2022.pdf (accessed on 5 June 2024).
4. Retting, R. *Pedestrian Traffic Fatalities by State: 2019 Preliminary Data, Spotlight on Highway Safety*; Governors Highway Safety Association: Washington, DC, USA, 2020; Available online: https://www.ghsa.org/resources/Pedestrians20 (accessed on 1 July 2024).
5. Tefft, B.C.; Arnold, L.S.; Horrey, W.J. *Examining the Increase in Pedestrian Fatalities in the United States, 2009–2018 (Research Brief)*; AAA Foundation for Traffic Safety: Washington, DC, USA, 2021.
6. Sullivan, J.M.; Flannagan, M.J. The role of ambient light level in fatal crashes: Inferences from daylight saving time transitions. *Accid. Anal. Prev.* **2002**, *34*, 487–498. [CrossRef]
7. Campbell, B.J.; Zegeer, C.V.; Huang, H.H.; Cynecki, M.J. A Review of Pedestrian Safety Research in the United States and Abroad. Federal Highway Administration. Report No. FHWA-RD-03-042. 2004. Available online: https://www.fhwa.dot.gov/publications/research/safety/pedbike/03042/03042.pdf (accessed on 18 July 2024).
8. Zegeer, C.; Bushell, M. Pedestrian crash trends and potential countermeasures from around the world. *Accid. Anal. Prev.* **2012**, *44*, 3–11. [CrossRef]

9. Schneider, R.J. United States Pedestrian Fatality Trends, 1977 to 2016. *Transp. Res. Rec.* **2020**, *2674*, 1–15. [CrossRef]
10. Chang, D. *National Pedestrian Crash Report*; Report, No. DOT HS 810 968; National Highway Traffic Safety Administration: Washington, DC, USA, 2008.
11. Siddiqui, N.A.; Chu, X.; Guttenplan, M. Crossing locations, light conditions, and pedestrian injury severity. *Transp. Res. Rec.* **2006**, *1982*, 141–149. [CrossRef]
12. Sanders, R.L.; Schneider, R.J.; Proulx, F.R. Pedestrian fatalities in darkness: What do we know, and what can be done? *Transp. Policy* **2022**, *120*, 23–39. [CrossRef]
13. Subramanian, L.; O'Neal, E.E.; Mallaro, S.; Williams, B.; Sherony, R.; Plumert, J.M.; Kearney, J.K. A Comparison of Daytime and Nighttime Pedestrian Road-Crossing Behavior Using an Immersive Virtual Environment. Traffic Injury Prevention. 2022. Available online: https://www.tandfonline.com/action/journalInformation?journalCode=gcpi20 (accessed on 5 June 2024).
14. Sullivan, J.M.; Flannagan, M.J. *Characteristics of Pedestrian Risk in the Darkness*; UMTRI-2001-33; The University of Michigan Transportation Research Institute: Ann Arbor, MI, USA, 2001.
15. Tyrrell, R.A.; Wood, J.M.; Owens, D.A.; Borzendowski, S.W.; Sewall, A.S. The conspicuity of pedestrians at night: A review. *Clin. Exp. Optom.* **2016**, *99*, 425–434. [CrossRef]
16. Wood, J.A.; Lacherez, P.; Tyrrell, R.A. Seeing pedestrians at night: Effect of driver age and visual abilities. *Ophthal. Physiol. Opt.* **2014**, *34*, 452–458. [CrossRef]
17. Ferenchak, N.N.; Abadi, M.G. Nighttime pedestrian fatalities: A comprehensive examination of infrastructure, user, vehicle, and situational factors. *J. Saf. Res.* **2021**, *79*, 14–25. [CrossRef]
18. Bella, F.; Silvestri, M. Effects of safety measures on driver's speed behavior at pedestrian crossings. *Accid. Anal. Prev.* **2015**, *83*, 111–124. [CrossRef]
19. Behnood, A.; Mannering, F. Time-of-day variations and temporal instability of factors affecting injury severities in large-truck crashes. *Anal. Methods Accid. Res.* **2019**, *23*, 100102. [CrossRef]
20. Song, L.; Li, Y.; Fan, W.; Liy, P. Modeling pedestrian-injury severities in pedestrian-vehicle crashes considering spatiotemporal patterns: Insights from different hierarchical Bayesian random-effects models. *Traffic Inj. Prev.* **2021**, *22*, 524–529. [CrossRef]
21. Alogaili, A.; Mannering, F. Differences between day and night pedestrian-injury severities: Accounting for temporal and unobserved effects in prediction. *Anal. Methods Accid. Res.* **2022**, *33*, 100201. [CrossRef]
22. Rosén, E.; Stigson, H.; Sander, U. Literature review of pedestrian fatality risk as a function of car impact speed. *Accid. Anal. Prev.* **2011**, *43*, 25–33. [CrossRef]
23. Green, M. *Roadway Human Factors: From Science to Application*; Lawyers and Judges Publishing Company: Tucson, AZ, USA, 2020.
24. Tyrrell, R.; Brooks, J.; Wood, J.; Carberry, T. Nighttime Conspicuity from the Pedestrian's Perspective. In Proceedings of the 83rd Annual Meeting of the Transportation Research Board, Washington, DC, USA, 11–15 January 2004.
25. Schwebel, D.C.; Stavrinos, D.; Byington, K.W.; Davis, T.; O'Neal, E.E.; de Jong, D. Distraction and pedestrian safety: How talking on the phone, texting, and listening to music impact crossing the street. *Accid. Anal. Prev.* **2012**, *45*, 266–271. [CrossRef]
26. Hatfield, J.; Murphy, S. The effect of mobile phone use on pedestrian crossing behavior at signalized and signalized intersections. *Accid. Anal. Prev.* **2007**, *39*, 197–205. [CrossRef]
27. Wanvik, O. Effects of road lighting: An analysis based on Dutch accident statistics 1987–2006. *Accid. Anal. Prev.* **2009**, *41*, 123–128. [CrossRef]
28. Fitzpatrick, K.; Park, E. Nighttime effectiveness of the pedestrian hybrid beacon, rectangular rapid flashing beacon, and LED-embedded crossing sign. *J. Saf. Res.* **2021**, *79*, 273–286. [CrossRef]
29. Fhwa PedSAFE. *Pedestrian Safety Guide and Countermeasure Selection System*; Federal Highway Administration: Washington, DC, USA, 2013; Available online: http://www.pedbikesafe.org/pedsafe/countermeasures_detail.cfm?CM_NUM=8. (accessed on 5 June 2024).
30. Hakkert, A.S.; Gitelman, V.; Ben-Shabat, E. An evaluation of crosswalk warning systems: Effects on pedestrian and vehicle behaviour. *Transp. Res. Part F* **2002**, *5*, 275–292. [CrossRef]
31. Hussain, Q.; Alhajyaseen, W.K.M.; Kharbeche, M.; Almallah, M. Safer pedestrian crossing facilities on low-speed roads: Comparison of innovative treatments. *Accid. Anal. Prev.* **2023**, *180*, 106908. [CrossRef]
32. Vignali, V.; Bichicchi, A.; Simone, A.; Lantieri, C.; Dondi, G.; Costa, M. Road sign vision and driver behaviour in work zones. *Transp. Res. Part F Traffic Psychol. Behav.* **2019**, *60*, 474–484. [CrossRef]
33. Høye, A.; Laureshyn, A. SeeMe at the crosswalk: Before-after study of a pedestrian crosswalk warning system. *Transp. Res. Part F Traffic Psychol. Behav.* **2019**, *60*, 723–733. [CrossRef]
34. Shurbutt, J.; Van Houten, R.; Turner, S.; Huitema, B. An analysis of the effects of stutter flash LED beacons to increase yielding to pedestrians using multilane crosswalks. *Transp. Res. Rec.* **2009**, *2073*, 69–78.
35. Patella, S.; Sportiello, S.; Carrese, S.; Bella, F.; Asdrubali, F. The effect of a LED lighting crosswalk on pedestrian safety: Some experimental results. *Safety* **2020**, *6*, 20. [CrossRef]
36. Porter, B.E.; Neto, I.; Balk, I.; Jenkins, J.K. Investigating the effects of Rectangular Rapid Flash Beacons on pedestrian behavior and driver yielding on 25mph streets: A quasi-experimental field study on a university campus. *Transp. Res. Part F Traffic Psychol. Behav.* **2016**, *42*, 509–521. [CrossRef]

37. Shurbutt, J.; Van Houten, R. Effects of Yellow Rectangular Rapid-flashing Beacons on Yielding at Multilane Uncontrolled Crosswalks. Federal Highway Administration. 2010. Available online: http://www.fhwa.dot.gov/publications/research/safety/pedbike/10043/10043 (accessed on 5 June 2024).
38. Goswamy, A.; Abdel-Aty, M.; Islam, Z. Factors affecting injury severity at pedestrian crossing locations with Rectangular RAPID Flashing Beacons (RRFB) using XGBoost and random parameters discrete outcome models. *Accid. Anal. Prev.* **2023**, *181*, 106937. [CrossRef]
39. Costa, M.; Lantieri, C.; Vignali, V.; Ghasemi, N.; Simone, A. Evaluation of an integrated lighting-warning system on motorists yielding at unsignalized crosswalks during nighttime. *Transp. Res. Part F Traffic Psychol. Behav.* **2020**, *68*, 132–143. [CrossRef]
40. cf. fig. 303 art. 135 of the Regulation implementing the Highway Code. Available online: https://www.gazzettaufficiale.it/eli/id/1992/12/28/092G0531/sg (accessed on 5 June 2024).
41. Ministero Delle Infrastrutture e dei Trasporti. *Norme Funzionali e Geometriche per la Costruzione delle Strade*; Italian Ministry of Transportation Decreto Ministeriale: Rome, Italy, 2001.
42. Hussain, Q.; Alhajyaseen, W.K.M.; Pirdavani, A.; Brijs, K.; Shaaban, K.; Brijs, T. Do detection-based warning strategies improve vehicle-yielding behavior at uncontrolled midblock crosswalks? *Accid. Anal. Prev.* **2021**, *157*, 106166. [CrossRef]
43. Edvardsson Björnberg, K. Vision Zero and Other Road Safety Targets. In *The Vision Zero Handbook*; Edvardsson Björnberg, K. Hansson, S.O., Belin, M.Å., Tingvall, C., Eds.; Springer: Cham, Switzerland, 2023. [CrossRef]
44. Carrese, S.; Pallante, L.; Patella, S.M.; Sportiello, S. Assessing The Impact of Led-Illuminated Crosswalks on Pedestrian Safety *Transp. Res. Procedia* **2023**, *69*, 719–726. [CrossRef]
45. Peña-García, A.; Hurtado, A.; Aguilar-Luzón, M.C. Impact of public lighting on pedestrians' perception of safety and well-being *Saf. Sci.* **2015**, *78*, 142–148. [CrossRef]
46. Ministero delle infrastrutture e della Mobilità sostenibile, Piano Nazionale della Sicurezza 2030. Delibera CIPESS 14 April 2022, n 13. Available online: https://www.mit.gov.it/nfsmitgov/files/media/notizia/2022-09/Delibera%2014%20aprile%202022%20-%20Piano%20nazionale%20sicurezza%20stradale%202030.pdf (accessed on 5 June 2024).
47. Lantieri, C.; Costa, M.; Vignali, V.; Acerra, E.M.; Marchetti, P.; Simone, A. Flashing in-curb LEDs and beacons at unsignalized crosswalks and driver's visual attention to pedestrians during nighttime. *Ergonomics* **2021**, *64*, 330–341. [CrossRef]
48. Vignali, V.; Cuppi, F.; Acerra, E.; Bichicchi, A.; Lantieri, C.; Simone, A.; Costa, M. Effects of median refuge island and flashing vertical sign on conspicuity and safety of unsignalized crosswalks. *Transp. Res. Part F Traffic Psychol. Behav.* **2019**, *60*, 427–439 [CrossRef]
49. Metropolitan Agency of Bologna, Municipality of Bologna. *Urban Road Safety Plan (PSSU) 2016–2018, Delibera di Consiglio P.G 540417/2019*; Metropolitan Agency of Bologna: Bologna, Italy, 2019.

smart cities

MDPI

Article

Accessibility Measures to Evaluate Public Transport Competitiveness: The Case of Rome and Turin

Alessandro Zini [1], Roberta Roberto [2], Patrizia Corrias [3], Bruna Felici [3] and Michel Noussan [4,5,*]

1 Studies, Analysis and Evaluations Unit, ENEA, 00044 Frascati, Italy; alessandro.zini@enea.it
2 Department of Energy Technologies and Renewable Sources, ENEA, 13040 Saluggia, Italy; roberta.roberto@enea.it
3 Studies, Analysis and Evaluations Unit, ENEA, 00196 Rome, Italy; patrizia.corrias@enea.it (P.C.); bruna.felici@enea.it (B.F.)
4 Department of Energy, Politecnico di Torino, 10129 Turin, Italy
5 Paris School of International Affairs, SciencesPo, 75007 Paris , France
* Correspondence: michel.noussan@polito.it

Abstract: The transport sector worldwide relies heavily on oil products, and private cars account for the largest share of passenger mobility in several countries. Public transport could represent an interesting alternative under many perspectives, including a decrease in traffic, pollutants, and climate emissions. However, for public transport to succeed, it should be attractive for final users, representing a viable alternative to private mobility. In this work, we analyse the spatial distribution of public transport service provision within two metropolitan cities, considering the three key dimensions of mobility, competitiveness, and accessibility of public transport. The results show that private car performs better than public transport in all scopes considered, and that performance indicators are highly variable among city areas, indicating inequalities in social and environmental sustainability in urban systems. The outcomes of the analysis provide interesting insights for policy makers and researchers that deal with similar topics, and can also be extended to other cities and countries.

Keywords: urban mobility; public transport; accessibility; equity

1. Introduction

Globally, there is a clear trend towards increasing concentration of population in major cities [1]. Historically, cities are created also because the geographic concentration of productive activities allows an economic advantage through economies of scale and networks [2]. After reaching an optimal level of concentration, however, typical dis-economies of concentration emerge, including congestion and slower travel speeds. Urbanisation-related phenomena often include spatial expansion in the peri-urban area, also driven by rising real estate costs in the metropolitan centre, which may, in turn, be accompanied by increased land consumption, urban sprawl, and increased travel distances for a large share of the population. When these phenomena occur together, the conditions for increased car dependency are created. Among the possible consequences of urbanisation, the aggravation of phenomena of social stratification, between 'rich' and 'poor' neighbourhoods, is also significant.

The complexity of the objectives that contemporary societies should pursue and the interdependence of the variables at stake is evident: harmonising the city's conurbation economies with the reduction in congestion dis-economies, reducing the environmental impact, and reconciling principles of efficiency and economy in the management of public affairs with those of social equity.

Defining operational concepts such as 'mobility', 'competitiveness' of public transport (PT) services and 'accessibility' in cities, as well as measurable and repeatable indicators, thus provides important understandings. It can be useful both in the assessment of investment projects and in the analysis of the degree of success in the pursuit of environmental sustainability and social equity objectives.

This evaluation requires a benchmarking of PT against private car-based mobility, which is currently a default choice for users in several cities worldwide. The performance of PT services, and any future implementation, must confront the problem of car dependency, along with the underlying political-economic factors of car-dependent societies [3].

The objective of this study is to analyse the spatial distribution of PT service provision within two Italian metropolitan cities considering the three key dimensions of mobility, competitiveness, and accessibility of PT. Specifically, the paper focuses on the mobility during the morning peak hour in Rome and Turin to: (1) compare the local PT system to car mobility in terms of competitiveness and accessibility, and (2) identify possible conditions of territorial inequality with respect to mobility (see Section 3, Research Questions and Methodology).

Among the main results, the analyses show that, both in Rome and Turin, private car performs better than PT in all scopes considered (mobility, competitiveness, and accessibility). In all urban areas, a gap is identified, and in some of them it is significantly high. The spatial distributions of mobility by private car are mostly specular to those of mobility by public transport, and show a centre-periphery pattern in the spatial distribution of service efficiency. Significant variability in performance indicators among city areas are also identified, indicating inequalities in social and environmental sustainability in urban systems.

The innovative contribution of this work is the combined implementation of the above-mentioned key dimensions to present a comprehensive analysis of PT performance in the different areas of a metropolitan city by processing available data with a good level of spatial detail. The research on mobility is typically conducted through sectoral studies, which specifically analyse the sociological, economic, or technological aspects. In this work, we have instead adopted an interdisciplinary approach. The mapping of the cities of Turin and Rome was carried out by analysing, through the use of combined indices, the levels of mobility, competitiveness, and accessibility by private car and PT. Regarding the measure of accessibility, in particular, a new empirical metric was used through the combination of public transport availability and habitual trips, i.e., between supply and demand, considering travel time as a relevant parameter. The study, conducted using freely accessible databases, is the first to have used the approach proposed in the cities of Rome and Turin.

2. Literature Review on Mobility and Accessibility in Urban Areas

The study of mobility, understood in a more general sense, involves the evaluation of a complex combination of elements, ranging from individual factors (such as attitudes, values, intentions, available time and financial resources) to contingent factors (such as the cost and availability of different modes of transport, location of services, physical attributes of the built environment, social norms and culturally dominant values) [4–6].

The main reference models [4] are represented by different approaches:

1. Utility or rational choice theory, based on the evaluation of costs and benefits [7–12];
2. Accessibility theory, which evaluates spatial and temporal factors [13–18];
3. Socio-psychological theories, which consider the impact of attitudes, norms, values and emotions [19–23];
4. 'New mobility perspective' approach, which is largely based on the method of analysing social practices [24–27].

Urban mobility studies also include a strand that focuses its analysis on the idea of designing cities in which the car becomes less necessary, such as New Urbanism, Smart Growth, Transit-Oriented Development, Compact City, and the 15 min City [13]. This strand, on the one hand, seems to share the approach of the accessibility theory mentioned above, but on the other hand appears to refer specifically to the concept of avoidance of travelling by car, incorporating the "avoid" component of the A-S-I approach (which classifies and prioritise measures in "avoid", "shift", and "improve") [28,29]. Similarly, the issue of functional polycentrism of cities needs to be considered. Cities with more places with higher densities of economic and institutional activity would be able to generate lower overall travel demand and also greater preference for PT use. Polycentricity has become a key concept in regional studies [30], which proposes that cities develop a spatial distribution of population and economic activity around multiple urban subcentres. Its opposite is the monocentric city, with a metropolitan core with a high concentration of jobs that becomes increasingly dispersed as one moves away from it. A polycentric city is potentially characterised by shorter distances between home and work and less congestion [31]. Moreover, polycentricity, especially in large cities, appears to promote lower operating costs and greater PT efficiency because it ensures a minimum size scale at different points in the city [32].

A distinction must be made between the terms "mobility" and "accessibility". Mobility refers to the degree of ease of movement, i.e., the ease with which one can move through space using a transport mean [33]. Accessibility, on the other hand, refers to the ability of an individual to move around a given area according to his or her needs and desires. In particular, accessibility can be defined as the ability of an individual in a given place to participate in an activity or set of activities [34], or as the number of opportunities, such as work, shopping, etc., that can be reached from a given place in a given time by car, PT or non-motorised modes of transport [35]. It follows that the mobility and accessibility levels of a city or neighbourhood are not necessarily correlated [33].

In this perspective, accessibility corresponds to 'mobility capital': an extension of the classical concept of capital (physical, economic, cultural, and social), where mobility emerges as one of the resources for action [36], directly linked to the concept of life opportunities [37] and capability [38,39]. Accessibility as understood thusly directly involves equity and justice in transport, especially for PT. Moreover, accessibility can be seen as the key to an inclusive transport system [40].

The social dimension of transport is addressed in several documents and studies outlining the priority principles of the European Union [41,42].

The European Union pays special attention to accessibility of transport for specific socio-demographic segments (young people, people at risk of poverty, people with a low level of education, people with reduced mobility, people living in isolated areas [41]). Equal attention is paid to horizontal equity principles, such as reducing geographical disparities, ensuring that all regions can benefit from infrastructure development [42]).

Three different ways of understanding the term equity can be traced in humanistic theory [43]: (1) the egalitarian principle, (2) the utilitarian principle, and (3) Rawl's principle. The choice between one or the other, which falls within the sphere of the political evaluation, is full of implications for the implementation of PT plans. Under the egalitarian principle, each individual has equal rights. This implies that the provision of transport services should not differ from region to region. Under the utilitarian principle, on the other hand, the 'good' to be pursued is the maximisation of total aggregate utility. This can result in a focus on territorial areas with higher added value, more attractive journeys, and higher population and employment density. In the framework of Rawl's principle [44] inequalities are allowed only if they lead to benefits for each individual, while restorative justice intervenes to balance unsustainable situations.

On a practical level, following Rawl, a plan is to be chosen if it is able to ensure the best condition, among all plans, for a group or an area that is in the worst position. In the context of PT plans, however, Rawl's principle has had little application. Better success has been achieved in the application of the utilitarian principle [43], especially in the practice of Cost–Benefit Analysis (CBA), or also in the variant of Social Cost–Benefit Analysis (SCBA). SCBA shares with CBA the approach of assigning a monetary value to the effects of the planned intervention. Usually, the benefits estimated in the CBA include time savings, vehicle operating costs, journey costs, impacts on accidents, journey quality, greenhouse gases, and indirect taxes [45].

In this context, some scholars [18,46] observe that the uncritical use of CBA in the evaluation of investment projects, which makes use of indicators such as travel time savings and property value by attributing a monetised value to them, can lead to unintended consequences. A focus on time value alone inevitably ends up favouring the fastest means of transport, the car, while a focus on economic value can lead to favouring regions with higher added value. A Social CBA could improve the evaluation process by analysing additional aspects, including safety, health and environmental and social impacts [18]. However, given the broad range of effects to be accounted for, this is rarely applied during the planning of mobility plans and infrastructure.

One of the risks of including CBA in mobility planning is associated with the possibility that the focus on the time-saving indicator becomes exclusive, overlooking other considerations. The risk is that the fastest mode of transport, i.e., the car, or high-speed railway lines serving commuting flows will be favoured [45].

An efficient PT system is one of the keys to success in improving overall mobility in modern cities. According to Thomson [47], 'all efforts to improve rush-hour car travel will fail unless public transport is also improved'. Efficiency levels must also not be excessively heterogeneous across urban areas.

In addition, the monetisation of achievable benefits, estimated by using territorial added value, could lead to increased territorial socio-economic gaps in investment [48].

The guiding rule is that CBA is to be regarded as a tool and not as a principle, instrumental in assessing the goodness of one plan against a similar competing one. It therefore should not be used to evaluate the best solution between investment projects in new private mobility versus public mobility [45]. Other evaluation tools also seem capable of broadening the horizon to non-monetary or non-monetisable benefits, such as landscape impact or reducing urban inequalities. These include, for example, Multi-Criteria Analysis, seen as a means for establishing trade-offs between different stakeholders involved in the project [49]. On a practical level, the criteria of economy and efficiency in the management of the PT systems may conflict with the criteria of justice and inclusiveness.

From a methodological point of view, the studies of mobility plans and investment projects consider PT accessibility according to three distinct notions: (1) "how easy it is to walk to a PT stop" from a given point in space, i.e., accessibility to PT [50]; (2) "how many people can be served by the PT service" [51]; and (3) "how many destinations people seek could be reached with the current PT offer", i.e., accessibility by means of PT [52].

The last notion appears to be the most valid for the purposes of this paper, having defined accessibility as the ability to visit places of activity (such as shops, workplaces, services, businesses, public spaces, etc.) using a particular transport system in an acceptable time or cost [53]. It also aims to intersect supply and demand for mobility [45], whereas the first two concepts of accessibility do not allow to go beyond the analysis of service supply. In some studies, the qualitative components of mobility demand are also captured with opinion surveys [52]. Another issue is how to measure service levels and/or accessibility of PT services within cities. Two main approaches emerge in the literature. The first one is the isochrones method [54], also complemented by the estimation of the number of opportunities reachable in a 30 or 40 min trip (e.g., [55]). The second is the gravity-based model method, aimed at measuring the travel costs (travel time and/or travel cost) between one zone and all others, weighted by the attractiveness of the respective zones (e.g., [56,57]). A third method is based on measures of transport network connectivity, although still relatively less used [58,59].

Travel time is often considered an intuitive measure that fits well with people's perception of distance friction [60–62]. Time, however, is not the only factor influencing modal choice decisions. Other elements include reliability, punctuality, comfort, vehicle/station characteristics and, considering PT, frequency, among others [5,63]. As an example, in a study conducted on metro users, Raveau et al. [64] indicate that users, given the same travel time, tend to favour linear and direct routes and routes that are better known or more heavily travelled. It follows that a PT trip involving more complex routes or the use of more vehicles, even if of shorter duration, may not be very competitive compared to the intrinsically "linear" route with one's own vehicle (car or motorbike). Empirical evidence in the research literature on the spatial distribution of PT supply shows a centripetal dynamic, with higher service levels in central areas. This evidence is valid for both European cities (such as Gothenburg, Paris, Madrid [65]), and non-European cities (such as Melbourne, San Francisco, Montreal, New York, Sydney and Boston [54,62,65]). Population density and demand levels to a good extent account for this dynamic. Available studies on the competition of PT versus the private vehicle in US and European cities show that the latter is able to provide a higher level of accessibility [54,66–72]. The spatial distribution of the level of accessibility is rather more complicated according to the empirical literature. Suburbs connected to public mobility corridors capable of satisfying part of the travel demand of its residents, for example, may exhibit a fair level of accessibility [73]. Mobility in Rome and Turin has been object of discussion and study in recent years. Greenpeace [74] has focused on urban development and mobility in Rome, amongst other things. They propose supply side measures: enhancing the supply of PT services, particularly rail transport; focusing on peripheral and more densely populated areas of residents; encouraging walking and bicycling; and disincentivising private car mobility (e.g., with traffic-restricted zones and 30 km/h speed limit zones). Lelo and Risi [75] focuses more on structural and morphological factors related to the history of urban development and socio-economic factors. He highlights, for example, how people residing in more suburban areas, where housing costs, population density and workplace are lower, are forced to make long commutes due to inefficient PT services. Within the framework of research on the 15 min city concept and its variants, such as the 20 min district, Staricco [76] developed a methodological framework to analyse levels of accessibility on foot to 20 types of local services from each census section in the city of Turin. Ceccato et al. [77] calculated accessibility in Turin using travel speed for car and PT.

Rome and Turin have already been the subject of analysis by this research group. In [78], the mobility habits and related energy consumption and emissions of remote workers during 2015–2019 were studied. A new survey of the mobility habits of a sample of remote workers (about 2200) who are based in Rome is underway.

3. Research Questions and Methodology

This work is part of a broader research activity aimed at assessing the adequacy of the PT system to meet the needs of users to travel for work and to reach services and places of interest (culture and art, sport, leisure and recreation, shopping, etc.) and at identifying the conditions of mobility inequalities in the main Italian cities. The paper focuses on the mobility during the morning peak hour in two Italian metropolitan cities: Rome and Turin. The choice of the two cities is due to the attention recently given to them in numerous debates and studies, and to their specific physical and historical-urban characteristics, which lead to different impacts on urban circulation and represent a challenge for initiatives aimed at improving transport sustainability (see Section 4).

This paper focuses on the following two main research questions:

1. How is the local PT system in the cities of Turin and Rome compared to the car with reference to the competitiveness indicator and the accessibility indicator of journeys made during the morning rush hour on working days?
2. Are there conditions of territorial inequality with respect to mobility during the morning rush hour on working days in the two cities?

The analysis was conducted by examining three main indicators (calculated from the isochrones of PT and car trips and origin/destination (O/D) trip matrices):

(1) The mobility index for PT and cars, seen as a measure of the degree of fluidity of the respective carriers in the urban territory (Equation (1));

(2) The index of competitiveness of PT with respect to the car, defined as the ratio between the respective mobility indexes (Equation (2));
(3) The degree of accessibility of PT, here considered as a measure of the possibility of travelling with PT the typical trips of the residents of the individual neighbourhoods (Equation (3)).

An isochrone is defined as the line joining points located at equal "distances" in terms of travel time from a given location. The isochrones, shown as areas overlapping territorial maps, enable us to visualise the proportion of territory that can be reached in a given time from the central point. They are typically computed by means of simulations and models based on the available infrastructure and the schedule of PT services [79]. In this work, isochrones are calculated through the analysis of location data provided by TravelTime [80] based on the PT schedules planned by the respective transport operators [81]. Since they do not take into account the inherent daily irregularity of travel times due to accidents, breakdowns, delays, and unusual traffic conditions, the degree of statistical uncertainty cannot be estimated. The isochrones are centred on the PT service stops, obtained from the Open Street Map repository [82,83], 2759 and 8659 PT stops in the cities of Turin and of Rome, respectively.

By centring the isochrones on the PT stops, the time due to multimodality, i.e., the time required to reach the stop on foot, by car, or by any other means, is not counted. This assumption, while not limiting the scope of the analysis or the significance of the results, avoids the choice of multimodality criteria. Such criteria, while used in some studies [84,85], may be overly arbitrary or complex, making the final results more difficult to be interpreted and compared.

The territorial unit of analysis is the sub-municipal zone. The municipalities of Turin and Rome are subdivided into 94 and 153 urban zones, respectively, which approximately reflect the size of the city's 'neighbourhood'. The sub-municipal zones of Turin are statistical zones [86] resulting from the aggregations of several census sections as defined by the Italian National Institute of Statistics—ISTAT [87]. The sub-municipal zones of Rome derive from the subdivision of the municipal territory for statistical and spatial planning and management purposes [88].

TP includes all of the TP means operating in Rome and Turin: train, metro, bus and tram, while travels with car include private cars, as in [89].

All analyses are conducted by considering the 30 min outbound trips at 7 a.m., representing the peak-hour for morning commuting.

The isochrones are calculated on Wednesdays in May 2023, i.e., on a weekday during the school attendance period for all school orders. The month of May 2023 does not fall within the COVID and post-COVID period and, therefore, its effect on car travel mode is considered negligible. The choice of considering Wednesday as representative for all the working days is a common practice in the field, as the mobility demand during working days generally show little deviations from a day to another, as also confirmed by literature study on mobility patterns [90].

The O/D matrices between sub-municipal zones are calculated from TomTom [89] data, based on average weekdays (Monday through Friday) of October 2022. This period, although it does not coincide with the period for which the location data from Travel time are available, is used for the calculations without invalidating the results. A previous study on vehicle mobility profiles in Turin [90] shows that hourly mobility patterns have a very similar behaviour in May and October, especially during the morning peak hour (although considering pre-COVID trends).

The mobility index (MI), expressed in km^2, is calculated as the average area of the isochrones generated in each urban zone by the 30 min trips at 07:00 from each PT stop. The average area is calculated as the sum of the isochrones divided by the number of stops in each zone,

$$MI_z = \frac{\sum_i A_{i,z}}{S_z} \tag{1}$$

where z is the indicator of each zone (ranging from 1 to 94 in Turin and from 1 to 153 in Rome); S is the number of PT stops in z; $A_{i,z}$ is the area of the ith isochrone in zone z.

The competitiveness index CI is calculated as

$$CI_z = \frac{MI_{z(PT)}}{MI_{z(car)}} = \frac{\sum_i A_{i,z(PT)}}{\sum_i A_{i,z(car)}} \tag{2}$$

where MI_z is calculated for both PT and private cars. In the latter case, to perform a consistent analysis, the isochrones' areas are calculated in each PT stop considering the trip in private car.

The accessibility index AI is calculated as

$$AI_z = \frac{n_z\left\{N, A_{i,(TP)}\right\}}{N_z} \tag{3}$$

where N is the number of trips made by car in each zone; $n_z\left\{N, A_{i,(TP)}\right\}$ is the subset of N that falls within the area of the PT isochrone.

The parameter CI has a minimum of 0, and an undefined theoretical maximum value. A value greater than 1 indicates that moving by PT is faster than moving by private means. In fact, it is rather unlikely that MI reaches and exceeds a value of 1. In practice, by comparing mobility by PT with that by private means, we assume that the latter constitutes a benchmark.

The parameter AI has a minimum of 0, and a purely theoretical maximum of 1. An hypothetical value of 1 would indicate that all habitual movements of residents could be fulfilled through a 30 min journey with PT.

The implementation of index AI is intended to enrich the isochrone concept. In this way, the isochrone loses its typical characteristic of anisotropy, in which all points in the space around the starting point have the same utility. Using the index proposed in this article, an isochrone is more 'relevant' the more it envelopes habitual movements of residents. As a consequence, this analysis implicitly allows different weights to be assigned to zones based on their importance, resulting in indicators that are more closely aligned with real mobility dynamics.

Moran's spatial mobility autocorrelation index has been used to estimate the significance of the PT mobility spatial trend. It indicates the cross-product statistic between the PT mobility index for a neighbourhood and its spatial lag, expressed in deviations from its mean.

4. Rome and Turin: Territorial Context and Characteristics

Rome is the capital of Italy, and its biggest city, both in terms of population, with 2.8 million inhabitants, and size, being almost 1300 square km (its density is around 2150 inhabitants/km^2). In an European context, Rome is also the eighth-largest city in terms of population. Rome shows a very important level of congestion, especially on the ring roads and consular radial roads, as the concentration of work opportunities in certain areas of the city has a strong impact on daily commuting patterns. Considering the demographic data for 2022 [87], the citizens of Rome are, on average, 46.4 years old, with 23.3% of the population being at least 65 years old (these figures are very similar to the national trend, with an average age of 46.2 and 23.8% of people being at least 65 years old). The total income per capita in Rome in 2022 is EUR 29,155 [91], with a 4.3% increase over 2021, much higher than the national average of EUR 20,039, but lower than the figure of Milan (EUR 37,734).

Rome is also one of the Italian cities with the highest motorisation rate (considering the ten largest cities), with 628 cars per 1000 inhabitants in 2021 [92]. The use of public transport in Rome remains limited, as in an average working day, the total trips are 0.72 million, which is less than 15% of the total trips [93].

Rome's public transport network includes 37 km of tram tracks, 32 km of trolleybuses, and 58 km of subway (the second in Italy after Milan), as well as 1960 buses (of which 25 are electric or hybrid, and 500 run on natural gas) [94]. The total offer of Rome's public transport stands at 8159 pkm per inhabitant, the third-highest value in Italy, after Milan and Venice. However, this indicator should be analysed by considering the fact that many non-residents use the public transport services, including tourists and commuters. The largest part of the pkm offer is from buses, 51% of the total, and from subway, 46% of the total.

Turin is the fourth-largest city in Italy, with 850,000 inhabitants, comparable to other European cities like Amsterdam, Marseille, Kraków, and Valencia. Turin is also in the top list for population density in Italy (considering cities with more than 50,000 inhabitants) with around 6525 inhabitants/km^2. As of 2022 [87], the citizens of Turin are, on average, 47.6 years old, with 26.0% of the population being at least 65 years old (these figures are both higher than the Italian ones). The total income per capita in Turin in 2022 is EUR 26,971 [91], with a 3.4% increase over 2021, which is higher than the national average of EUR 20,039, but lower than other large cities in central or northern Italy, including Florence (EUR 28,409) and Bologna (EUR 29,978). Turin has a motorisation rate of 581 cars per 1000 inhabitants in 2021, with an 11% decrease compared to 2018 [92].

The modal share of public transport in Turin is around 9% of the total daily trips, with an estimated value of 260 thousand trips per day in the city in 2022 [95]. These figures remain much lower compared to available data before the pandemic, with the modal share being around half of the 2013 level.

Turin's public transport is based on 73 km of tram tracks (the second city in Italy after Milan) and 14 km of subway, as well as 750 buses (of which 96 are electric or hybrid and 213 run on natural gas) [94]. The annual public transport offer is 4325 pkm per inhabitant, which is lower than average figures for provincial capitals in Italy (4748 pkm/inhab.) and Northern Italy (6048 pkm/inhab.). The pkm offer is mostly from buses, 58% of the total, followed by 22% from subway and 20% from tram.

In terms of land consumption, Rome and Turin are, respectively, in first and second place in the ranking of Italian provinces by area consumed. The metropolitan city of Rome has the largest area consumed in Italy in 2022, with more than 70,300 hectares, increased by a further 235 hectares in 2023. The province of Turin occupies about 58,500 hectares, with an increase of 168 hectares in 2023. The urban areas of Rome and Turin are reported in Figure 1.

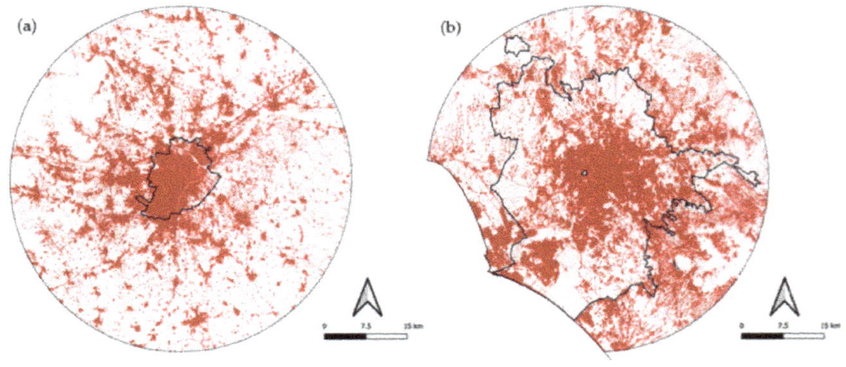

Figure 1. Urban areas in Turin (**case a, left**) and Rome (**case b, right**).

5. Results

Figures 2 and 3 show the PT mobility index (PT-MI) and car mobility index (car-MI) in Turin and Rome on weekdays at peak hour (7 a.m.). In both cases, the degree of car mobility in all urban zones appears to be higher than that of PT (see Figure 4), confirming the results from studies conducted in other cities [54,63,66–71]. In none of the urban zones of the two cities is PT actually giving a better mobility service when compared to private cars at morning peak hours. Moreover, in absolute terms, the maximum PT-MI is lower than the minimum car-MI, with values of 56.2 km^2 against 62 km^2 in Turin and 52.6 km^2 against 67 km^2 in Rome.

PT mobility is higher in Turin than in Rome. On average, in Turin, a journey of 30 min at 7 a.m. is equivalent to an isochrone of 34 km^2, while in Rome, the same is 18 km^2. In parallel, mobility by car is greater in Rome than in Turin (233 km^2 and 189 km^2, respectively).

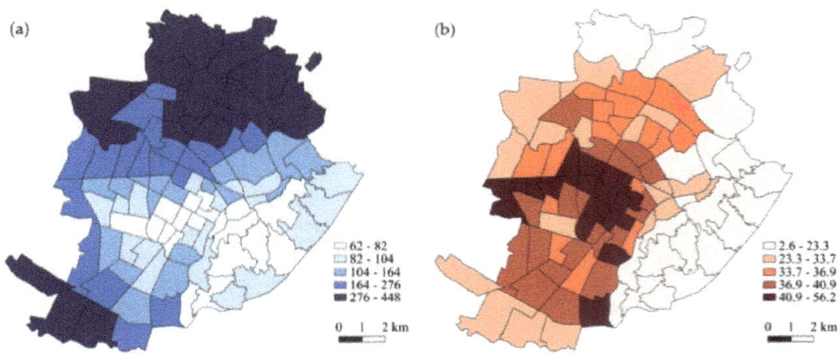

Figure 2. Mobility index (km^2) of (**a**) car and (**b**) PT trips in the city of Turin—average isochrone area (at 07:00 on weekdays).

More specifically, in Turin (Figure 2) the maximum PT isochrone does not exceed 56.2 km^2, while the maximum isochrone for car travels is about eight times higher. The zones characterised by the highest car-MI are those close to the motorway network and ring roads. However, also without taking these areas into account, a degree of mobility by car that is almost five times higher than the one by PT is observed. Large portions of the city territory (corresponding to the hill areas—in the East—and peripheral areas) show values of less than 23.3 km^2 for PT. Some of them are industrial or green/forested areas, while others are low-density residential zones in the hilly part of the city.

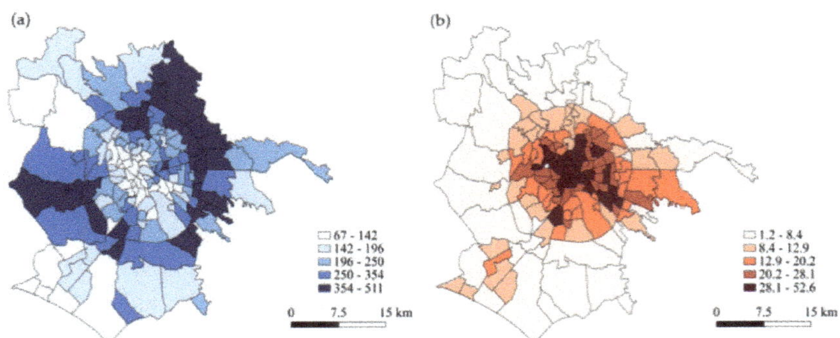

Figure 3. Mobility index (km^2) of (**a**) car and (**b**) PT trips by (**a**) car and (**b**) PT in the city of Rome (at 07:00 on weekdays).

In the case of Rome (Figure 3), the gap between the mobility index of PT and cars is even more evident. Again, the areas with maximum car mobility are close to the ring road and highway connections. Large peripheral urbanised areas show very low PT mobility degrees, with a 30 min coverage area of at most 8.4 km^2.

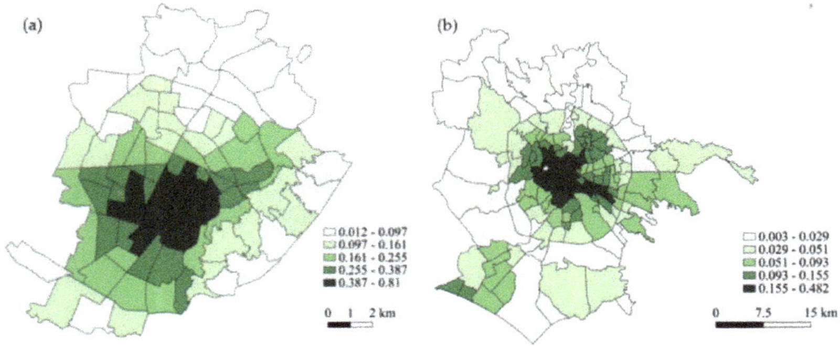

Figure 4. Competitiveness index of public versus car transport in (**a**) Turin and (**b**) Rome.

Analysing the spatial distribution in the two cities, a clear trend emerges with respect to the centre/periphery dimension. The peripheral urban zones show a lower PT-MI than the central ones. At the same time, peripheral areas appear to be more efficient in private transport. Both findings agree with other studies in the literature [54,63,66–71]. The degree of statistical dispersion of mobility in neighbourhoods is high, the coefficient of variation relative to PT isochrones being 0.36 in Turin and 0.60 in Rome, while for car isochrones the values are 0.61 and 0.44, respectively.

Moran's spatial mobility autocorrelation index for both cities is positive (0.640 for Turin and 0.478 for Rome) and significant ($p < 0.01$ after 999 permutations), suggesting a geographical distribution not attributable to a random spatial process. A more detailed representation of the mobility index of PT and car trips for each city is provided in Appendix A.

The competitiveness index (CI), calculated as the ratio between isochrones (Equation (2)), is shown in Figure 4. Overall, the average ratio between the degree of competitiveness of PT and that of car (calculated on all PT stops) is 0.18 for Turin and 0.08 for Rome. In other words, for the two cities, the average isochrone of PT has an area equal, respectively, to 18% and 8% of that relative to travels by car. A more detailed representation of the CI is provided in Appendix A.

Finally, Figure 5 shows the accessibility index (AI) in relation to 30 min trips made at 7 a.m. It represents a measure of the amount of everyday trips that can theoretically be satisfied by PT. On average, a 30 min PT trip in Turin would satisfy 74% of regular commuters in the morning rush hour. In Rome, the percentage is considerably lower, around 41%. The centre/periphery dynamic of variation still seems to be confirmed, but with some attenuation with respect to the MI. This evidence appears to be in line with other studies (e.g., [73]). Some peripheral areas appear to be less disadvantaged. For example, those in the eastern districts in the city of Rome, as well as a sort of ridge in the city of Turin identified by rail transport and by the subway. Moreover, Moran's accessibility autocorrelation index, when compared with Moran's mobility autocorrelation index, shows an even greater gap between the two cities (0.412 for Turin and 0.234 for Rome). This figure seems to suggest that, in a larger city such as Rome, the attenuation of the spatial trend in terms of accessibility is even greater. A more detailed representation of the AI in Rome and Turin is available in Appendix A.

The distribution of the PT indicators (MI, CI, and AI) is variable between the different areas of the two cities. This is indicated by high values of the coefficient of variation, calculated as the ratio between the standard deviation and the mean after weighting the administrative areas by their respective resident population. This dispersion indicator is particularly high for the PT competitiveness (the coefficient of variation of CI is 0.93 for Turin and 1.25 for Rome), and slightly lower for the accessibility indicator (the coefficient of variation of AI is 0.81 for Turin and 1.04 for Rome). The statistical results indicate the presence of significant territorial inequalities in the attractiveness of PT supply.

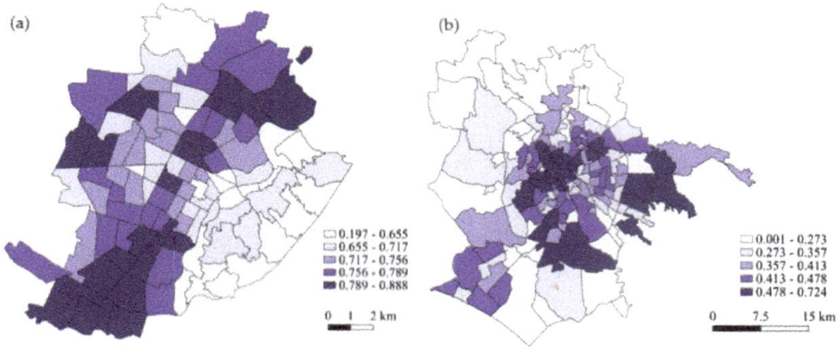

Figure 5. Accessibility index of public transport in (**a**) Turin and (**b**) Rome.

6. Discussion

The results of the analysis show that, in all of the urban zones of the cities examined, the degree of mobility allowed by car is higher than the one by PT.

With respect to the competitiveness indicator proposed (see Equation (2)), we observe that, in Turin and Rome, the PT is never more competitive than travelling by car, with CI remaining always far lower than 1. In both cities, there is also a large gap in PT competitiveness between central city areas and peripheral areas (Figures 2–4), in line with some empirical evidence in the literature for Melbourne, San Francisco, Gothenburg, Paris, Montreal, New York, Madrid, Sydney, and Boston [54,62,65,96]. The centripetal pattern in the spatial distribution of PT competitiveness is the result of economic drivers, urban policy choices and long-standing historical events. First, historically, central neighbourhoods are also those with the highest population density. A service typically based on density, such as PT, is certainly more efficient when provided to the most highly populated neighbourhoods. In addition, central districts are those that concentrate most economic activity and are sites of institutional and governmental offices. They generate inbound work and study trips in the morning peak and, therefore, receive more attention in PT planning. Spatial disparities in the distribution of PT services are influenced by the morphology of the cities' road network, which often has a radial structure, as in the case of Rome. This has negative effects especially for road PT, which is forced to proceed on highly congested road axes. In both Turin and Rome, the aforementioned conditions of disadvantage are added to other forms of social inequality, such as the lack of services in peripheral areas and lower average incomes. Moreover, road congestion in urban areas can have a negative impact on measures to alleviate PT saturation on roads. In fact, the expansion of PT services on roads, unless the lane network is increased, would lead to less improvements than expected.

Rome's post-war demographic growth was characterised by illegal or unapproved housing developments sector that led to the growth of neighbourhoods that lacked services and infrastructure [97–100]. In the case of Turin, urban development was conditioned by the presence of Fiat, Italy's largest industrial company at the time, around which grew sleeping neighbourhoods for workers employed by the company and its suppliers. Although with different causes and in different ways, the development of the two cities has led to high land use without adequate infrastructure and quality services. In suburban areas, where the results of the analysis show large gaps in PT competitiveness (Figure 4), the lowest socio-economic indicators of quality of life (such as the percentage of college graduates, the unemployment rate, the percentage of households with potential economic hardship and the social hardship index) are also found [75].

In newly urbanised areas, PT development is often forced to 'follow' building construction, mostly delayed, prioritising road transport, which is easier to implement, but has low competitiveness. In such cases, it is difficult for the people to avoid using private vehicles. The problem is accentuated when new neighbourhoods have a low population density, are residential, and are far from central areas.

Regarding the PT mobility index, some peripheral areas seem to be less disadvantaged. For example, those in the eastern districts of the city of Rome and a kind of backbone of the city of Turin identified by rail and metro transport. In these cases, the rail PT system makes a positive contribution. This is true for the eastern area of Rome, where discrete levels of mobility, PT competitiveness, and accessibility are noted. Lelo and Risi indicate [101], in these same neighbourhoods, some of the highest indices for unemployment, the lowest number of graduates and a high socio-economic disadvantage, and suggest a trend linking illegal construction practices in the city of Rome with social and economic marginality. In this context, the importance of an integrated interpretation of the various dimensions of analysis is therefore evident, including the type of PT (road or rail).

Looking at the AI, considered as a percentage of habitual journeys that in theory can be satisfied by PT (see Equation (3)), the trend dominated by the centre/periphery dimension observed for CI appears more nuanced for both Rome and Turin. This is due to the presence of metro and train stops and probably a combination of other behavioural and choice factors. At the same time, the lower degree of accessibility that characterises the city of Rome compared to that of Turin seems likely to be attributable to the higher incidence of daily commuting and the greater distances to be travelled.

When analysing the competitiveness of PT in different areas of the same city with reference to travel time, the empirical literature mostly seems to suggest that it is greater in central areas. This has been highlighted for Melbourne [54], San Francisco [67], Gothenburg [96], Paris, Montreal, New York, Madrid, Sydney, and Boston [65]. Chia and Lee [73], in a study on the Australian city of Brisbane, find that when the accessibility measure considers not only travel time, but also the potential effect of transfer location, the previous trend is not confirmed, some outer-city suburbs located along major bus corridors have a relatively higher level of accessibility. This also holds true for our map of Rome, when considering TomTom trips: some suburbs, especially those served by metro lines, see an improved level of accessibility through PT.

The results indicate the presence of significant territorial inequalities in the attractiveness of PT supply. The inequalities seem to be less significant when considering accessibility, which may incorporate other factors such as proximity to public rail transport or the role of people's housing choices in favouring shorter commuting times. The findings also show that he policy actions need to be tailored to the specific area of the city based on quantitative evidence.

In urban planning policies, the issue of mobility should not be a sectoral policy involving only experts. We believe that an integrated approach is needed, pursuing an integration between urban development and mobility plans and including socio-economic sustainability and environmental sustainability as indispensable elements in the definition and implementation of effective spatial planning and urban regeneration policies. Urban regeneration policies (in particular combining a set of actions in different fields of intervention), in order to be effective and contribute to the concrete reduction in inequalities, shall combine innovative practices of governance, active citizenship participation, methods, and approaches to improve the access to common goods and public spaces and services.

PT should be planned and operated in a way that makes it more accessible, and provide a competitive service compared to private mobility, not only on the economic dimension. To make PT service more effective, it might be appropriate to introduce some forms of incentives for PT companies so that they are motivated to transport more passengers, and not just to provide a service. Analysis of available data and information on users' mobility should be encouraged, also by making big data publicly available.

7. Conclusions

In this work, we analysed the spatial distribution of PT service provision in two metropolitan cities in Italy according to mobility, competitiveness, and accessibility indices calculated during the morning rush hour. In this context, accessibility is intended as a metric of the extent to which PT can be used for the typical commute of residents in different neighbourhoods.

The results show that private car use in the two cities offers higher levels of mobility, competitiveness, and accessibility than PT. In all urban areas, the car mobility index is higher than the PT mobility index in both Rome and Turin and, in absolute terms, the maximum value PT-MI is lower than the minimum value of car-MI. The gap is very high, even excluding areas close to the motorway network and ring roads, and is higher in Rome than in Turin.

The spatial distributions of mobility by private car are mostly specular to those of mobility by public transport. People use the car more because they are close to high traffic roads and/or the public offer is not satisfactory. There is also a clear centre-periphery trend in the spatial distribution of service efficiency: PT is more efficient in the centre, and decreases moving towards the periphery; the opposite trend occurs for cars. The PT mobility index is higher in Turin than in Rome, while the car mobility index is higher in Rome than in Turin. In terms of competitiveness, the average PT isochrone is 18% and 8% of that for car travel in Turin and Rome, respectively. On average, a 30 min PT trip in Turin would satisfy 74% of the morning rush-hour commute. In Rome, the percentage is considerably lower, around 41%.

The joint analysis of regular commuting by residents and isochrone areas shows a better assessment of PT accessibility in some peripheral areas of cities. For example, for Rome, in particular those served by the metro lines (eastern suburbs of Rome) and those with a high level of self-contained commuting (the 'city within a city' districts of the Roman coastline). The results also show significant variability in performance indicators between city areas, implying issues of social equity and environmental sustainability of urban systems. The distribution of PT indicators is variable between different areas of the two cities. The distribution of PT indicators is variable between different areas of the two cities, especially with regard to PT competitiveness, which indicates the presence of significant spatial inequalities in the attractiveness of PT supply. Critical aspects in mobility in the cities of Turin and Rome, in part associated with the specific historical and socio-economic context in which the cities have developed in recent decades, impact the levels of equity and quality of life of residents.

The indicators presented in this paper are based on figures related to the morning peak hour, which is the most critical for congestion in most Italian cities. However, the analysis of the evening peak hour may lead to different results and insights. For this reason, a future application of this model will address other hours of operation to shed light on the effect of different times over the day. The proposed methodology can be usefully applied to other cities, so as to also identify any common patterns or elements that suggest correlations. Integrating the analysis with big data and more comprehensive sets of information also regarding the other dimensions involved (social, economic, structural aspects, etc.) is an interesting development scenario and a useful tool to support urban policies.

The indicators presented in this paper are based on figures related to the morning peak hour, which is the most critical for congestion in most Italian cities. However, the analysis of the evening peak hour may lead to different results and insights. For this reason, a future application of this model will address other hours of operation to shed light on the effect of different times over the day.

Author Contributions: Conceptualisation, All; methodology, A.Z., R.R. and M.N.; software, A.Z.; formal analysis, A.Z., M.N. and R.R.; data curation, A.Z.; writing—original draft preparation, All; writing—review and editing, All; visualisation, A.Z.; supervision, R.R. All authors have read and agreed to the published version of the manuscript.

Funding: This research received no external funding.

Data Availability Statement: Data are available upon request.

Conflicts of Interest: The authors declare no conflicts of interest.

Abbreviations

The following abbreviations are used in this manuscript:

AI	accessibility index
car-MI	car mobility index
CBA	cost–benefit analysis
CI	competitiveness index
MI	mobility index
PT	public transport
PT-MI	public transport mobility index
SCBA	social cost–benefit analysis
SUMP	Sustainable Urban Mobility Plan

Appendix A

Appendix A.1

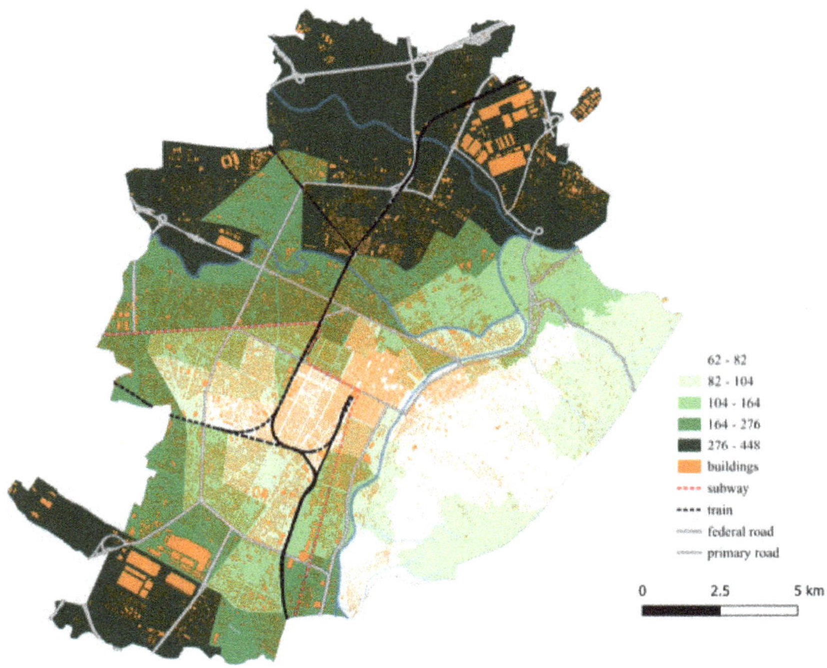

Figure A1. Mobility index (km^2) of car trips in the city of Turin (at 07:00 on weekdays).

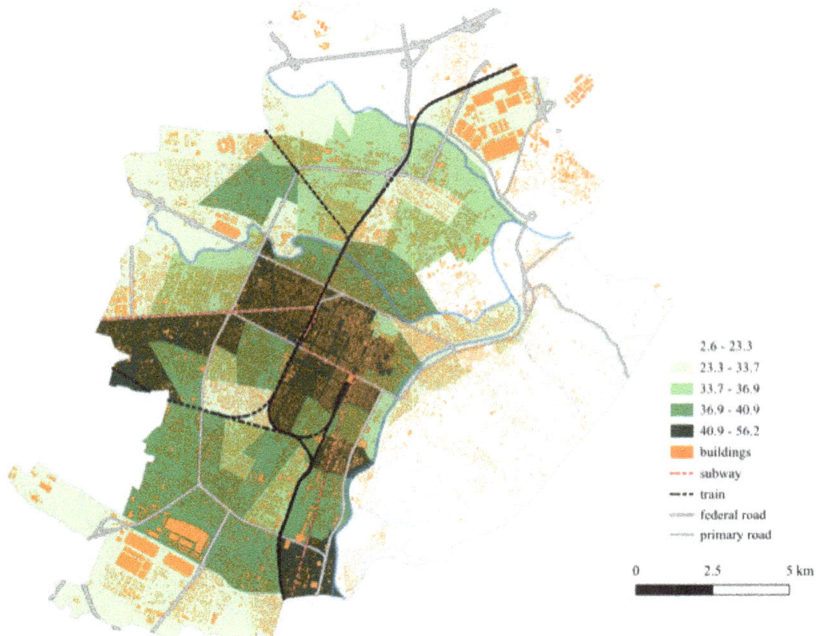

Figure A2. Mobility index (km^2) of TP trips in the city of Turin (at 07:00 on weekdays).

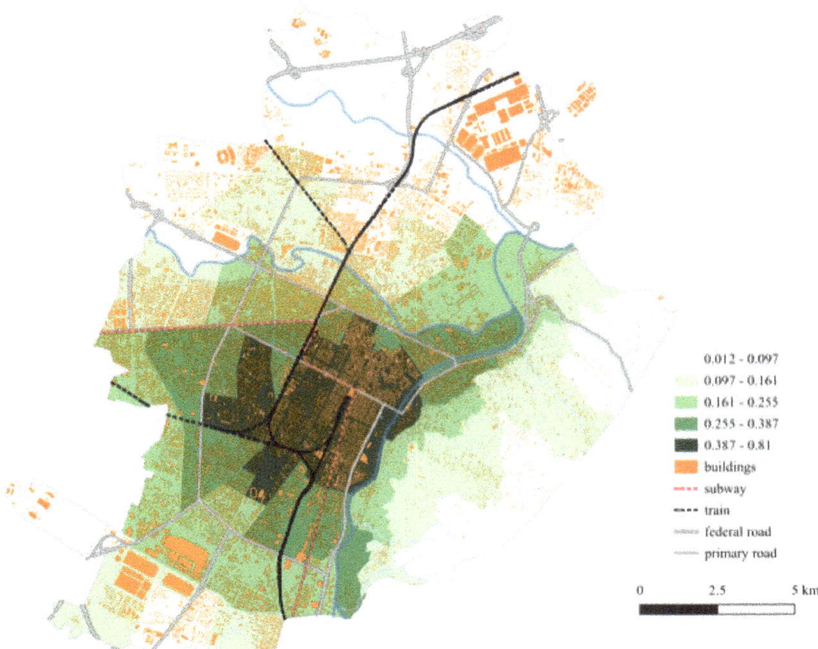

Figure A3. Competitiveness index in the city of Turin (at 07:00 on weekdays).

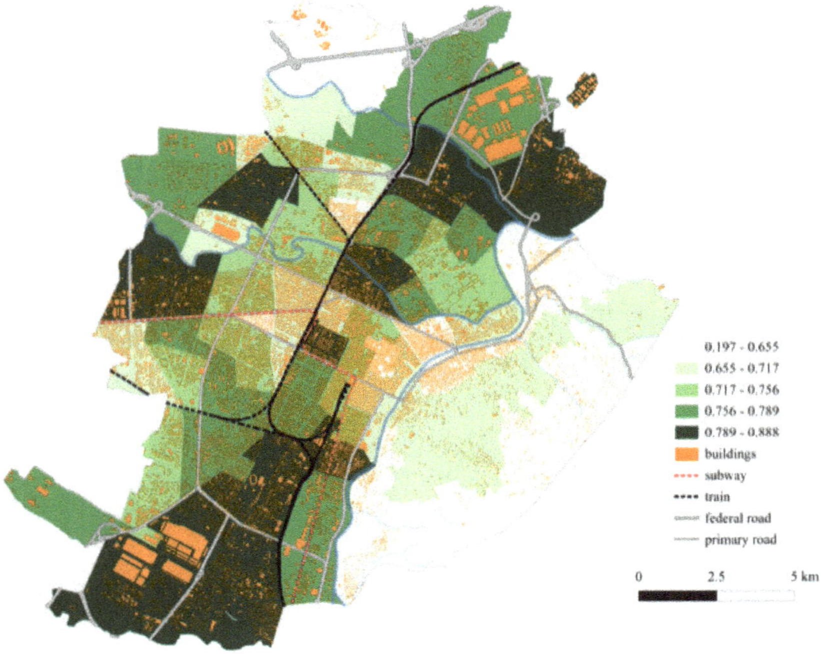

Figure A4. Accessibility index in the city of Turin (at 07:00 on weekdays).

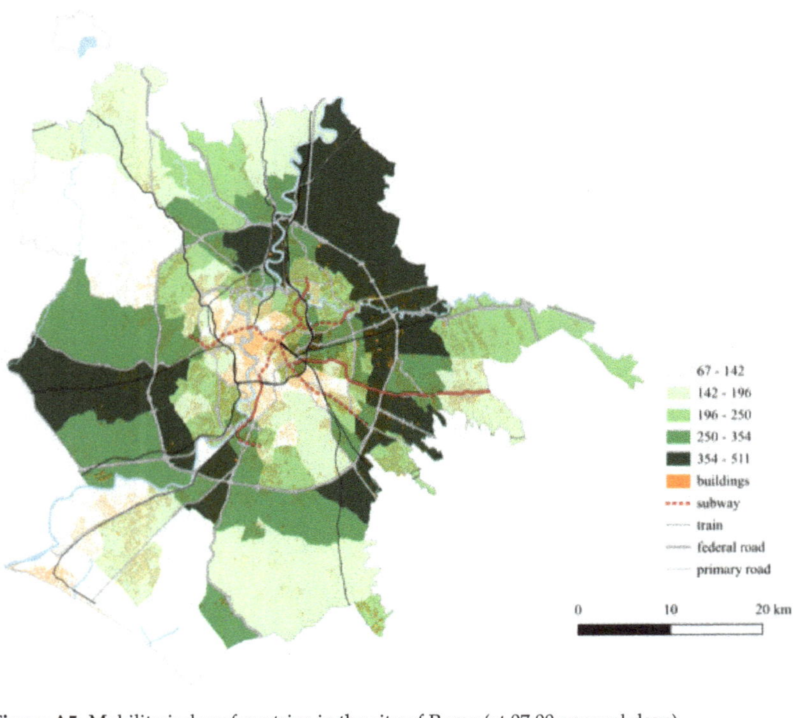

Figure A5. Mobility index of car trips in the city of Rome (at 07:00 on weekdays).

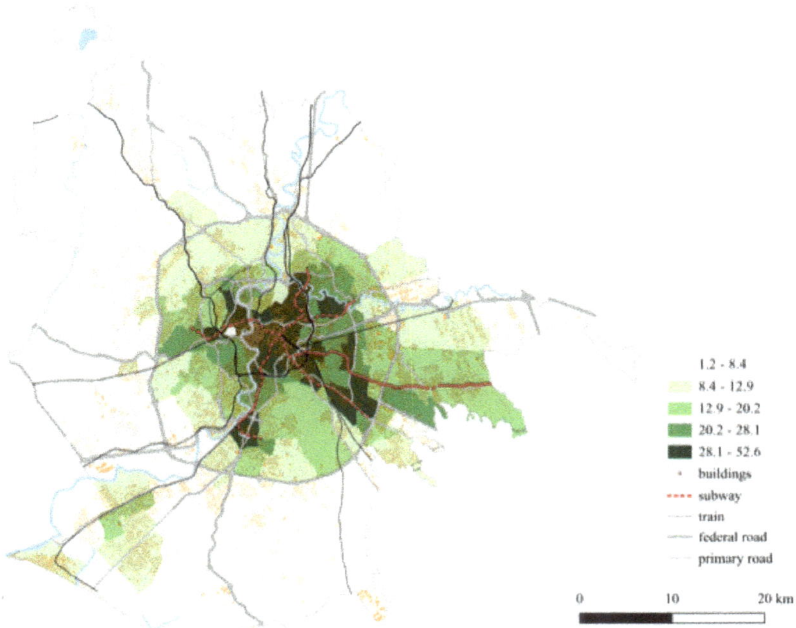

Figure A6. Mobility index of TP trips in the city of Rome (at 07:00 on weekdays).

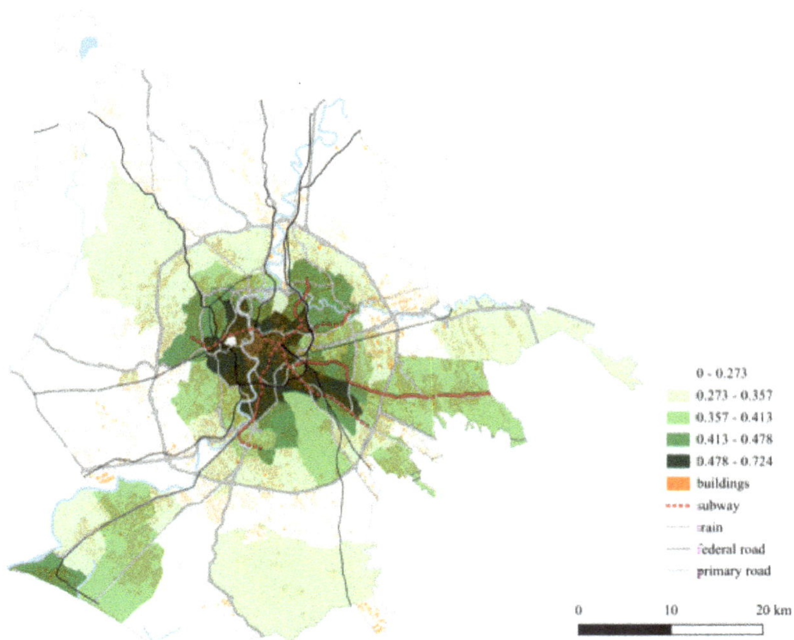

Figure A7. Competitiveness index in the city of Rome (at 07:00 on weekdays).

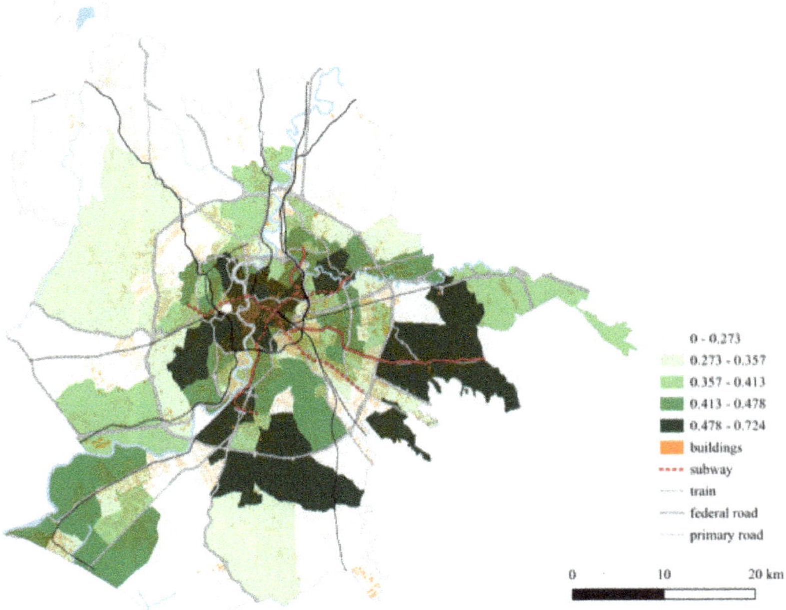

Figure A8. Accessibility index in the city of Rome (at 07:00 on weekdays).

References

1. Swilling, M.; Hajer, M.; Baynes, T.; Bergesen, J.; Labbe, F.; Musango, J.; Ramaswami, A.; Robinson, B.; Salat, S.; Suh, S.; et al. The Weight of Cities: Resource Requirements of Future Urbanization. 2018. Available online: https://www.resourcepanel.org/sites/default/files/documents/document/media/the_weight_of_cities_full_report_english.pdf (accessed on 6 September 2023).
2. Mashall, A. *Principles of Economics*, 8th ed.; Macmillan: London, UK, 1920.
3. Mattioli, G.; Roberts, C.; Steinberger, J.K.; Brown, A. The political economy of car dependence: A systems of provision approach. *Energy Res. Soc. Sci.* **2020**, *66*, 101486. [CrossRef]
4. Lucas, K.; Schwanen, T. Auto Motives: Understanding Car Use Behaviors. In *Governing Cities on the Move: Functional and Management Perspectives on Transformations of European Urban Infrastructures*; Lucas K., Blumenberg, E., Weinberger, R., Eds.; Emerald Group Publishing: Bradford, UK, 2011; pp. 3–38.
5. Lunke, E.B.; Fearnley, N.; Aarhaug, J. Public transport competitiveness vs. the car: Impact of relative journey time and service attributes. *Res. Transp. Econ.* **2021**, *90*, 101098. [CrossRef]
6. STEG, L. Can public transport compete with the private car? *IATSS Res.* **2003**, *27*, 27–35. [CrossRef]
7. Domencich, T.A.; McFadden, D. Urban travel Demand: A Behavioral Analysis: A Charles River Associates Research Study. 1975. Available online: https://eml.berkeley.edu/~mcfadden/travel/front.pdf. (accessed on 24 May 2024).
8. Ben-Akiva, M.; Lerman, S.R. *Discrete Choice Analysis: Theory and Application to Travel Demand*; The MIT Press: Cambridge, MA, USA, 2018.
9. Foerster, J.F. Mode choice decision process models: A comparison of compensatory and non-compensatory structures. *Transp. Res. Part A Gen.* **1979**, *13*, 17–28. [CrossRef]
10. Gärling, T.; Eek, D.; Loukopoulos, P.; Fujii, S.; Johansson-Stenman, O.; Kitamura, R.; Pendyala, R.; Vilhelmson, B. A conceptual analysis of the impact of travel demand management on private car use. *Transp. Policy* **2002**, *9*, 59–70. [CrossRef]
11. Becker, G.S. A Theory of the Allocation of Time. *Econ. J.* **1965**, *75*, 493–517. [CrossRef]
12. Zax, J.; Kain, J. Commutes, quits, and moves. *J. Urban Econ.* **1991**, *29*, 153–165. [CrossRef]
13. de Coevering P.P., V. *The Interplay Between Land Use, Travel Behaviour and Attitudes: A Quest for Causality*; Delft University of Technology: Delft, The Netherlands, 2021. [CrossRef]
14. Hägerstrand, T. What about people in regional science? *Reg. Sci. Assoc. Pap.* **1970**, *24*, 7–21. [CrossRef]
15. Lenntorp, B. *Paths in Space-Time Environment: A Time-Geographic Study of Movement Possibilities of Individuals*; Department of Geography, The Royal University of Lund: Lund, Sweden, 1976.
16. Pirie, G.H. Measuring accessibility: A review and proposal. *Environ. Plan. A* **1979**, *11*, 299–312. [CrossRef]
17. Kwan, M.P. Gender differences in space-time constraints. *Area* **2000**, *32*, 145–156. [CrossRef]

8. Lucas, K.; van Wee, B.; Maat, K. A method to evaluate equitable accessibility: Combining ethical theories and accessibility-based approaches. *Transportation* **2016**, *43*, 473–490. [CrossRef]
9. Schwartz, S.H. Normative influences on altruism. *Adv. Exp. Soc. Psychol.* **1977**, *10*, 221–279.
0. Triandis, H.C. *Interpersonal Behavior*; Brooks/Cole: Baltimore, MD, USA, 1977.
1. Ajzen, I. The theory of planned behavior. *Organ. Behav. Hum. Decis. Process.* **1991**, *50*, 179–211. [CrossRef]
2. Bamberg, S.; Schmidt, P. Incentives, morality, or habit? Predicting students' car use for university routes with the models of Ajzen, Schwartz and Triandis. *Environ. Behav.* **2003**, *35*, 264–285. [CrossRef]
3. Hunecke, M.; Haustein, S.; Grischkat, S.; Bohler, S. Psychological, socio- demographic, and infrastructural factors as determinants of ecological impact caused by mobility behavior. *J. Environ. Psychol.* **2007**, *27*, 277–292. [CrossRef]
4. Reckwitz, A. Toward a theory of social practices: A development in culturalist theorizing. *Eur. J. Soc. Theory* **2002**, *5*, 243–263. [CrossRef]
5. Sheller, M. Automotive emotions: Feeling the car. *Theory Cult. Soc.* **2004**, *21*, 221–245. [CrossRef]
6. Urry, J. *Mobilities*; Polity: Cambridge, UK, 2007.
7. Merriman, P. Automobility and the geographies of the car. *Geogr. Compass* **2009**, *3*, 586–599. [CrossRef]
8. TUMI—Transformative Urban Mobility Initiative. Sustainable Urban Transport: Avoid-Shift-Improve (A-S-I). 2019. Available online: https://transformative-mobility.org/multimedia/avoid-shift-improve-instruments/ (accessed on 31 July 2023).
9. Jarre, M.; Noussan, M.; Campisi, E. Avoid–Shift–Improve: Are Demand Reduction Strategies Under-Represented in Current Energy Policies? *Energies* **2024**, *17*, 4955. [CrossRef]
0. Derudder, B.; Meijers, E.; Harrison, J.; Hoyler, M.; Liu, X. Polycentric urban regions: Conceptualization, identification and implications. *Reg. Stud.* **2022**, *56*, 1–6. [CrossRef]
1. Parr, J. The Polycentric Urban Region: A Closer Inspection. *Reg. Stud.* **2004**, *38*, 231–240. [CrossRef]
2. Susilo, Y.O.; Maat, K. The influence of built environment to the trends in commuting journeys in the Netherlands. *Transportation* **2007**, *34*, 589–609. [CrossRef]
3. Handy, S. *Accessibility-vs. Mobility-Enhancing Strategies for Addressing Automobile Dependence*; U.S. UC Davis: Institute of Transportation Studies: Davis, CA, USA 2002.
4. Hansen, W. How Accessibility Shapes Land Use. *J. Am. Plan. Assoc.* **1959**, *25*, 73–76. [CrossRef]
5. O'Toole, R. *Gridlock: Why We're Stuck in Traffic and What to Do About It*; Cato Institute Press: Washington, DC, USA, 2010.
6. Borlini, B.; Memo, F. *Ripensare L'accessibilità Urbana*; Cittalia: Rome, Italy, 2009; Paper 2/2009. Available online: https://www.cittalia.it/wp-content/uploads/2020/01/citt%C3%A0-P-2009-02-Ripensare-l%E2%80%99accessibilit%C3%A0-urbana.pdf (accessed on 31 July 2023).
7. Dahrendorf, R. *Per un Nuovo Liberalismo*; Laterza: Bari, Italy, 1993.
8. Sen, A. *Commodities and Capabilities*; North-Holland: Amsterdam, The Netherlands, 1985.
9. Nussbaum, M. Nature, Function, and Capability: Aristotle on Political Distribution. In *Oxford Studies in Ancient Philosophy: Supplementary Volume*; Oxford University Press: Oxford, UK, 1988; pp. 73–76.
0. Bruzzone, F.; Cavallaro, F.; Nocera, S. The definition of equity in transport. *Transp. Res. Procedia* **2023**, *69*, 440–447. [CrossRef]
1. European Commission, Study on the Social Dimension of the Future EU Transport System Regarding Users and Passengers. 2021. Available online: https://transport.ec.europa.eu/transport-themes/social-issues-equality-and-attractiveness-transport-sector/studies/study-social-dimension-future-eu-transport-system-regarding-users-and-passengers_en (accessed on 6 September 2023).
2. European Commission, Cohesion Policy Investments on Track for Rail Transport. 2021. Available online: https://ec.europa.eu/regional_policy/whats-new/panorama/2021/03/03-12-2021-cohesion-policy-investments-on-track-for-rail-transport_en (accessed on 6 September 2023).
3. Thomopoulos, N.; Grant-Muller, S.; Tight, M. Incorporating equity considerations in transport infrastructure evaluation: Current practice and a proposed methodology. *Eval. Program Plan.* **2009**, *32*, 351–359. [CrossRef]
4. Rawls, J. *A Theory of Justice: Original Edition*; Harvard University Press: Cambridge, MA, USA, 1971.
5. *International Transport Forum, Improving Transport Planning and Investment Through the Use of Accessibility Indicators*; OECD: Paris, France, 2019. [CrossRef]
6. Jones, P.; Lucas, K. The social consequences of transport decision-making: Clarifying concepts, synthesising knowledge and assessing implications. *J. Transp. Geogr.* **2012**, *21*, 4–16. [CrossRef]
7. Thomson, J.M. *Great Cities and Their Traffic*; Gollancz: London, UK, 1977.
8. Martens, K.; Ciommo, F.D. Travel time savings, accessibility gains and equity effects in cost–benefit analysis. *Transp. Rev.* **2017**, *37*, 152–169. [CrossRef]
9. Tsamboulas, D.; Yiotis, G.S.; Panou, K.D. Use of Multicriteria Methods for Assessment of Transport Projects. *J. Transp. Eng.* **1999**, *125*, 407–414. [CrossRef]
0. Inayathusein, A.; Cooper, S. *London's Accessibility Indicators*; OECD: Paris, France, 2018. [CrossRef]
1. Dolega, L.; Pavlis, M.; Singleton, A. Estimating attractiveness, hierarchy and catchment area extents for a national set of retail centre agglomerations. *J. Retail. Consum. Serv.* **2016**, *28*, 78–90. [CrossRef]
2. Di Ciommo, F. *How the Inaccessibility Index Can Improve Transport Planning and Investment*; OECD: Paris, France, 2018. [CrossRef]

53. Frank, L.; Bradley, M.; Kavage, S.; Chapman, J.; Lawton, T. Urban form, travel time, and cost relationships with tour complexity and mode choice. *Transportation* **2008**, *35*, 37–54. [CrossRef]
54. Dovey, K.; Woodcock, I.; Pike, L. Isochrone Mapping of Urban Transport: Car-dependency, Mode-choice and Design Research *Plan. Pract. Res.* **2017**, *32*, 402–416. [CrossRef]
55. Bocarejo S., J.P.; Oviedo H., D.R. Transport accessibility and social inequities: A tool for identification of mobility needs and evaluation of transport investments. *J. Transp. Geogr.* **2012**, *24*, 142–154. [CrossRef]
56. Hull, A.; Silva, C.; Bertolini, L. (Eds.) Gravity-based accessibility measures for integrated transport-land use planning (GraBAM) In *Accessibility Instruments for Planning Practice*; COST: Bruxelles, Belgium, 2012; pp. 117–124.
57. Delmelle, E.C.; Casas, I. Evaluating the spatial equity of bus rapid transit-based accessibility patterns in a developing country The case of Cali, Colombia. *Transp. Policy* **2012**, *20*, 36–46. [CrossRef]
58. Mishra, S.; Welch, T.F.; Jha, M.K. Performance indicators for public transit connectivity in multi-modal transportation networks *Transp. Res. Part A Policy Pract.* **2012**, *46*, 1066–1085. [CrossRef]
59. Kaplan, S.; Popoks, D.; Prato, C.G.; Ceder, A.A. Using connectivity for measuring equity in transit provision. *J. Transp. Geogr* **2014**, *37*, 82–92. [CrossRef]
60. Mavoa, S.; Witten, K.; McCreanor, T.; O'Sullivan, D. GIS based destination accessibility via public transit and walking in Auckland New Zealand. *J. Transp. Geogr.* **2012**, *20*, 15–22. [CrossRef]
61. Salonen, M.; Toivonen, T. Modelling travel time in urban networks: Comparable measures for private car and public transport *J. Transp. Geogr.* **2013**, *31*, 143–153. [CrossRef]
62. Kawabata, M. Spatiotemporal Dimensions of Modal Accessibility Disparity in Boston and San Francisco. *Environ. Plan. A Econ Space* **2009**, *41*, 183–198. [CrossRef]
63. Hess, D.B. Access to Employment for Adults in Poverty in the Buffalo-Niagara Region. *Urban Stud.* **2005**, *42*, 1177–1200 [CrossRef]
64. Raveau, S.; Muñoz, J.; de Grange, L. A topological route choice model for metro. *Transp. Res. Part A Policy Pract.* **2011**, *45*, 138–147 [CrossRef]
65. Biazzo, I.; Monechi, B.; Loreto, V. General scores for accessibility and inequality measures in urban areas. *R. Soc. Open Sci.* **2019** *6*, 190979. [CrossRef] [PubMed]
66. Kawabata, M. Job Access and Employment among Low-Skilled Autoless Workers in US Metropolitan Areas. *Environ. Plan. A Econ. Space* **2003**, *35*, 1651–1668. [CrossRef]
67. Kawabata, M.; Shen, Q. Commuting Inequality between Cars and Public Transit: The Case of the San Francisco Bay Area 1990–2000. *Urban Stud.* **2007**, *44*, 1759–1780. [CrossRef]
68. Levinson, D.M. Accessibility and the journey to work. *J. Transp. Geogr.* **1998**, *6*, 11–21. [CrossRef]
69. Silva, C.; Pinho, P. The Structural Accessibility Layer (SAL): Revealing how Urban Structure Constrains Travel Choice. *Environ Plan. A Econ. Space* **2010**, *42*, 2735–2752. [CrossRef]
70. Shen, Q. A Spatial Analysis of Job Openings and Access in a U.S. Metropolitan Area. *J. Am. Plan. Assoc.* **2001**, *67*, 53–68 [CrossRef]
71. Niedzielski, M.; Śleszyński, P. Analyzing accessibility by transport mode in Warsaw, Poland. *Geogr. Pol.* **2008**, *81*, 61–78.
72. Kwok, R.C.W.; Yeh, A.G.O. The Use of Modal Accessibility Gap as an Indicator for Sustainable Transport Development. *Environ Plan. A Econ. Space* **2004**, *36*, 921–936. [CrossRef]
73. Chia, J.; Lee, J.B. Extending public transit accessibility models to recognise transfer location. *J. Transp. Geogr.* **2020**, *82*, 102618 [CrossRef]
74. Onlus, G. L'insostenibile Mobilità di Roma. 2021. Available online: https://www.greenpeace.org/italy/rapporto/14279/linsostenibile-mobilita-di-roma/ (accessed on 6 September 2023).
75. Lelo, K.; Monni, S.; Tomassi, F. *Le Sette Rome. La Capitale Delle Disuguaglianze Raccontata in 29 Mappe*; Donzelli Editore: Roma, Italy, 2021
76. Staricco, L. 15-, 10- or 5-minute city? A focus on accessibility to services in Turin, Italy. *J. Urban Mobil.* **2022**, *2*, 100030. [CrossRef]
77. Ceccato, R.; Deflorio, F.; Diana, M.; Pirra, M. Measure of urban accessibility provided by transport services in Turin: A traveller perspective through a mobility survey. *Transp. Res. Procedia* **2020**, *45*, 301–308. [CrossRef]
78. Roberto, R.; Zini, A.; Felici, B.; Rao, M.; Noussan, M. Potential Benefits of Remote Working on Urban Mobility and Related Environmental Impacts: Results from a Case Study in Italy. *Appl. Sci.* **2023**, *13*, 607. [CrossRef]
79. David O'Sullivan, A.M.; Shearer, J. Using desktop GIS for the investigation of accessibility by public transport: An isochrone approach. *Int. J. Geogr. Inf. Sci.* **2000**, *14*, 85–104. [CrossRef]
80. Travel Time. What is TravelTime? Available online: https://docs.traveltime.com/api/overview/introduction (accessed on 24 July 2023).
81. TravelTime. Public Transport Model Overview. Available online: https://help.traveltime.com/en/articles/8523661-public-transport-model-overview (accessed on 24 September 2024).
82. Geofabrik. OpenStreetMap. Available online: https://www.geofabrik.de/geofabrik/openstreetmap.html (accessed on 24 July 2023)
83. OpenStreetMap. OpenStreetMap. Available online: https://www.openstreetmap.org/#map=5/42.088/12.564 (accessed on 24 July 2023).

84. Gamper, J.; Böhlen, M.; Cometti, W.; Innerebner, M. Defining Isochrones in Multimodal Spatial Networks. In *Proceedings of the 20th ACM International Conference on Information and Knowledge Management, Glasgow Scotland, UK, 24–28 October 2011*; ACM Digital Library: New York, NY, USA; pp. 2381–2384 [CrossRef]
85. Forsch, A.; Dehbi, Y.; Niedermann, B.; Oehrlein, J.; Rottmann, P.; Haunert, J.H. Multimodal travel-time maps with formally correct and schematic isochrones. *Trans. GIS* **2021**, *25*, 3233–3256. [CrossRef]
86. di Torino, C. Open Data of the City on Turin, Statistical Zones. Available online: http://aperto.comune.torino.it/sv/dataset/zone-statistiche (accessed on 6 September 2023).
87. ISTAT. Italian National Institute of Statistics. Available online: https://www.istat.it/en/ (accessed on 6 September 2023).
88. OpenStreetMap. Open Data of the City on Rome, Urbanistic Zones. Available online: https://wiki.openstreetmap.org/wiki/Rome (accessed on 6 September 2023).
89. TomTom. Origin-Destination Matrix. Available online: https://www.tomtom.com/products/origin-destination-matrix-analysis/ (accessed on 24 July 2023).
90. Noussan, M.; Carioni, G.; Sanvito, F.D.; Colombo, E. Urban Mobility Demand Profiles: Time Series for Cars and Bike-Sharing Use as a Resource for Transport and Energy Modeling. *Data* **2019**, *4*, 108. [CrossRef]
91. Il Sole 24 Ore. Da Portofino a Cavargna: Ecco, Comune per Comune, Dove i Redditi Sono Cresciuti Di Più. 2024. Available online: https://www.ilsole24ore.com/art/da-portofino-cavargna-ecco-comune-comune-dove-redditi-sono-cresciuti-piu--slittano-misure-irpef-ires-AFAmvJiD?refresh_ce (accessed on 7 October 2024).
92. ISFORT. Rapporto Mobilità 2022. 2022. Available online: https://www.isfort.it/wp-content/uploads/2023/07/RapportoMobilita2022.pdf (accessed on 6 September 2023).
93. Roma Servizi per la Mobilità Srl. Rapporto Roma Mobilità 2022. 2022. Available online: https://drive.google.com/file/d/1lqPT1wUBSq_3ix48yKt7dxcqu4XAquBv/view (accessed on 6 September 2023).
94. ISTAT. Ambiente Urbano—2021. 2023. Available online: https://www.istat.it/it/archivio/286822 (accessed on 6 September 2023).
95. Agenzia della Mobilità Piemontese. Indagine IMQ 2022—Rapproto di Sintesi. 2023. Available online: https://mtm.torino.it/wp-content/uploads/dati-statistiche/indagine-imq-2022/IMQ_2022_Risultati_27-07-2023.pdf (accessed on 6 September 2023).
96. Erik, E.; Solá, A.G.; Larsson, A. Featured Graphic. Spatial Inequality and Workplace Accessibility: The Case of a Major Hospital in Göteborg, Sweden. *Environ. Plan. A Econ. Space* **2012**, *44*, 2295–2297. [CrossRef]
97. Berdini, P. *Breve Storia Dell'abuso Edilizio in Italia dal Ventennio Fascista al Prossimo Futuro*; Donzelli Editore: Roma, Italy, 2010.
98. Cellamare, C. *Le Diverse Periferie di Roma e le Forme di Autorganizzazione*; Working Papers.; Rivista online di Urban@it—2/2016; Centro nazionale di studi per le politiche urbane: Bologna, Italy 2016.
99. Cerasoli, M. *Periferie Urbane Degradate. Regole Insediative e Forme Dell'abitare. Come Intervenire?* Cittalia: Rome, Italy, 2008; Paper 2/2008.
100. Cini, L.; Colloca, P.; Maggini, F.; Tomassi, F.; Valbruzzi, M. Inchiesta su periferie urbane, disagio socio-economico e voto. I casi di Bologna, Firenze e Roma. *Ance Quad. Sci. Politica* **2021**, *28*, 137–177. [CrossRef]
101. Lelo, K.; Risi, G. Sviluppo urbano tra abusivismo e disuguaglianze: Il caso di Roma. *DITE Riv. Studio Delle Din. Territ. Assoc. Ital. Sci. Reg.* **2022** . Available online: https://www.dite-aisre.it/sviluppo-urbano-tra-abusivismo-e-disuguaglianze-il-caso-di-roma/ (accessed on 6 September 2023).

smart cities

MDPI

Article

Optimizing Smart City Street Design with Interval-Fuzzy Multi-Criteria Decision Making and Game Theory for Autonomous Vehicles and Cyclists

Maryam Fayyaz [1], Gaetano Fusco [2], Chiara Colombaroni [2], Esther González-González [1,*] and Soledad Nogués [1]

[1] GEURBAN Research Group, School of Civil Engineering, University of Cantabria, 39005 Santander, Spain; maryam.fayyaz@alumnos.unican.es (M.F.); soledad.nogues@unican.es (S.N.)

[2] Department of Civil, Constructional and Environmental Engineering (DICEA), Sapienza University of Rome, 00184 Rome, Italy; gaetano.fusco@uniroma1.it (G.F.); chiara.colombaroni@uniroma1.it (C.C.)

* Correspondence: gonzalezme@unican.es

Highlights:

What are the main findings?

- Safety is the most critical factor in designing urban streets that integrate cyclists and autonomous vehicles (AVs);
- Green infrastructure and smart technology adoption are the optimal integration strategies.

What are the implications of the main findings?

- These strategies foster a balanced coexistence of cyclists and AVs, leading to a more efficient transport system and a more sustainable urban environment in the driverless era.
- This research provides valuable guidance for urban planners and decision makers on the implementation of AVs on our streets, while advocating for sustainable and active mobility.

Abstract: Encouraging older and newer mobility alternatives to standard privately owned cars, such as cycling and autonomous vehicles, is necessary to reduce pollution, enhance safety, increase transportation efficiency, and create a more sustainable urban environment. Implementing mobility plans that identify the use of different transport modes in their confidence intervals can lead to the development of smarter and more efficient cities, where all citizens can benefit from safe and environmentally friendly streets. This research aims to provide insights into designing urban streets that seamlessly integrate autonomous vehicles and cyclists, promoting sustainable mobility while ensuring urban transport efficiency. With this aim, the research identifies and prioritizes the factors that are relevant to street design as well as the appropriate strategies to address them. Our methodology combines Multi-Criteria Decision-Making (MCDM) with Game theory to identify and realize the most convenient conditions for this integration. Initially, the basic factors were identified using the value-interval fuzzy Delphi method. Following this, the factors were weighted with the interval-fuzzy Analytic Network Process (ANP), and the cause-and-effect variables were evaluated using the interval-fuzzy Decision-Making Trial and Evaluation Laboratory ANP (DANP). Finally, Game theory was employed to determine the optimal model for addressing these challenges. The results indicate that safety emerged as the most significant factor and two optimal strategies were identified; the integration of green infrastructure and smart technology.

Keywords: autonomous vehicles; smart city; street design; game theory; interval-fuzzy MCDM

1. Introduction

In recent years, there has been a growing interest in transport alternatives that can offer more sustainable, efficient, and inclusive solutions for urban mobility, such as bicycles, or that enhance technological performance, such as self-driving or Autonomous Vehicles (AVs).

Achieving dynamic, inclusive, and environmentally friendly urban environments that accommodate the varied requirements and desires of city dwellers is linked to the prioritizing non-motorized modes of transport [1,2], such as cycling. Bicycles are low-cost, zero-emission, and healthy modes of transport that can improve physical and mental well-being and reduce the demand for car trips and parking spaces [3]. By providing secure and easily accessible pathways for cyclists, citizens' health is improved, car dependency is decreased, and both traffic congestion and atmospheric contamination are alleviated [4,5]. However, urban cyclists face various challenges, such as safety concerns, a lack of dedicated infrastructure, traffic congestion, and conflicts with other road users [6], which result in their minority use in most cities.

The imminent arrival of AVs has generated an increasing need to plan and design roadway spaces that can accommodate both traditional and new transportation modes [7], which can make cycling even more difficult. Emerging AVs can operate without human intervention, using sensors, cameras, and artificial intelligence to comprehend and navigate the traffic environment [8,9], and it is expected that AVs can potentially enhance the safety, convenience, and accessibility of urban transport, as well as optimize the use of road space and energy [10]. However, to fully realize the benefits of AVs, it is not enough to simply introduce them into existing street spaces designed for conventional cars [10,11]. Cyclists are vulnerable road users who require adequate infrastructure and protection from motorized traffic, while AVs are intelligent agents who need technological devices to communicate and cooperate with other vehicles and streets, requiring new traffic regulations and formal rules [12,13].

Autonomous vehicles have distinct capacities to tackle certain difficulties. They are primarily designed to strictly comply with traffic restrictions, therefore minimizing the likelihood of accidents caused by human mistakes, which is a major issue for cyclists [14]. Furthermore, AVs' capacity to effectively communicate with one another and the surrounding infrastructure shows a potential for improving bicycle safety and preventing accidents through precise anticipation of their actions. By sharing real-time data with other AVs and traffic management systems, they can adjust their speed, direction, and braking patterns to avoid collisions [15]. For instance, if an AV detects a cyclist approaching an intersection, it can communicate this information to other AVs nearby, prompting them to slow down or stop to ensure the cyclist's safety. Additionally, AVs can interact with smart traffic signals and road signs, which can provide warnings to both drivers and cyclists about potential hazards ahead. This interconnected network enhances the overall safety of urban environments by minimizing human error and ensuring a coordinated response to dynamic traffic situations [16]. Moreover, AVs can potentially mitigate traffic congestion by optimizing routes and decreasing the number of vehicles on the road [17], therefore improving the practicality and effectiveness of cycling as a means of urban transportation.

While there are positive aspects to consider, designing streets to accommodate a variety of transportation modes, including bicycles and self-driving cars, poses challenges that require thorough evaluation and prioritization, based on their importance and level of uncertainty [16,18]. Although previous studies analyzed street design factors and aspects of the interaction between cyclists and AVs, no comprehensive research was found that simultaneously addresses the identification of lane design factors and the proposal of strategies to address them. This research aims to provide valuable insights into the issues and guidelines for redesigning streets to integrate autonomous vehicles and cyclists harmoniously and efficiently, creating a more balanced relationship between different road users and promoting sustainable mobility patterns. This research proposes a novel mixed-method approach, combining Multi-Criteria Decision Making methods (MCDM), fuzzy Delphi, and Analytical Network Process (ANP) with DEMATEL (DANP) with Game theory. The Delphi method involves consulting experts on the most relevant factors affecting street design for autonomous vehicles and cyclists, as extracted from the literature. Then, the DANP is applied to evaluate and rank these factors. Later, Game theory is used to address the issues with several technological, behavioral, spatial, and policy strategies. This combination introduces several advantages, namely, comprehensive evaluation of conflicting factors, flexibility in decision making, providing multiple optimal strategies, and practicality, ensuring robustness against uncertainties.

Based on the above, the present study is organized as follows: Section 2 reviews the existing literature to understand the factors involved in designing street spaces for both bicycles and AVs and methods used to assess and estimate urban design and planning factors and policies. Section 3 details the combined methodology of fuzzy MDCM and Game theory, outlining how these methods collectively identify and prioritize key issues in smart city street design and strategies to overcome them. Section 4 presents the results obtained, pointing to the most effective strategies to address the integration of cyclists and AVs. Finally, Section 5 provides the main conclusions of the research, discusses the results, and introduces further lines of research.

2. Literature Review

2.1. Street Design for Cyclists and AVs

Bicycle use is directly linked to several factors that relate not only to the formal configuration of cycling infrastructures, i.e., the existence and width of cycling lanes, but also to the urban environment in which they are located [19,20]. These factors can be classified into barriers that discourage or facilitators that enhance the activity [21,22].

In a study of the relationship between urban planning and physical activity, Lee and Moudon [23] established four categories of barriers that discourage activities such as walking and cycling, namely, opportunity, distance, access, and safety. These barriers include unsafe route conditions, poor maintenance, or a lack of high-quality route-related facilities. Based on this classification, Wang et al. [19] conducted an extensive review of the literature, organizing the barriers as follows: opportunity barriers related to the lack of cycling or recreational facilities; access and distance barriers associated with unconnected networks or large travel distances; safety barriers regarding insecure environments or crossing points and high crime and accident rates; and physical setting barriers involving pleasant landscapes, high-quality environments, and weather conditions. Well-maintained paths and short and flattened trails were identified as the most effective attributes to encourage cycling.

Traffic and interaction with motorized vehicles are highlighted as major concerns. Streets with heavy traffic and/or high traffic speeds put cyclists at risk and create a sense of insecurity [24,25], which particularly affects women [26]. The provision of appropriate cycling infrastructures/facilities, such as cycling lanes or trails and bike racks, is obviously the first requirement to enhance this mode of transportation [20,21]. However, the physical separation of theses cycle lanes increases objective and perceived levels of safety and comfort, as pointed out by several studies [27,28].

A relevant factor when talking about cycling infrastructure is accessibility. To make cycling accessible and usable for all people, including people with disabilities, inclusive street design should be promoted, and it is a topic on which there is still little research [29, 30]. Design requirements should consider, among other technical aspects, a wider lane width and turning radius, adapted to the specific characteristics of the type of bicycle used by people with disabilities [29]. In addition, the continuity of networks and on-street bicycle lanes is considered a major facilitator to cycling activity [20,31].

Intersections, where the interaction between cyclists and other road users is higher, are a particularly sensitive element [31–33]. In their study of cyclists' risk perceptions, Parkin et al. [27] demonstrated that roundabouts were associated with higher perceived risk than traffic signal-controlled intersections. Roundabouts can be designed as high-capacity and high-speed intersections or as intersections with wide approaches and traffic lanes that help calm traffic while providing connections between different links in the network [34].

Intersections not only increase the perceived risk for cyclists, who are some of the most vulnerable road users, but also account for a significant proportion of all crashes [34].

A number of accidents are due to drivers failing to see cyclists in time, demonstrating that visibility is also an important factor. Improving the visibility of cyclists is particularly relevant in adverse weather conditions, on streets with poor or no lighting and at night [35,36]. Solving such problems can also help improve the sustainability of street conditions. Sustainable practices, including the use of renewable materials and energy-efficient designs, are key to improving safety while reducing environmental impacts and ensuring long-term sustainability. For example, adaptive street lighting that adjusts intensity based on the detection of vehicles, pedestrians, and cyclists, is a priority when dealing with visibility issues, as well as the promotion of energy-efficient street design [37]. Similarly, new pavements can prevent falls and injuries and also mitigate the heat island effect and water runoff problems associated with asphalt [38]. In fact, road surface conditions are also relevant factors in cycling infrastructures: deteriorated pavement, the presence of obstacles, glass/debris, and especially slopes have a negative influence [19,21,24,25]. Although it may be attractive to some experienced cyclists [39], in general the presence of steep inclines is a factor negatively associated with cycling for transportation [40]. This is because unfavorable slopes can cause accidents and slow traffic flow.

Atmospheric environmental factors, such as meteorological conditions and air quality, also exert influences on individuals' cycling activity levels. Bad weather, ice and snow, and the presence of air pollutants have been found to have deterrent effects for cyclists [19,25]. These findings concur with those of Helbich et al. [41], who confirmed significant weather effects on cycling, highlighting that trips for leisure and in surrounding areas are more sensitive to weather conditions than trips for commuting and in more densely populated central areas.

Urban design features of travel lanes and intersections will be affected by the introduction of AVs [42,43], which may either emphasize or reduce some of the above-mentioned negative factors that affect cycling. Research on cyclist autonomous vehicle interaction has focused on similar key issues, often linked to safety, such as communication technology requirements and the design of traffic infrastructure. For example, several authors highlight the difficulties in detecting and predicting the trajectory and behavior of cyclists [44–47]. To address these issues, authors have proposed improving the interaction through communication systems. Using intelligent sensors for AVs, vehicles can better communicate the intention to merge with or overtake cyclists [48]. Solutions to facilitate safe interactions include equipping not only AVs but also the cyclists. Berge et al. [44] propose a support system for cyclists based on a passive beacon or chip system that connects them with vehicles and other road users as well as with infrastructure. On the other hand, in addition to improving current technologies, such as adaptive driving algorithms [45], advances in artificial intelligence are expected to help increase the detection capabilities of AVs [49].

In general, authors suggest that to improve the safety of cyclists it is necessary to segregate lanes by mode and multi-level crossing with AVs and active travel modes [50]. In a recent study, Ngwu et al. [51] evaluated adolescent cyclists' perceptions of traffic infrastructure designs, pointing to spacious bike lanes, separated lanes for cyclists and AVs, most preferably with physical barriers, and clear markings and signage for AVs and cyclists as the most important design elements.

However, AV-only lanes may also pose a challenge to trips undertaken by other modes of transport [52]. In contrast to suburban areas, AVs may negatively affect street compatibility with cyclists in city centers due to increased traffic volume [53]. An efficient approach involves assessing the effects of lane design on both traffic congestion and environmental considerations. When designing lanes that allow bicycles and autonomous cars to coexist peacefully, it is important to consider different route choices and the interactions between human-driven vehicles (HVs) and AVs. Allocating exclusive lanes for connected and autonomous vehicles (CAVs) can significantly enhance and decrease emissions traffic efficiency [54,55] and reduce the likelihood of accidents [56]. In addition, the process of incorporating AVs into the current road infrastructure necessitates substantial financial resources and a considerable amount of time [57].

Another feasible design option to enhance network efficiency is the implementation of a multi-lane AV zone, dedicated to both AVs and HVs, as suggested by Roy et al. [58]. In this case, AVs must possess the ability to reliably perceive and respond appropriately to interactions with cyclists and drivers. This includes being able to interpret visual cues such as eye contact and manual signals, as emphasized by Park and Sohn [56].

Despite its significance, research on improving bicycle AV interactions has been scarce. This scarcity is attributed by Wang et al. [59] to the quick progress of AV technology, which surpasses academic research endeavors, posing challenges for academics in staying updated with advancements and conducting thorough studies. In addition, the complexity of urban settings and the ever-changing interactions between street users present difficulties in creating feasible simulation scenarios and gathering pertinent data [2].

These contributions serve as examples of research that emphasizes the importance of carefully planning and designing streets that improve cycling experiences and interactions between cyclists and AVs. Even while cutting-edge AV technology has the potential to dramatically lower accident rates and increase bike safety, challenges involving the requirements for suitable infrastructure and intricate urban interactions still exist. These types of complex decisions, where diverse conflicting factors should be considered simultaneously, need decision support tools that can help planners and designers identify the most optimal design and planning strategies, as explained below.

2.2. Methods Used to Assess and Estimate Urban Design and Planning Challenges and Policies

The design of street spaces affects the mobility, safety, and well-being of various road users, and is always associated with several factors that seem even more complex with the introduction of new modes of transport, such as AVs. To assess and identify these factors, various methods and criteria can be used, e.g., Multi-Criteria Decision Making (MCDM) or Game theory.

MCDM methods assist decision makers in making well-informed choices in intricate scenarios with numerous objectives and constraints [60]. The fuzzy approach to MCMD has the relevant advantage that it enables a better representation of the vagueness of different criteria perceptions inherent in the human decision-making process involving quantitative and qualitative attributes.

For example, the fuzzy Delphi method (FDM) is an MCDM technique that uses expert opinions and linguistic variables to reach a consensus on the importance and uncertainty of the factors [61]. Liu et al. [62] showed that the FDM can effectively prioritize the infrastructure needs for connected and autonomous vehicles (CAVs), such as safe harbors and charging facilities, which are crucial for their integration into urban environments. Additionally, Liu et al. [63] demonstrated the application of this method in optimizing dedicated lanes for CAVs, considering factors like traffic flow and safety, which are pertinent to urban design and transportation planning. Using the MCDM technique like Electre III and AHP, Kiciński et al. [64] assessed various scenarios for the urban public transportation system in Cracow, Poland. Ten criteria encompassing economic, technical, social, and environmental elements formed the basis of the assessment of several combinations of high-speed rail, tram, bus, and subterranean transportation systems.

Game theory is a branch of mathematics and economics that studies the strategic behavior of rational agents in situations of conflict and co-operation [65–67], and can be used to model and analyze various phenomena in urban planning, design, and transport studies [68,69]. For example, Cortés-Berrueco et al. [70] applied Game theory to analyze driver strategies and their impact on traffic flow. Their study highlights a method of optimizing mobility by modeling driver interactions and lane-changing behaviors, with significant implications for both urban design and traffic control strategies. In a similar vein, Zhu et al. [71] propose a Game theory-based lane-changing conflict management model for automated vehicles. It addresses conflicts of interest between vehicles, such as sacrificing speed for lane-changing vehicles. This method has also been used in autonomous-driving-related studies, such as by Liu et al. [62]. By integrating autonomous driving technologies, cities can optimize transportation systems, promote sustainability, and enhance the urban experience. Urban planners can focus on optimizing traffic flow, reducing congestion, and promoting alternative modes of transportation. The framework also emphasizes safety, reducing accidents, and promoting pedestrian-friendly environments. Also related to AVs, Wang and colleagues [59] introduced a novel approach to determine lane changes in communal spaces, which could influence urban planning. They utilized the overtaking expectation parameter and a unique approach to determine the optimal option in situations where no driver is willing to switch lanes.

3. Research Methodology

This research proposes to create a thorough three-step model by combining Multi-Criteria Decision Making (MCDM)—fuzzy Delphi and fuzzy DANP—with Game theory to facilitate the design of road space for the peaceful and effective cohabitation of cyclists and AVs. It utilizes a range of tactics, including technical breakthroughs, behavioral insights, spatial considerations, and legislative frameworks, to tackle the issues related to their interplay.

In this study, MCDM techniques were employed to assess and rank the different factors that impact cyclists' AV interactions in lane design, like safety, efficiency, and user preferences [72]. The fuzzy Delphi method (FDM) is an enhancement of the traditional Delphi method, integrating fuzzy logic to handle the inherent uncertainty and subjectivity in expert opinions. The traditional Delphi method involves a series of iterative rounds of questioning among a panel of experts to reach a consensus on a particular issue. The FDM could offer significant advantages in identifying factors (i.e., barriers and facilitators) in smart city street design for AVs and cyclists by systematically aggregating expert opinions with uncertainty handling, ensuring robust consensus. This method enhances decision-making accuracy by incorporating diverse perspectives and addressing the inherent vagueness in expert judgments. By incorporating fuzzy logic, FDM allows for a more flexible and nuanced expression of expert judgments, capturing the degree of confidence or uncertainty in their responses.

The combination of the fuzzy ANP and the Decision-Making Trial and Evaluation Laboratory (DEMATEL) was then used to determine the final relative importance of designing factors and calculate the internal relationships among them, given that this hybrid method may manage complicated situations with interdependencies and feedback among criteria (i.e., factors) and alternatives [73,74]. Combining fuzzy ANP with DEMATEL (DANP) has the potential to simplify and standardize decision making, while also offering a more thorough and practical examination of the issue at hand [75].

Lastly, Game theory under fuzzy sets also provides a significant framework for comprehending and simulating the strategic interactions between cyclists and AVs. Game theory models can accurately represent the motivations and actions of both parties involved, taking into account elements such as the perception of risk, the willingness to co-operate, and the drive to compete. Researchers can utilize Game theory to analyze cyclist AV interactions and devise techniques to enhance safety and optimize co-operation on the road.

By integrating these three methodologies, a thorough and resilient framework was constructed to facilitate decision making in the design of lanes for bicycles and AVs, while also considering various factors and uncertainties (Figure 1). Specifically, combining MCDM and Game theory enables a comprehensive evaluation that optimizes conflicting design factors, allows for a flexible decision making by offering multiple optimal strategies, and ensures the robustness of solutions against uncertainties. To achieve this, this methodology was grounded in the following assumptions:

- Decision makers are rational and aim to optimize the model's defined objectives;

- The methods applied (fuzzy Delphi, fuzzy DANP, and Game theory) effectively manage uncertainties and complex interdependencies among the factors;
- Interactions between factors and strategies are modeled linearly using normalized weights derived from the fuzzy ANP;
- The data collected from expert surveys are reliable and accurately represent real-world priorities.

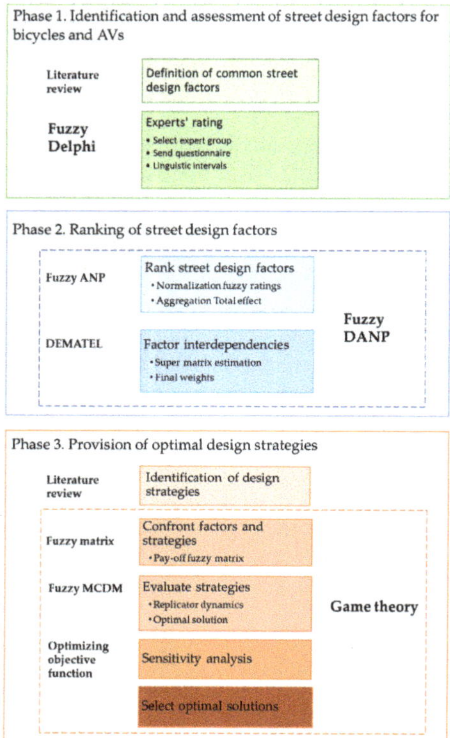

Figure 1. Proposed methodology for street design for bicycle AV interaction.

3.1. *Identifying Factors for Bicycles and AVs Street Design with Interval Fuzzy Delphi Technique*

The first step in designing street spaces is to identify and prioritize the most relevant factors or criteria (barriers and facilitators) in lane design for bicycles and AVs. To validate and ensure the selection of factors, the interval-fuzzy Delphi technique was used.

First, the desired factors were extracted from the literature review. We conducted a comprehensive literature search using electronic databases such as Google Scholar, Scopus, Science Direct, and others to identify relevant articles related to cycling infrastructure design, barriers to cycling, smart city street design, AV street design, and interactions between cyclists and AVs. As a result, seven key factors were identified in this study.

Once the factors were selected, the fuzzy Delphi technique was used to rate the importance of the factors, with a group of five experts in urban planning. Each member rated the importance independently and confidentially using a questionnaire (Figure 2) and the valuation ratings provided in Table 1. Then, the interval average of the importance of each index was calculated. Finally, the average of these numbers was diffused. Any index with a score above 0.5 was used as the final index.

Introduction:

Thank you for participating in this Delphi survey. This survey aims to identify and evaluate the factors (barriers and facilitators) in the design of smart city streets that affect the integration of autonomous vehicles and cyclists. Your expertise is crucial in assessing the significance of these factors using interval judgments. Please provide your responses considering the current and future scenarios of smart city development.

Instructions:

For each factor listed below, please indicate the range of importance (on a scale from 1 to 10) based on your professional judgment. The lower bound of the interval should represent the minimum level of importance, while the upper bound should represent the maximum level of importance.

1. **Structure:** Refers to the physical design and engineering of the street infrastructure that impacts the movement and interaction of autonomous vehicles and cyclists.
   Importance Interval: [___, ___]
2. **Sustainability**: Involves the ability to maintain street functionality with minimal environmental impact, incorporating renewable materials and energy-efficient designs.
   Importance Interval: [___, ___]
3. **Atmospheric Environmental conditions**: Concerns the influence of weather conditions, air quality, and other atmospheric factors on the safety and efficiency of autonomous vehicles and cyclists.
   Importance Interval: [___, ___]
4. **Visual Aspect**: Includes the design aesthetics, visibility, and signage that affect both the usability and appeal of the street for autonomous vehicles and cyclists.
   Importance Interval: [___, ___]
5. **Safety:** Covers all aspects of street design that protect the safety of autonomous vehicles and cyclists, including traffic management, lighting, and emergency response systems.
   Importance Interval: [___, ___]
6. **Slope**: Relates to the gradient or incline of the street, which may influence the movement, control, and energy consumption of both autonomous vehicles and cyclists.
   Importance Interval: [___, ___]
7. **Accessibility**: Refers to the ease of access and usability of the streets for all users, including those with disabilities, ensuring inclusive design for both autonomous vehicles and cyclists.
   Importance Interval: [___, ___]

Figure 2. Expert's Delphi questionnaire.

Table 1. Interval fuzzy Likert ratings normalized.

Linguistic Variables	Very Low	Low	Medium Low	Medium	Medium High	High	Very High
Equivalent interval numbers	[0.0–0.15]	[0.15–0.3]	[0.3–0.45]	[0.45–0.6]	[0.6–0.75]	[0.75–0.9]	[0.9–1.0]

3.2. Ranking and Importance of Street Design Factors with Interval-Fuzzy DANP

The subsequent step involves the application of the fuzzy DANP. The ANP expands on the Analytic Hierarchy Process (AHP) by accommodating network structures instead of only hierarchical ones [74,76]. DEMATEL is an MCDM technique that uses a communication link matrix to analyze interdependencies and causal linkages within a system [77]. It computes an ANP super-matrix to determine criterion and sub-criterion weights by using the same total effect numbers, which are then balanced to achieve unlimited power for the weights [74]. This is crucial to methodologically evaluate interrelationships and interdependencies between criteria, i.e., street design factors. In order to achieve reliable results in our study, this combined method offered a thorough framework for prioritizing factors based on their global relevance.

Various integration approaches between ANP and DEMATEL exist, each suited to specific goals or structural issues. Typically, DEMATEL identifies critical aspects and their relationships, while ANP assigns weights to factors and prioritizes alternatives. Another approach involves constructing interdependent matrices for ANP using DEMATEL, illustrating how each cluster and node influences others. In this study, we applied the combined DEMATEL and ANP algorithms described by Nematkhan et al. [78].

3.2.1. Fuzzy ANP

The fuzzy ANP is a method used for decision making, where criteria weights are provided as intervals, as in the present study. This process involves multiple steps, including the normalization of intervals and their aggregation to determine final weights, which help in handling the inherent uncertainties in the data.

An Interval Fuzzy Element (IVFE) is an advanced concept in fuzzy set theory that combines the principles of fuzzy sets with interval numbers to better handle uncertainty and imprecision. In traditional fuzzy sets, each element has a membership degree between 0 and 1, indicating its degree of belonging to the set. IVFE extends this by allowing the membership degree itself to be an interval, denoted as $[\mu_{min}, \mu_{max}]$, thus capturing a range of possible membership values. This is particularly useful in scenarios where the exact degree of membership is uncertain or varies within a range. Additionally, the factor's value can also be represented as an interval, [a, b], reflecting the inherent uncertainty in the data. This dual interval representation—one for the membership degree and one for the factor's value—provides a more flexible and robust framework for modeling and analyzing uncertain information.

The steps in this algorithm are as follows [74,79–81]:

1. Initial Computation: Experts' assessment on the mutual influence of the n factors selected is derived from IVFE according to Equation (1).

$$\widetilde{G} = \begin{bmatrix} \widetilde{g}^{11} & \cdots & \widetilde{g}^{1j} & \cdots & \widetilde{g}^{1n} \\ \vdots & & \vdots & & \vdots \\ \widetilde{g}^{i1} & \cdots & \widetilde{g}^{ij} & \cdots & \widetilde{g}^{in} \\ \vdots & & \vdots & & \vdots \\ \widetilde{g}^{n1} & \cdots & \widetilde{g}^{nj} & \cdots & \widetilde{g}^{nn} \end{bmatrix} \tag{1}$$

where $\widetilde{g}^{ij} = \left(\widetilde{\gamma}_1^{ij}, \ldots, \widetilde{\gamma}_t^{ij}, \ldots, \widetilde{\gamma}_s^{ij}\right)$ such that $\widetilde{\gamma}_t^{ij} = \left[\widetilde{\gamma}_t^{ijL}, \widetilde{\gamma}_t^{ijR}\right]$.

The decision matrix G represents pairwise relationships among criteria (factors).

$\widetilde{g}$: this represents the elements of the decision matrix, which were provided by the s experts and then converted into the decision matrix.

$\widetilde{\gamma}$: this refers to the assessment of the t-th expert and is expressed as fuzzy numbers, as denoted by the tilde sign, with the upper and lower bounds denoted by the letters R and L, respectively.

2. Standardization and Aggregation: The direct impact matrix D is standardized and then used to obtain the comprehensive impact matrix $\widetilde{T}$ using Equation (2), which provides an absorbing state of a Markov chain process as the limit of matrices D_1, D_2, ... D_m [82]:

$$\widetilde{T} = \lim_{m\to\infty}\left(\widetilde{D} + \widetilde{D}^2 + \widetilde{D}^3 + \cdots + \widetilde{D}^m\right) = \widetilde{D}\left(\widetilde{I} - \widetilde{D}\right)^{-1} \quad (2)$$

where $\widetilde{T}$ denotes the comprehensive impact matrix and the I is the identity matrix, corresponding to Equations (3) and (4).

$$\left(I - \widetilde{D}\right) = \begin{bmatrix} 1 & \cdots & 0 & \cdots & 0 \\ \vdots & 1 & \vdots & & \vdots \\ 0 & \cdots & 1 & \cdots & 0 \\ \vdots & & \vdots & 1 & \vdots \\ 0 & \cdots & 0 & \cdots & 1 \end{bmatrix} - \begin{bmatrix} 0 & \cdots & \widetilde{d}_{1j} & \cdots & \widetilde{d}_{1n} \\ \vdots & & \vdots & & \vdots \\ \widetilde{d}_{i1} & \cdots & \widetilde{d}_{ij} & \cdots & \widetilde{d}_{in} \\ \vdots & & \vdots & & \vdots \\ \widetilde{d}_{n1} & \cdots & \widetilde{d}_{nj} & \cdots & \widetilde{d}_{nn} \end{bmatrix} = \begin{bmatrix} 1 & \cdots & 0 & \cdots & 0 \\ \vdots & 1 & \vdots & & \vdots \\ 0 & \cdots & 1 & \cdots & 0 \\ \vdots & & \vdots & 1 & \vdots \\ 0 & \cdots & 0 & \cdots & 1 \end{bmatrix} \quad (3)$$

Also:

$$\widetilde{D}\left(I - \widetilde{D}\right)^{-1} = \widetilde{D}I = \begin{bmatrix} 0 & \cdots & \widetilde{d}_{1j} & \cdots & \widetilde{d}_{1n} \\ \vdots & & \vdots & & \vdots \\ \widetilde{d}_{i1} & \cdots & \widetilde{d}_{ij} & \cdots & \widetilde{d}_{in} \\ \vdots & & \vdots & & \vdots \\ \widetilde{d}_{n1} & \cdots & \widetilde{d}_{nj} & \cdots & \widetilde{d}_{nn} \end{bmatrix} \cdot \begin{bmatrix} 1 & \cdots & 0 & \cdots & 0 \\ \vdots & 1 & \vdots & & \vdots \\ 0 & \cdots & 1 & \cdots & 0 \\ \vdots & & \vdots & 1 & \vdots \\ 0 & \cdots & 0 & \cdots & 1 \end{bmatrix} \quad (4)$$

3. Computation Using Coefficient of Variation: In order to compute the normalized effect matrix, we used the Variable Homogeneity Factor of the Coefficient of Variation (VHFCV), a measure of the relative variability. To achieve this objective, we first apply the VHFCV operator to the direct effect matrix Φ in Equations (5) and (6).

$$\Phi = \begin{bmatrix} \varphi_{11} & \cdots & \varphi_{1j} & \cdots & \varphi_{1n} \\ \vdots & & \vdots & & \vdots \\ \varphi_{i1} & \cdots & \varphi_{ij} & \cdots & \varphi_{in} \\ \vdots & & \vdots & & \vdots \\ \varphi_{n1} & \cdots & \varphi_{nj} & \cdots & \varphi_{nn} \end{bmatrix} \quad (5)$$

$$\varphi_{ij} = VHFCV\left(\widetilde{g}^{ij}\right) \quad (6)$$

4. Normalization: The direct impact matrix Φ is then normalized by using the following Equations (7) and (8):

$$H = \frac{\Phi}{s} \quad (7)$$

$$s = max\left(\underset{1\le i\le n}{max}\sum_{j=1}^{n}\varphi_{ij}, \underset{1\le j\le n}{max}\sum_{i=1}^{n}\varphi_{ij}\right) \quad (8)$$

5. Total Effect Determination: The last step is to determine the total influential matrix by using the relation in Equation (9):

$$Z = \lim_{m\to\infty}\left(H + H^2 + H^3 + \cdots + H^m\right) = H(I - H)^{-1} \quad (9)$$

6. Calculation of r and c: Values r and c are typically the row and column sums of the relation matrix, used to determine the prominence and relation of each factor in the decision-making process. These values are calculated based on Equation (10):

$$r = \left[\sum_{j=1}^{n} t_{ij}\right]_{n\times 1}, c = \left[\sum_{j=1}^{n} t_{ij}\right]'_{n\times 1} \tag{10}$$

The row sums (r) represent the total influence exerted by each factor, indicating its prominence as a driver within the network. The column sums (c) reflect the total influence received by each factor, showcasing its dependency. The difference ($r - c$) identifies whether a criterion is a net influencer or influenced, while the sum ($r + c$) indicates the overall significance of the criterion within the system.

Once the Delphi method was completed, we considered interval-based elements and applied normalization to the decision matrix. Using a decision matrix, which is a methodical instrument for analyzing and contrasting possibilities according to particular standards, decision-makers can appraise options in an organized way. The options are represented by rows, the criteria by columns, and the matrix values show how well each choice fits with each condition. This procedure allows for the optimal choice to be made by considering both quantitative and qualitative information, usually from reliable sources or experts. Crucially, this is not an isolated action, rather, it is a stage in a series of related processes, each of which builds on the one before it.

1. Normalization: Normalize each interval weight by the sum of all interval weights. For an interval $[\mathrm{a_i}, \mathrm{b_j}]$, the normalized interval is given as follows:

$$\left[\frac{\boldsymbol{\alpha}_i}{\sum \boldsymbol{\alpha}_i}, \frac{\mathbf{b}_i}{\sum \mathbf{b}_i}\right]$$

2. Summing the Intervals: Compute the sum of the lower bounds and the upper bounds of all interval weights.
3. Normalization Calculation: Normalize each interval weight by dividing each lower and upper bound by the corresponding sums computed in the previous step.

3.2.2. DEMATEL

To depict the internal relationships among the main factors, the interval fuzzy-DEMATEL technique was employed. This technique allows experts to articulate their opinions on the factors, including their direction and intensity with greater precision.

In this part, the threshold value was not used in the calculation of the total communication matrix (allowing for the preservation of all internal connections). After obtaining the r and c, to evaluate the influence and interdependence among factors, we used the combined DEMATEL and ANP algorithms to compute the super-matrix and the weighted super-matrix based on Equation (11):

$$T_C = \begin{array}{cc} & \begin{array}{ccccc} \mathrm{D_1} & \cdots & \mathrm{D_j} & \cdots & \mathrm{D_m} \\ \mathrm{c_{11}} \cdots \mathrm{c_{1n_1}} & \cdots & \mathrm{c_{j1}} \cdots \mathrm{c_{jn_j}} & \cdots & \mathrm{c_{m1}} \cdots \mathrm{c_{mn_m}} \end{array} \\ \begin{array}{cc} \mathrm{D_i} & \begin{array}{c} \mathrm{c_{11}} \\ \mathrm{c_{12}} \\ \vdots \\ \mathrm{c_{1n_1}} \\ \vdots \end{array} \\ \vdots & \\ \mathrm{D_i} & \begin{array}{c} \mathrm{c_{i1}} \\ \mathrm{c_{i2}} \\ \vdots \\ \mathrm{c_{1n_1}} \\ \vdots \end{array} \\ \vdots & \\ \mathrm{D_m} & \begin{array}{c} \mathrm{c_{m1}} \\ \mathrm{c_{m2}} \\ \vdots \\ \mathrm{c_{mn_m}} \end{array} \end{array} & \begin{bmatrix} \mathrm{g_c^{11}} & \cdots & \mathrm{g_c^{1j}} & \cdots & \mathrm{g_c^{1n}} \\ \vdots & \vdots & \vdots & \vdots & \vdots \\ \mathrm{g_c^{i1}} & \cdots & \mathrm{g_c^{ij}} & \cdots & \mathrm{g_c^{in}} \\ \vdots & \vdots & \vdots & \vdots & \vdots \\ \mathrm{g_c^{n1}} & \cdots & \mathrm{g_c^{nj}} & \cdots & \mathrm{g_c^{nn}} \end{bmatrix} \end{array} \tag{11}$$

Subsequently, the normalization of matrix Tc was performed as per Equation (12):

$$T_C^{nor} = \begin{bmatrix} T_C^{nor_{11}} & \cdots & T_C^{nor_{1j}} & \cdots & T_C^{nor_{1n}} \\ \vdots & & \vdots & & \vdots \\ T_C^{nor_{i1}} & & T_C^{nor_{ij}} & \cdots & T_C^{nor_{in}} \\ \vdots & & \vdots & & \vdots \\ T_C^{nor_{n1}} & \cdots & T_C^{nor_{nj}} & \cdots & T_C^{nor_{nn}} \end{bmatrix} \quad (12)$$

To derive the un-weighted super-matrix, we transformed the comprehensive impact matrix T based on the interdependence of dimension relations and associated clusters, guided by Equation (13):

$$W_C = (T_C^{nor})' = \begin{matrix} & & \\ D_1 & \begin{matrix} c_{11} \\ c_{12} \\ \vdots \\ c_{i1} \\ c_{i2} \end{matrix} \\ \vdots & \vdots \\ D_i & c_{in_1} \\ \vdots & \vdots \\ D_m & \begin{matrix} c_{m1} \\ c_{m2} \\ \vdots \\ c_{mn_m} \end{matrix} \end{matrix} \overset{\begin{matrix} D_1 & \cdots & D_j & \cdots & D_m \\ c_{11} \cdots c_{1n_1} & \cdots & c_{m1} \cdots c_{mn_m} & \cdots & c_{j1} \cdots c_{jn_j} \end{matrix}}{\begin{bmatrix} W_c^{11} & \cdots & W_c^{i1} & \cdots & W_c^{m1} \\ \vdots & \vdots & \vdots & \vdots & \vdots \\ W_c^{1j} & \cdots & W_c^{ij} & \cdots & W_c^{mj} \\ \vdots & \vdots & \vdots & \vdots & \vdots \\ W_c^{1m} & \cdots & W_c^{im} & \cdots & W_c^{mm} \end{bmatrix}} \quad (13)$$

The weighted super-matrix W_c^* was derived based on Equation (14)

$$W_C^* = T_D^{nor} \odot W_C = \begin{bmatrix} t_D^{nor_{11}} \odot W_c^{11} & \cdots & t_D^{nor_{i1}} \odot W_c^{i1} & \cdots & t_D^{nor_{m1}} \odot W_c^{m1} \\ \vdots & & \vdots & & \vdots \\ t_D^{nor_{1j}} \odot W_c^{1j} & \cdots & t_D^{nor_{ij}} \odot W_c^{ij} & \cdots & t_D^{nor_{mj}} \odot W_c^{mj} \\ \vdots & & & & \vdots \\ t_D^{nor_{1m}} \odot W_c^{1m} & \cdots & t_D^{nor_{im}} \odot W_c^{im} & \cdots & t_D^{nor_{mm}} \odot W_c^{mm} \end{bmatrix} \quad (14)$$

Finally, the weighted super-matrix was raised to the power Φ until convergence, resulting in a stable super-matrix term. This allows for acquiring the global priority vector, which specifies the weights that are influential $w = (w_1, \ldots w_e, \ldots, w_{on})$ from $lim\varphi \to \infty^- (W_c^*)^\varphi$ for the factors. The symbol Φ represents the converged power of the weighted super-matrix, stabilizing to indicate the global priorities of the factors. In Equation (14), Φ is computed iteratively until the matrix converges, ensuring all factors' weights are normalized. Equation (14) uses the Hadamard product ($\odot$) to denote element-wise multiplication, differentiating it from standard matrix multiplication ($\times$).

The weights (W) derived from DANP indicate the relative importance of each factor in the context of smart city street design for AVs and cyclists. Additionally, DANP allowed us to assess the influence each factor exerts on others (influence) and the extent to which it is affected by them (influence received). This dual analysis helps us to identify not only which factors are most critical but also how they interact within the decision-making framework, guiding more effective urban design strategies.

3.3. *Identifying Strategies with Fuzzy Game Theory*

The third step in designing street spaces for cyclists and AVs is to apply Game theory to determine the optimal strategy, considering the interaction between cyclists and AVs. To apply Game theory, the factors can be framed as the first player and the strategies as the second player. The objective is to establish an optimal model where the second player, representing the strategies in the context of AV integration, emerges victorious.

Game theory can help us understand how rational or irrational agents behave in strategic situations, and how to design mechanisms or algorithms that achieve desirable objectives [67,77,83]. When all players keep their current tactics and there is no way for any of them to gain an advantage over the others, this condition is called a Nash equilibrium, and it is a key idea in Game theory [69,84], given that all finite games include a Nash equilibrium. Using the minimax algorithm, which is a recursive method that determines the best course of action for a player in a zero-sum game and suppose that the other player likewise plays optimally, is one approach to locating a Nash equilibrium. In a zero-sum game, one player's success is another's failure, and no player's gain is more than zero [77,81,84] (Table 2).

Table 2. Pay-off matrix.

		Player 2		
		Strategy 1	**Strategy 2**	**Strategy n**
Player 1	**Challenge 1**	(a_{11}, b_{11})	(a_{12}, b_{12})	(a_{1n}, b_{1n})
	Challenge 2	(a_{21}, b_{21})	(a_{22}, b_{22})	(a_{31}, b_{31})
	Challenge n	(a_{n1}, b_{n1})	(a_{n2}, b_{n2})	(a_{nn}, b_{nn})

3.3.1. Identification of Strategies

The first player in this framework stands for the factors that must be reduced, i.e., the weights derived from DANP, normalized using a [0, 1] interval to ensure compatibility; the second player is made up of the approaches meant to deal with these factors, which must be maximized in order to produce the best possible results. This methodology entails giving these players precise definitions and evaluating how well different mitigation techniques work for the concerns that were discovered.

Four primary strategies were chosen for evaluation: climate-responsive design strategies, green infrastructure integration, advanced structural design and engineering, and smart technology integration. These selections were based on a thorough assessment of the literature. These tactics were selected because they have a major effect on passing the stated factors. Their proven efficacy and applicability served as a basis for selection, guaranteeing that the tactics selected offered a comprehensive and doable method for accomplishing the goals of the research. The approach may be reviewed to improve the study's depth and applicability if different strategies—or a different number of strategies—are determined to be required.

3.3.2. Game Theory with Fuzzy Matrix

Within the field of evolutionary Game theory, scholars examine diverse behavioral patterns displayed by game agents with the goal of identifying behaviors that are most likely to become dominant over time [84]. Central to this investigation are enduring players' behavioral trends described by Equation (15):

$$\begin{gathered} \frac{dx_i}{dt} = x_i[f_i(x) - \varphi(x)] \\ \text{that } \varphi(x) = \sum_{i=1}^{n} x_i f_i(x) \end{gathered} \tag{15}$$

where x_i stands for the percentage of the ith strategy type in the players' population and the vector $x = (x_1, x_2, \ldots, x_n)$ reflects the distribution of n kinds. As for the fitness of the i type and the average population fitness, they are represented by $f_i(x)$ and $\varphi(x)$, respectively [69]. The equation is defined on the n-dimensional simplex because each component x_i represents a proportion, and the sum of all components of the vector x equals one. In mathematical terms, this means that the vector x lies within a space where each element x_i is a non-negative number, and the total sum of these elements is constrained to be exactly one. This condition ensures that x adheres to the properties of a simplex, a geometric concept that generalizes the idea of a triangle or tetrahedron to n dimensions. The simplex constraint is often used in optimization problems and probability distributions to represent feasible solutions or probabilities that must sum to a whole. It is assumed in the replicator equation that the population is distributed uniformly. The same idea can be applied to choose among different strategies to face the factors identified and assessed as explained in the previous subsections. By substituting A for the evolutionary game's reward matrix, we obtain Equation (16) from (15):

$$\frac{dx_i}{dt} = x_i\left[\left(Ax\right)_i - x^T Ax\right], \tag{16}$$

The term $(Ax)_i$ represents the anticipated outcome, whereas $x^T Ax$ stands for the population's average fitness. The current proportions of each population use dictate the state of the evolutionary game. The current state of the population at time t may be described by the probabilities of employing the first mode, which is p, and the second mode, which is $1 - p$ [77,83].

The input pay-off matrix of the evolutionary game viewpoint of the decision-making issue represents the evolvi6ng strategies for street design, denoted as component $x_i (x_i \in X)$. Consider M as the pay-off matrix that represents the anticipated result of the MCDM issue. Let x_i represent the expected benefit of the ith strategy of the street design and y_j represent the probability of the jth strategies to deal with barriers and facilitators [69] (Table 3). The super-matrix derived from the DEMATEL process feeds into the Game theory model by influencing the structure of the pay-off matrix (M). Specifically, the weighted relationships in the super-matrix guide the prioritization of strategies in the game-theoretic framework, ensuring alignment with the factors' causal influences.

Table 3. Linguistic variables to the fuzzy scale value.

Linguistic Variables *	Likert Scale	Fuzzy Scale
EH	9	(7, 9, 9)
VH	7	(5, 7, 9)
M	5	(3, 5, 7)
VL	3	(1, 3, 5)
EL	1	(1, 1, 3)

* EH (Extremely High), VH (Very High), M (Medium), VL (Very Low), EL (Extremely Low).

The replicator dynamics of the evolutionary game may be described as Equation (17). The replicator dynamics, which govern the evolution of strategies in a population, are represented by differential equations, where $M_{ij} \times x_i$ indicates the rate of change in strategies over time. In the context of this model, Equation (17) refers to the pay-off functions of a street design. Street design strategies are implemented as a player from zero- and one-sum Game theory. The symbol Z represents the value of the objective function in the optimization problem. Minimizing $Z = x_1 + x_2 + x_3 + x_4$ reflects the goal of optimizing the allocation of resources or strategies under a set of linear constraints, i.e., to achieve the optimal distribution of strategies addressing the identified factors. These constraints represent the limitations or interactions within the system, ensuring that the strategies adhere to the evolutionary dynamics described by the replicator equations. Thus, the linear optimization problem is a specific case of the broader evolutionary game framework, aiming to find the optimal strategy distribution that minimizes the total resource allocation while satisfying the dynamic constraints.

$$\text{Minimize } Z\,\frac{dx_i}{dt} \quad \begin{cases} \frac{dx_i}{dt} = x_i(G_i - F) \leq 1 \\ \frac{dy_j}{dt} = y_j(H_j - F) > 1, \end{cases} \tag{17}$$

where $G_i = \sum_{i=1}^{n} M_{ij} \times y_j$; $H_j = \sum_{i=1}^{m} M_{ij} \times x_i$ and $F = \sum_{i=1}^{m} \sum_{j=1}^{n} M_{ij} \times x_i \times y_j$.

The replicator dynamics of the evolutionary game may be solved numerically, leading to the evolutionarily stable and mixed stable strategy or Nash equilibrium stable fixed points. The linear profile of the plots showing the evolution of strategies over time and the confirmation of their stability by local stability analysis provide support for these fixed points [66,67,77,83,84].

4. Results

4.1. Identification and Ranking of Final Factors

As mentioned previously, we identified seven categories of factors that may be encountered in the smart city street design for AVs and cyclists from the literature review: Structure, Sustainability, Atmospheric environmental conditions, Visual aspect, Safety, Slope, and Accessibility. The fuzzy Delphi method confirmed that all identified factors are relevant and have an impact on the main problem. The factors were evaluated based on their importance, with scores indicating how critical each factor is in the context of urban planning and design.

Safety (C5) emerged as the most relevant factor, with a score of 0.970, suggesting that it is critical for ensuring the success of smart city street designs (Table 4). This high relevance is likely because safety is a fundamental requirement for the adoption and operation of AVs and the protection of cyclists. Any design that compromises safety could lead to serious consequences, hence its high weight and influence. Visual aspect (C4) and Accessibility (C7) both shared a significant level of importance, with scores of 0.857. These factors are crucial because they directly affect user experience and inclusivity. Visual appeal can greatly influence public acceptance, while accessibility ensures that the design meets the needs of all citizens, including those with disabilities. Structure (C1) and Slope (C6) both had moderate importance, each with a score of 0.685. These factors are vital for the physical feasibility and functionality of the street design. The structure determines how well the infrastructure can accommodate traffic flows, while the slope impacts vehicle control and cyclist safety. Sustainability (C2) had a slightly lower relevance with a score of 0.634, reflecting its importance in addressing long-term environmental impacts. Although crucial, it may be seen as less immediate compared to safety or accessibility in the context of day-to-day operations. Atmospheric environmental conditions (C3) was the least relevant factor with a score of 0.500. This lower relevance could be due to the perception that, while environmental factors are important, they may not directly impact the operational efficiency or safety of AVs and cyclists as immediately as the other factors.

In conclusion, and in line with other studies [85,86], Safety (C5) was the most critical factor due to its fundamental role in protecting lives and ensuring the smooth operation of autonomous systems, while Atmospheric environmental conditions (C3) was less critical, reflecting its more indirect impact on the immediate functionality of the system.

Table 4. Final intervals for each street design factor from Delphi technique.

Factors in Smart City Street Design for Autonomous Vehicles and Cyclists	Code	Interval Average	Score	Result
Structure	C1	[2.3, 2.9]	0.685	Acceptable
Sustainability	C2	[2.2, 2.8]	0.634	Acceptable
Atmospheric environmental conditions	C3	[1.8, 2.6]	0.500	Acceptable
Visual aspect	C4	[2.5, 3.3]	0.857	Acceptable
Safety	C5	[2.8, 3.4]	0.970	Acceptable
Slope	C6	[2.3, 2.9]	0.685	Acceptable
Accessibility	C7	[2.5, 3.3]	0.857	Acceptable

4.2. Relevance of Street Design Factors

The relative importance of the factors was estimated by means of the DANP approach using a Python/PuLP solver. The sum of the weights values is equal to 1, indicating the correctness of the weighting operation. The analysis of the factors' weights reveals that 'C5: Safety', with a weight of 0.205 (Figure 3), is the most crucial for optimizing smart city street design for AVs and cyclists. This underscores the paramount importance of ensuring the safety of both AVs and cyclists, likely driven by the need to minimize accidents and enhance public trust in autonomous transportation systems.

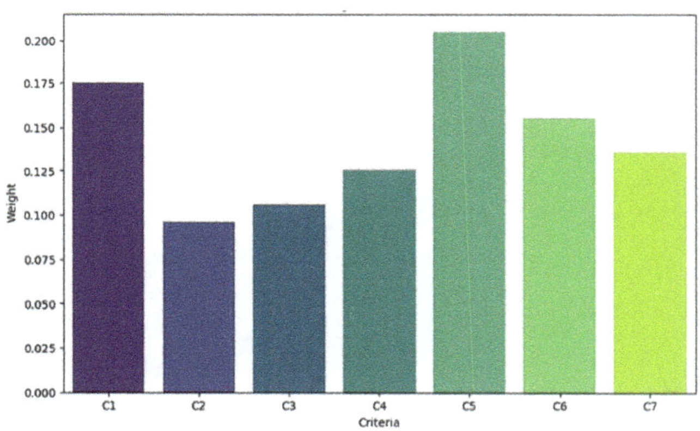

Figure 3. Weights of the street design factors obtained using interval-fuzzy ANP.

'C1: Structure' follows closely with a weight of 0.175, highlighting the necessity for robust and well-engineered street infrastructure that can support the technological requirements of smart city designs. The 'C6: Slope' factor, with a significant weight of 0.156, indicates that the gradient of streets is a vital consideration, likely affecting both vehicle performance and cyclist safety. 'C4: Visual aspect' also holds considerable importance with a weight of 0.126, reflecting the need for esthetically pleasing urban environments that can enhance the overall user experience and public acceptance of smart city initiatives. Conversely, 'C2: Sustainability' and 'C3: Atmospheric environment conditions' have lower weights of 0.096 and 0.106, respectively. While still important, their relatively lower significance suggests that, in this context, immediate functional and safety concerns take precedence over long-term environmental considerations. Notably, 'C7: Accessibility' with a weight of 0.136 may indicate a need for further clarity in distinguishing between the two structure-related criteria (C1 and C7) to avoid redundancy and improve the precision of the analysis.

Overall, the prioritization reflects a balanced approach where safety and structural integrity are deemed most critical, ensuring that smart city streets are both functional and secure for autonomous vehicles and cyclists.

In order to determine the threshold value for relationships, it sufficed to compute the mean values of matrix T according to the fuzzy DEMATEL technique. Once the threshold intensity was established, all values in the T matrix that were below the threshold were set to zero, indicating that the causal link was disregarded (Table 5).

Table 5. The pattern of causal relationships of the main factors.

	R	D	D − R	D + R
C1	[0.109 0.182]	[0.109 0.184]	[−1.387 2.044]	[0.219 0.366]
C2	[0.100 0.169]	[0.081 0.147]	[−1.907 −2.249]	[0.181 0.316]
C3	[0.114 0.188]	[0.095 0.165]	[−1.907 −2.249]	[0.209 0.353]
C4	[0.114 0.188]	[0.100 0.171]	[−1.430 −1.635]	[0.214 0.359]
C5	[0.109 0.182]	[0.200 0.286]	[9.062 1.042]	[0.310 0.468]
C6	[0.114 0.188]	[0.090 0.159]	[−2.384 −2.862]	[0.205 0.347]
C7	[0.114 0.188]	[0.100 0.171]	[−1.430 −1.635]	[0.214 0.359]

The summation of the components in each row in Table 6 represents the impact of a particular factor on the others within the system. 'C1: Structure' and 'C5: Safety' are the cause, and the rest represent the effect. Factors C1 and C5 positively influence the other variables, meaning they drive or cause changes in the rest of the variables. In contrast, the rest of the factors are described as the "effect" because they are negatively influenced or impacted on D − R by C1 and C5. Essentially, C1 and C5 are the driving forces (cause) that exert a positive impact in D − R, while the remaining factors are responsive or reactive (effect), experiencing a negative influence as a result.

Table 6. The pattern of causal relationships of the main factors.

	D − R (Cause)	D + R (Effect)
C1	0.328	0.292
C2	−2.078	0.249
C3	−2.078	0.281
C4	−1.533	0.287
C5	5.052	0.389
C6	−2.623	0.276
C7	−1.533	0.287

The analysis shows that 'Safety' (C5) is the most influential cause, with the highest positive D − R value of 5.052 (Figure 4). This indicates that safety plays a critical role in impacting other factors, emphasizing its paramount importance in the overall design and functionality of smart city streets. Additionally, 'Structure' (C1) also emerges as a significant cause with a positive D − R value of 0.328, highlighting the necessity for robust infrastructure to support the technological requirements of smart city designs. On the other hand, factors such as 'Sustainability' (C2), 'Atmospheric environment conditions' (C3), 'Visual aspect' (C4), 'Slope' (C6), and 'Accessibility' (C7) are primarily effects, as indicated by their negative D − R values. This suggests that these factors are more influenced by others rather than being primary influencers themselves.

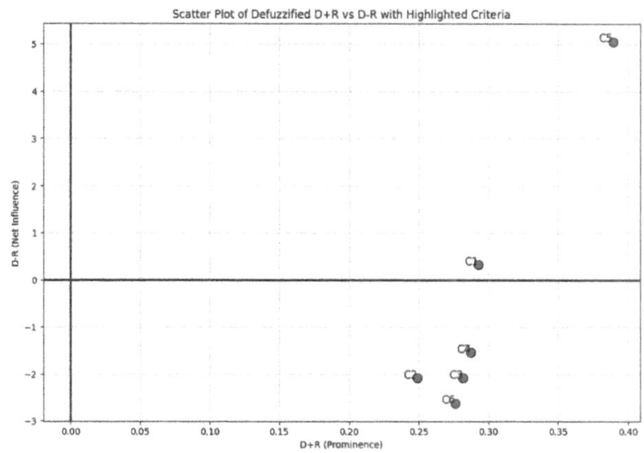

Figure 4. Cartesian co-ordinate diagram of interval fuzzy DEMATEL output for main factors.

4.3. Strategies

In this step, we define the first and second players. The first player includes factors that should be minimized. The strategies to address these are the second player, which should be maximized, because the goal of our problem is the most optimal way to deal with these barriers and facilitators, so the solutions to our problem are as follows:

Advanced Structural Design and Engineering (S1): Employing innovative structural design techniques and engineering practices enhances the structural characteristics, accessibility, and safety of streets. Traffic signs and road marks ensure the provision of clear information and thus safe navigation for all road users; dedicated lanes can increase capacity and ensure safety but mixed lanes could reduce the number of carriageways, and smart parking lots could release on-street parking space to be reconverted to new cycling lanes [63,87]. Moreover, new pavements can improve the surface conditions, strength, and load-bearing capacity of streets, solving the most demanding problems related to structural performance (due to platooning and reduced wheel wander) [88].

Green Infrastructure Integration (S2): Integrating green infrastructure elements, such as rain gardens and permeable pavements, as well as tree strips, into street designs enhance sustainability and visual aspect. Flood-resistant design features mitigate risks posed by natural disasters and extreme weather events, ensuring the longevity and reliability of streets [89]. These solutions manage stormwater runoff, reduce urban heat island effects, improve air quality, and reduce noise. Additionally, the inclusion of vegetated surfaces and landscaping enhances the esthetic appeal, guides transit itineraries [19], creates wildlife habitats, and promotes biodiversity, fostering healthier and more vibrant urban environments.

Climate-Responsive Design Strategies (S3): Implementing climate-responsive design strategies adapts streets to changing environmental conditions and minimizes vulnerability to climate-related risks. Measures such as shade structures, wind barriers, and snowmelt systems address atmospheric environmental conditions, the visual aspect, and safety. The automation of transport systems can contribute to energy efficiency by optimizing routes and driving patterns [90].

Smart Technology Integration (S4): Integrating smart technology solutions such as sensors, monitoring systems, and predictive analytics enhances the safety, structure efficiency, and sustainability of streets. Real-time monitoring of environmental conditions, traffic flow, and infrastructure performance enables proactive management and optimization [69]. Intelligent traffic management systems optimize signal timing and speed limits, reducing congestion and improving safety [89]. Furthermore, data-driven decision-making informs proactive maintenance strategies and evidence-based policy-making, facilitating sustainable urban development and enhancing the overall quality of life for residents and visitors alike [91].

According to Table 7, the saddle point is MIN MAX = 5 and MAX MIN = 3. The objective function was transformed into a classical form. The problem model was solved twice, once using the normal method and once with uncertainty weights.

Table 7. The pattern of causal relationships of the main factors and strategies.

		Second Player				
		S1	**S2**	**S3**	**S4**	**MAX**
First Player	**C1**	6	8	7	9	9
	C2	6	8	8	7	8
	C3	5	2	2	3	5
	C4	5	3	5	6	6
	C5	8	8	5	8	8
	C6	5	8	3	8	8
	C7	1	2	2	6	6
	MIN	1	2	2	3	

The problem model presented was derived from the replicator dynamics equation of the evolutionary game, described in Equation (17). The classical objective function for four strategies is as follows:

$$\text{minimize } Z = x_1 + x_2 + x_3 + x_4$$

The limitations of the problem are as follows:

$$\begin{gathered} 6x_1 + 8x_2 + 7x_3 + 9x_4 \leq 1 \\ 7x_1 + 8x_2 + 8x_3 + 6x_4 \leq 1 \\ 3x_1 + 2x_2 + 2x_3 + 5x_4 \leq 1 \\ 6x_1 + 5x_2 + 3x_3 + 5x_4 \leq 1 \\ 8x_1 + 5x_2 + 8x_3 + 8x_4 \leq 1 \\ 8x_1 + 3x_2 + 8x_3 + 5x_4 \leq 1 \\ 6x_1 + 2x_2 + 2x_3 + x_4 \leq 1 \\ \text{That } x_i \geq 0 \end{gathered}$$

After solving the problem for the strategies in Python/PuLP solver, their importance is obtained as follows:

$$\text{Optimal Solution } (x_1,\ x_2, x_3,\ x_4) : [0, 0.058, 0, 0.088]$$

$$\text{Optimal Value of the Objective Function: } 0.147$$

Therefore, the optimal solution is to use the solutions x_1 = S2 = 0.058 and x_4 = S4 = 0.088, leading to an optimal value of the objective function at 0.147. In the context of optimizing smart city street design for AVs and cyclists using Game theory, the analysis reveals that the optimal strategies are 'Green Infrastructure Integration' (S2) and 'Smart Technology Integration' (S4). This indicates that these strategies are the most suitable for overcoming the identified factors in the smart city street design process.

'Green Infrastructure Integration' (S2) is highlighted as a key strategy, likely due to its ability to enhance sustainability and environmental friendliness, addressing the factors related to sustainability, atmospheric environment conditions, and visual aspect. 'Smart Technology Integration' (S4) also emerges as a critical strategy, emphasizing the importance of incorporating advanced technologies to improve the safety, efficiency, and overall functionality of the smart city streets. Together, these strategies provide a balanced approach, combining environmental considerations with technological advancements to create a comprehensive and effective solution for optimizing smart city street design for autonomous vehicles and cyclists.

The sensitivity analysis (Figure 5) demonstrates that changing the objective function for optimal solution values ensures the results are robust. The heat map visualizes the variations in the objective function's value due to these perturbations, showing a consistent range without significant deviations from the optimal value of 0.147. This indicates that the optimization strategy is not highly sensitive to small changes in the values of the strategies, affirming the correctness and stability of the results.

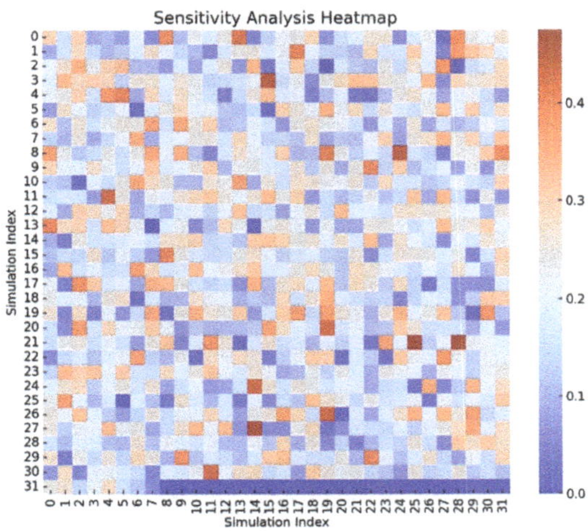

Figure 5. Sensitivity analysis with change in objective function z.

According to Figures 6 and 7, the majority of the resulting values cluster around the optimal value line, indicating that small perturbations in the strategies (S2: Green Infrastructure Integration and S4: Smart Technology Integration) do not significantly deviate the objective function from its optimal value. This suggests that the chosen optimal strategies are robust and reliable. While there are some resulting values above and below the optimal value, they are relatively close, mostly within the range of 0.1 to 0.3. This further reinforces the stability of the optimal solution. Given the scatter plot, it can be inferred that Green Infrastructure Integration and Smart Technology Integration as part of the smart city street design will yield consistent and reliable results. Even if there are slight changes or uncertainties in their implementation, the overall performance (objective function value) remains stable.

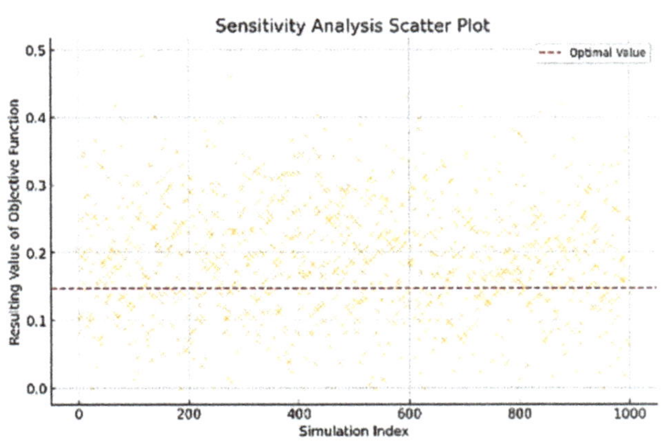

Figure 6. Sensitivity analysis with change in decision matrix.

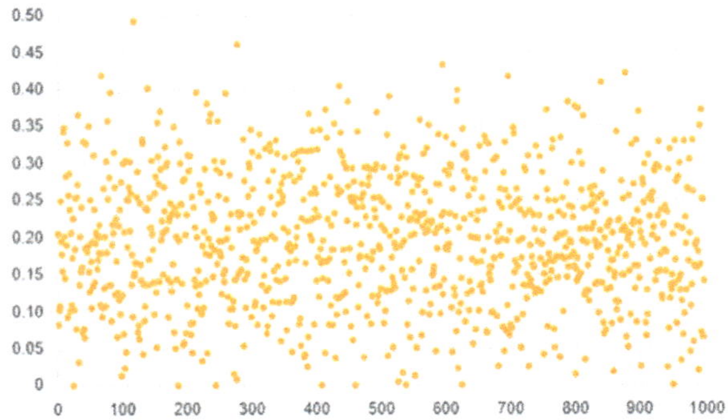

Figure 7. Sensitivity analysis with change in limitations.

The scatter plot also highlights that there are a few outliers where the resulting value deviates more significantly from the optimal value. These outliers can be investigated to understand specific conditions or variations that might cause a decrease in the performance. Addressing these could further enhance the robustness of the smart city street design.

5. Conclusions and Discussion

Designing streets for cyclists and self-driving vehicles in today's world is a vital and unavoidable urban necessity. With the increase in population and urban traffic, the need for transportation systems that can guarantee, safety, efficiency, and environmental sustainability is becoming increasingly evident. One of the main reasons for this demand is the increasing use of bicycles and the imminent need for the integration of autonomous vehicles. As a clean and economical vehicle, bicycles play an important role in reducing air pollution and traffic. On the other hand, by using advanced technologies, self-driving vehicles have the ability to reduce accidents, improve traffic flow, and increase transportation efficiency. Designing streets that can simultaneously meet the needs of cyclists and self-driving vehicles can help improve the quality of urban life. This type of design not only provides greater safety and comfort for cyclists, but also allows self-driving vehicles to move through cities with higher efficiency and fewer accidents. The harmonization of urban transport to accommodate both bicyclists and AVs presents a transformative opportunity to enhance urban mobility, making city streets safer, cleaner and more efficient environments.

Our research employed a comprehensive methodology, combining Multi-Criteria Decision Making and Game theory, to systematically identify and address the primary challenges to this integration. Through the value-interval FDM, we confirmed the basic factors, which were then weighted and analyzed for cause-and-effect interactions using the interval-fuzzy DANP. Game theory was instrumental in determining the optimal strategies to overcome these factors, in line with the studies by Dabiri et al. [92] and Guo et al. [93], which further support the effectiveness of Game-theory-based approaches in optimizing transportation systems and street design in smart cities.

Our findings highlight safety as the most crucial factor in smart city street design for autonomous vehicles and cyclists. These findings align with other studies published in recent years. For example, in their survey to evaluate the relative importance of potential motivators and deterrents to cycling among current and potential cyclists, Winters et al. [25] reported that safety was the factor that most influenced the likelihood of cycling. Similar results were obtained by Fishman et al. [22], who found that safety was a major concern for all focus group participants, whether they were regular cyclists or not. This was due to a perceived lack of adequate cycling infrastructure, in addition to the negative attitudes of some car drivers described by regular cyclists. Additionally, more recent research by Chen and Liu [85] also highlighted safety as the most critical factor in smart city transportation designs, emphasizing its non-negotiable role in successful implementation. This study also supported the significance of accessibility and sustainability in ensuring inclusive and long-term viable urban solutions. The lower importance assigned to atmospheric environmental conditions also aligns with findings from Zhou et al. [86], who noted that environmental considerations, while essential, are often secondary to more immediate operational concerns in the context of AVs.

To effectively address this and other challenges, the integration of 'Green Infrastructure' and 'Smart Technology' emerged as the most viable solutions, underscoring the importance of a balanced approach that combines environmental sustainability with technological innovation. The emphasis on 'Green Infrastructure Integration' highlights the need to address sustainability, atmospheric environment conditions, and the visual aspect, leading to the incorporation of green elements in streets through permeable pavements, rain gardens, vegetative curb areas, and sidewalk trees [19,38,94]. 'Smart Technology Integration', through technologies such as intelligent traffic management systems and smart street lightening or personal beacons [37,44,48], is crucial for enhancing safety, efficiency, and functionality within smart city street designs. Furthermore, Chen et al. [85] and Radakovic et al. [95] explored the integration of smart technologies in shared autonomous vehicle systems, highlighting the efficiency gains and sustainability benefits that parallel the findings related to 'Smart Technology Integration'.

To strengthen the applicability of this methodology some additional validation steps can be performed. The proposed model could be applied to a real-world urban setting, such as redesigning a street to integrate autonomous vehicles (AVs) and cyclists. Collecting real-world data on safety improvements, traffic flow, and public satisfaction, and then comparing these metrics before and after implementation, could also be very useful. In addition, the impacts of strategies like "smart technology integration" or "green infrastructure" could be tested by using simulations or field experiments. Finally, engaging stakeholders (e.g., urban planners, cyclists, AV manufacturers) to assess the feasibility and practicality of the proposed solutions would be very relevant in real-world urban design challenges.

This research provides valuable guidance for urban planners and decision-makers to identify key design factors and implement strategic policies that not only facilitate the coexistence of cyclists and AVs, making interaction easier and safer, but also contribute to the broader goals of reducing traffic congestion and improving urban mobility. The implementation of these strategies promises to create a more harmonious and efficient urban transport system, ultimately benefiting all road users.

Author Contributions: Conceptualization, M.F.; methodology, M.F.; validation, M.F., G.F. and C.C.; formal analysis, M.F. and G.F.; investigation, M.F., S.N. and E.G.-G.; writing—original draft preparation, M.F.; writing—review and editing, M.F., S.N., E.G.-G. and G.F.; software, M.F.; visualization, M.F. and E.G.-G.; supervision, S.N. and E.G.-G.; project administration, S.N.; funding acquisition, S.N and E.G-G. All authors have read and agreed to the published version of the manuscript.

Funding: This publication is part of the research project entitled "Planning and design recommendations to guide new Autonomous Vehicles in Cities"—AV Cities (2023–2026), under Grant PID2022-140649OB-I00 funded by MCIN/AEI/10.13039/501100011033/FEDER, UE.

Data Availability Statement: The raw data supporting the conclusions of this articles will be made available by the authors on request.

Conflicts of Interest: The authors declare no conflicts of interest.

Nomenclature

AVs	Autonomous vehicles
AHP	Analytical Hierarchical Process
ANP	Analytic Network Process
CAVs	Connected and autonomous vehicles
DANP	Decision-Making Trial and Evaluation Laboratory ANP
DEMATEL	Decision-Making Trial and Evaluation Laboratory

FDM	Fuzzy Delphi method
HVs	Human-driven vehicles
IVFE	Interval-Fuzzy Element
MCDM	Multi-Criteria Decision Making

References

1. Larsen, J. The Making of a Pro-Cycling City: Social Practices and Bicycle Mobilities. *Environ. Plan. A Econ. Space* **2017**, *49*, 876–892 [CrossRef]
2. Liu, S.; Shen, Z.-J.M.; Ji, X. Urban Bike Lane Planning with Bike Trajectories: Models, Algorithms, and a Real-World Case Study *Manuf. Serv. Oper. Manag.* **2022**, *24*, 2500–2515. [CrossRef]
3. Ferretto, L.; Bruzzone, F.; Nocera, S. Pathways to Active Mobility Planning. *Res. Transp. Econ.* **2021**, *86*, 101027. [CrossRef]
4. Khreis, H.; May, A.D.; Nieuwenhuijsen, M.J. Health Impacts of Urban Transport Policy Measures: A Guidance Note for Practice *J. Transp. Health* **2017**, *6*, 209–227. [CrossRef]
5. Rojas-Rueda, D.; Nieuwenhuijsen, M.J.; Khreis, H.; Frumkin, H. Autonomous Vehicles and Public Health. *Annu. Rev. Public Health* **2020**, *41*, 329–345. [CrossRef]
6. Freudendal-Pedersen, M. Cyclists as Part of the City's Organism: Structural Stories on Cycling in Copenhagen. *City Soc.* **2015**, *27*, 30–50. [CrossRef]
7. Freemark, Y.; Hudson, A.; Zhao, J. Are Cities Prepared for Autonomous Vehicles? *J. Am. Plan. Assoc.* **2019**, *85*, 133–151. [CrossRef]
8. Khayyam, H.; Javadi, B.; Jalili, M.; Jazar, R.N. Artificial Intelligence and Internet of Things for Autonomous Vehicles. In *Nonlinear Approaches in Engineering Applications*; Springer International Publishing: Cham, Switzerland, 2020; pp. 39–68. [CrossRef]
9. Yoganandhan, A.; Subhash, S.D.; Hebinson Jothi, J.; Mohanavel, V. Fundamentals and Development of Self-Driving Cars. *Mater. Today Proc.* **2020**, *33*, 3303–3310. [CrossRef]
10. Rahman, M.M.; Thill, J.-C. Impacts of Connected and Autonomous Vehicles on Urban Transportation and Environment: A Comprehensive Review. *Sustain. Cities Soc.* **2023**, *96*, 104649. [CrossRef]
11. Shafiei, S.; Gu, Z.; Grzybowska, H.; Cai, C. Impact of Self-Parking Autonomous Vehicles on Urban Traffic Congestion. *Transportation* **2023**, *50*, 183–203. [CrossRef]
12. Deng, Q.; Sun, H.; Chen, F.; Shu, Y.; Wang, H.; Ha, Y. An Optimized FPGA-Based Real-Time NDT for 3D-LiDAR Localization in Smart Vehicles. *IEEE Trans. Circuits Syst. II Express Briefs* **2021**, *68*, 3167–3171. [CrossRef]
13. Terzi, S.; Savvaidis, C.; Votis, K.; Tzovaras, D.; Stamelos, I. Securing Emission Data of Smart Vehicles with Blockchain and Self-Sovereign Identities. In Proceedings of the 2020 IEEE International Conference on Blockchain (Blockchain), Virtual, 2–6 November 2020; IEEE: Piscataway, NJ, USA, 2020; pp. 462–469. [CrossRef]
14. Chougule, A.; Chamola, V.; Sam, A.; Yu, F.R.; Sikdar, B. A Comprehensive Review on Limitations of Autonomous Driving and Its Impact on Accidents and Collisions. *IEEE Open J. Veh. Technol.* **2024**, *5*, 142–161. [CrossRef]
15. Nascimento, A.M.; Vismari, L.F.; Molina, C.B.S.T.; Cugnasca, P.S.; Camargo, J.B.; de Almeida, J.R.; Inam, R.; Fersman, E.; Marquezini, M.V.; Hata, A.Y. A Systematic Literature Review About the Impact of Artificial Intelligence on Autonomous Vehicle Safety. *IEEE Trans. Intell. Transp. Syst.* **2020**, *21*, 4928–4946. [CrossRef]
16. Pyrialakou, V.D.; Gkartzonikas, C.; Gatlin, J.D.; Gkritza, K. Perceptions of Safety on a Shared Road: Driving, Cycling, or Walking near an Autonomous Vehicle. *J. Saf. Res.* **2020**, *72*, 249–258. [CrossRef]
17. Duarte, F.; Ratti, C. The Impact of Autonomous Vehicles on Cities: A Review. *J. Urban Technol.* **2018**, *25*, 3–18. [CrossRef]
18. Barnett, J.; Gizinski, N.; Mondragon-Parra, E.; Siegel, J.; Morris, D.; Gates, T.; Kassens-Noor, E.; Savolainen, P. Automated Vehicles Sharing the Road: Surveying Detection and Localization of Pedalcyclists. *IEEE Trans. Intell. Veh.* **2021**, *6*, 649–664. [CrossRef]
19. Wang, Y.; Chau, C.K.; Ng, W.Y.; Leung, T.M. A Review on the Effects of Physical Built Environment Attributes on Enhancing Walking and Cycling Activity Levels within Residential Neighborhoods. *Cities* **2016**, *50*, 1–15. [CrossRef]
20. Hull, A.; O'Holleran, C. Bicycle Infrastructure: Can Good Design Encourage Cycling? *Urban Plan. Transp. Res.* **2014**, *2*, 369–406 [CrossRef]
21. Lee, C.; Moudon, A.V. Neighbourhood Design and Physical Activity. *Build. Res. Inf.* **2008**, *36*, 395–411. [CrossRef]
22. Fishman, E.; Washington, S.; Haworth, N. Barriers and Facilitators to Public Bicycle Scheme Use: A Qualitative Approach. *Transp. Res. Part F Traffic Psychol. Behav.* **2012**, *15*, 686–698. [CrossRef]
23. Lee, C.; Moudon, A.V. Physical Activity and Environment Research in the Health Field: Implications for Urban and Transportation Planning Practice and Research. *J. Plan. Lit.* **2004**, *19*, 147–181. [CrossRef]
24. Fraser, S.D.S.; Lock, K. Cycling for Transport and Public Health: A Systematic Review of the Effect of the Environment on Cycling *Eur. J. Public Health* **2011**, *21*, 738–743. [CrossRef]
25. Winters, M.; Davidson, G.; Kao, D.; Teschke, K. Motivators and Deterrents of Bicycling: Comparing Influences on Decisions to Ride. *Transportation* **2011**, *38*, 153–168. [CrossRef]
26. Garrard, J.H.S.D.J. Women and Cycling. In *City Cycling*; Pucher, J., Buehler, R., Eds.; The MIT Press: Cambridge, MA, USA, 2012; pp. 211–233.
27. Parkin, J.; Wardman, M.; Page, M. Models of Perceived Cycling Risk and Route Acceptability. *Accid. Anal. Prev.* **2007**, *39*, 364–371 [CrossRef] [PubMed]

28. Li, Z.; Wang, W.; Liu, P.; Ragland, D.R. Physical Environments Influencing Bicyclists' Perception of Comfort on Separated and on-Street Bicycle Facilities. *Transp. Res. D Transp. Environ.* **2012**, *17*, 256–261. [CrossRef]
29. Clayton, W.; Parkin, J.; Billington, C. Cycling and Disability: A Call for Further Research. *J. Transp. Health* **2017**, *6*, 452–462. [CrossRef]
30. Clayton, W.; Musselwhite, C. Exploring Changes to Cycle Infrastructure to Improve the Experience of Cycling for Families. *J. Transp. Geogr.* **2013**, *33*, 54–61. [CrossRef]
31. Krizek, K.J.; Roland, R.W. What Is at the End of the Road? Understanding Discontinuities of on-Street Bicycle Lanes in Urban Settings. *Transp. Res. D Transp. Environ.* **2005**, *10*, 55–68. [CrossRef]
32. Daniels, S.; Brijs, T.; Nuyts, E.; Wets, G. Injury Crashes with Bicyclists at Roundabouts: Influence of Some Location Characteristics and the Design of Cycle Facilities. *J. Saf. Res.* **2009**, *40*, 141–148. [CrossRef] [PubMed]
33. Wang, K.; Akar, G. Street Intersection Characteristics and Their Impacts on Perceived Bicycling Safety. *Transp. Res. Rec. J. Transp. Res. Board* **2018**, *2672*, 41–54. [CrossRef]
34. Parkin, J.; Clark, B.; Clayton, W.; Ricci, M.; Parkhurst, G. *Understanding Interactions Between Autonomous Vehicles and Other Road Users: A Literature Review*; UWE Bristol Research Repository: Bristol, UK, 2016.
35. Wood, J.M. Nighttime Driving: Visual, Lighting and Visibility Challenges. *Ophthalmic Physiol. Opt.* **2020**, *40*, 187–201. [CrossRef] [PubMed]
36. Wood, J.M.; Lacherez, P.F.; Marszalek, R.P.; King, M.J. Drivers' and Cyclists' Experiences of Sharing the Road: Incidents, Attitudes and Perceptions of Visibility. *Accid. Anal. Prev.* **2009**, *41*, 772–776. [CrossRef] [PubMed]
37. Gagliardi, G.; Lupia, M.; Cario, G.; Tedesco, F.; Cicchello Gaccio, F.; Lo Scudo, F.; Casavola, A. Advanced Adaptive Street Lighting Systems for Smart Cities. *Smart Cities* **2020**, *3*, 1495–1512. [CrossRef]
38. Li, H.; Harvey, J.T.; Holland, T.J.; Kayhanian, M. The Use of Reflective and Permeable Pavements as a Potential Practice for Heat Island Mitigation and Stormwater Management. *Environ. Res. Lett.* **2013**, *8*, 015023. [CrossRef]
39. Titze, S.; Stronegger, W.J.; Janschitz, S.; Oja, P. Association of Built-Environment, Social-Environment and Personal Factors with Bicycling as a Mode of Transportation among Austrian City Dwellers. *Prev. Med.* **2008**, *47*, 252–259. [CrossRef]
40. Rodríguez, D.A.; Joo, J. The Relationship between Non-Motorized Mode Choice and the Local Physical Environment. *Transp. Res. D Transp. Environ.* **2004**, *9*, 151–173. [CrossRef]
41. Helbich, M.; Böcker, L.; Dijst, M. Geographic Heterogeneity in Cycling under Various Weather Conditions: Evidence from Greater Rotterdam. *J. Transp. Geogr.* **2014**, *38*, 38–47. [CrossRef]
42. Blau, M.; Akar, G.; Nasar, J. Driverless Vehicles' Potential Influence on Bicyclist Facility Preferences. *Int. J. Sustain. Transp.* **2018**, *12*, 665–674. [CrossRef]
43. Eichholz, L.; Mellinger, N.; Manz, W. Urban Cycling and Automated Vehicles (Rad-Auto-Nom Project). In Proceedings of the 28th Annual European Real Estate Society Conference, Milan, Italy, 22–25 June 2022; European Real Estate Society: Amsterdam, The Netherlands, 2021. [CrossRef]
44. Berge, S.H.; Hagenzieker, M.; Farah, H.; de Winter, J. Do Cyclists Need HMIs in Future Automated Traffic? An Interview Study. *Transp. Res. Part F Traffic Psychol. Behav.* **2022**, *84*, 33–52. [CrossRef]
45. Elliott, D.; Keen, W.; Miao, L. Recent Advances in Connected and Automated Vehicles. *J. Traffic Transp. Eng. (Engl. Ed.)* **2019**, *6*, 109–131. [CrossRef]
46. Hagenzieker, M.P.; van der Kint, S.; Vissers, L.; van Schagen, I.N.L.G.; de Bruin, J.; van Gent, P.; Commandeur, J.J.F. Interactions between Cyclists and Automated Vehicles: Results of a Photo Experiment *. *J. Transp. Saf. Secur.* **2020**, *12*, 94–115. [CrossRef]
47. Vlakveld, W.; van der Kint, S.; Hagenzieker, M.P. Cyclists' Intentions to Yield for Automated Cars at Intersections When They Have Right of Way: Results of an Experiment Using High-Quality Video Animations. *Transp. Res. Part F Traffic Psychol. Behav.* **2020**, *71*, 288–307. [CrossRef]
48. Hou, M.; Mahadevan, K.; Somanath, S.; Sharlin, E.; Oehlberg, L. Autonomous Vehicle-Cyclist Interaction: Peril and Promise. In Proceedings of the 2020 CHI Conference on Human Factors in Computing Systems, Honolulu, HI, USA, 25–30 April 2020; ACM: New York, NY, USA, 2020; pp. 1–12. [CrossRef]
49. Gwak, J.; Jung, J.; Oh, R.; Park, M.; Rakhimov, M.A.K.; Ahn, J. A Review of Intelligent Self-Driving Vehicle Software Research. *KSII Trans. Internet Inf. Syst.* **2019**, *13*, 5299–5320. [CrossRef]
50. Carrese, S.; Nigro, M.; Patella, S.M.; Toniolo, E. A Preliminary Study of the Potential Impact of Autonomous Vehicles on Residential Location in Rome. *Res. Transp. Econ.* **2019**, *75*, 55–61. [CrossRef]
51. Ngwu, O.L.; Rimu, A.; Deb, S. How to Design Traffic Infrastructure to Support Cyclists' Interaction with Autonomous Vehicles: Teenage Cyclists' Perceptions. *Hum. Factors Syst. Interact.* **2022**, *52*. [CrossRef]
52. Botello, B.; Buehler, R.; Hankey, S.; Mondschein, A.; Jiang, Z. Planning for Walking and Cycling in an Autonomous-Vehicle Future. *Transp. Res. Interdiscip. Perspect.* **2019**, *1*, 100012. [CrossRef]
53. Soteropoulos, A.; Berger, M.; Mitteregger, M. Compatibility of Automated Vehicles in Street Spaces: Considerations for a Sustainable Implementation. *Sustainability* **2021**, *13*, 2732. [CrossRef]
54. Alves, G.V.; Schwammberger, M. Towards a Digital Highway Code Using Formal Modelling and Verification of Timed Automata. *Electron. Proc. Theor. Comput. Sci.* **2022**, *371*, 77–85. [CrossRef]
55. Seilabi, S.E.; Pourgholamali, M.; Homem de Almeida Correia, G.; Labi, S. Robust Design of CAV-Dedicated Lanes Considering CAV Demand Uncertainty and Lane Reallocation Policy. *Transp. Res. D Transp. Environ.* **2023**, *121*, 103827. [CrossRef]

56. Park, C.; Sohn, S.Y. An Optimization Approach for the Placement of Bicycle-Sharing Stations to Reduce Short Car Trips: An Application to the City of Seoul. *Transp. Res. Part A Policy Pract.* **2017**, *105*, 154–166. [CrossRef]
57. Shunmuga Perumal, P.; Wang, Y.; Sujasree, M.; Tulshain, S.; Bhutani, S.; Suriyah, M.K.; Kumar Raju, V.U. LaneScanNET: A Deep-Learning Approach for Simultaneous Detection of Obstacle-Lane States for Autonomous Driving Systems. *Expert. Syst. Appl.* **2023**, *233*, 120970. [CrossRef]
58. Roy, K.; Hoang, N.H.; Vu, H.L. Modeling Autonomous Vehicles Deployment in a Multilane AV Zone With Mixed Traffic. *IEEE Trans. Intell. Transp. Syst.* **2022**, *23*, 23708–23720. [CrossRef]
59. Wang, S.; Li, Z.; Wang, Y.; Zhao, W.; Liu, T. Evidence of Automated Vehicle Safety's Influence on People's Acceptance of the Automated Driving Technology. *Accid. Anal. Prev.* **2024**, *195*, 107381. [CrossRef] [PubMed]
60. Sahoo, S.K.; Goswami, S.S. A Comprehensive Review of Multiple Criteria Decision-Making (MCDM) Methods: Advancements, Applications, and Future Directions. *Decis. Mak. Adv.* **2023**, *1*, 25–48. [CrossRef]
61. Rejeb, A.; Rejeb, K.; Keogh, J.G.; Zailani, S. Barriers to Blockchain Adoption in the Circular Economy: A Fuzzy Delphi and Best-Worst Approach. *Sustainability* **2022**, *14*, 3611. [CrossRef]
62. Liu, M.; Wan, Y.; Lewis, F.L.; Nageshrao, S.; Filev, D. A Three-Level Game-Theoretic Decision-Making Framework for Autonomous Vehicles. *IEEE Trans. Intell. Transp. Syst.* **2022**, *23*, 20298–20308. [CrossRef]
63. Liu, H.; Yang, M.; Guan, C.; Chen, Y.S.; Keith, M.; You, M.; Menendez, M. Urban Infrastructure Design Principles for Connected and Autonomous Vehicles: A Case Study of Oxford, UK. *Comput. Urban Sci.* **2023**, *3*, 34. [CrossRef]
64. Kiciński, M.; Solecka, K. Application of MCDA/MCDM Methods for an Integrated Urban Public Transportation System—Case Study, City of Cracow. *Arch. Transp.* **2018**, *46*, 71–84. [CrossRef]
65. Ashofteh, P.-S.; Far, S.M.; Golfam, P. Application of Multi-Criteria Decision-Making of CODAS and SWARA in Reservoir Optimal Operation Using Marine Predator Algorithm Based on Game Theory. *Water Resour. Manag.* **2023**, *37*, 4385–4412. [CrossRef]
66. Collins, B.C.; Kumral, M. Examining Impact and Benefit Agreements in Mineral Extraction Using Game Theory and Multiple-Criteria Decision Making. *Extr. Ind. Soc.* **2022**, *10*, 101094. [CrossRef]
67. Punetha, N.; Jain, G. Game Theory and MCDM-Based Unsupervised Sentiment Analysis of Restaurant Reviews. *Appl. Intell.* **2023**, *53*, 20152–20173. [CrossRef] [PubMed]
68. Kaviari, F.; Mesgari, M.S.; Seidi, E.; Motieyan, H. Simulation of Urban Growth Using Agent-Based Modeling and Game Theory with Different Temporal Resolutions. *Cities* **2019**, *95*, 102387. [CrossRef]
69. Ibrahim, M.A.R.; Jaini, N.I.; Khalif, K.M.N.K. A Comprehensive Review of Hybrid Game Theory Techniques and Multi-Criteria Decision-Making Methods. *J. Phys. Conf. Ser.* **2021**, *1988*, 012056. [CrossRef]
70. Cortés-Berrueco, L.E.; Gershenson, C.; Stephens, C.R. Traffic Games: Modeling Freeway Traffic with Game Theory. *PLoS ONE* **2016**, *11*, e0165381. [CrossRef]
71. Zhu, L.; Yang, D.; Cheng, Z.; Yu, X.; Zheng, B. A Model to Manage the Lane-Changing Conflict for Automated Vehicles Based on Game Theory. *Sustainability* **2023**, *15*, 3063. [CrossRef]
72. Chellappa, V.; Ginda, G. Application of Multiple-Criteria Decision Making Methods for Construction Safety Research. *Proc. Inst. Civ. Eng.-Manag. Procure. Law* **2024**, *177*, 127–136. [CrossRef]
73. Alimohammadlou, M.; Alinejad, S. Challenges of Blockchain Implementation in SMEs' Supply Chains: An Integrated IT2F-BWM and IT2F-DEMATEL Method. *Electron. Commer. Res.* **2023**. [CrossRef]
74. Mubarik, M.S.; Kazmi, S.H.A.; Zaman, S.I. Application of Gray DEMATEL-ANP in Green-Strategic Sourcing. *Technol. Soc.* **2021**, *64*, 101524. [CrossRef]
75. Wu, H.; Zhong, W.; Zhong, B.; Li, H.; Guo, J.; Mehmood, I. Barrier Identification, Analysis and Solutions of Blockchain Adoption in Construction: A Fuzzy DEMATEL and TOE Integrated Method. *Eng. Constr. Archit. Manag.* **2023**. [CrossRef]
76. Li, Y.; Zhao, G.; Wu, P.; Qiu, J. An Integrated Gray DEMATEL and ANP Method for Evaluating the Green Mining Performance of Underground Gold Mines. *Sustainability* **2022**, *14*, 6812. [CrossRef]
77. Liu, T.; Deng, Y.; Chan, F. Evidential Supplier Selection Based on DEMATEL and Game Theory. *Int. J. Fuzzy Syst.* **2018**, *20*, 1321–1333. [CrossRef]
78. Nematkhah, F.; Raissi, S.; Ghezavati, V. An Integrated Fuzzy DEMATEL-Fuzzy ANP Approach to Nominate Diagnostic Method and Measuring Total Predictive Performance Score. *Saf. Reliab.* **2017**, *37*, 48–72. [CrossRef]
79. Chatterjee, K.; Kar, S. Multi-Criteria Analysis of Supply Chain Risk Management Using Interval Valued Fuzzy TOPSIS. *OPSEARCH* **2016**, *53*, 474–499. [CrossRef]
80. Mohammadzadeh, A.; Sabzalian, M.H.; Zhang, C.; Castillo, O.; Sakthivel, R.; El-Sousy, F.F.M. Training Interval Type-2 Fuzzy Systems Based on Error Backpropagation. In *Modern Adaptive Fuzzy Control Systems*; Springer: Cham, Switerland, 2023; pp. 49–93. [CrossRef]
81. Chaqooshi, A.F.; Rajabani, N.; Khalili Esbuei, S.; Hakimi, N. Identifying and Ranking Appropriate Resilience Supply Chain Strategies, Hybrid Approach of Game Theory and Fuzzy MCDM. *J. Ind. Manag. Perspect.* **2019**, *9*, 9–31. [CrossRef]
82. Zhou, X.; Hu, Y.; Deng, Y.; Chan, F.T.S.; Ishizaka, A. A DEMATEL-Based Completion Method for Incomplete Pairwise Comparison Matrix in AHP. *Ann. Oper. Res.* **2018**, *271*, 1045–1066. [CrossRef]
83. Debnath, A.; Bandyopadhyay, A.; Roy, J.; Kar, S. Game Theory Based Multi Criteria Decision Making Problem under Uncertainty: A Case Study on Indian Tea Industry. *J. Bus. Econ. Manag.* **2018**, *19*, 154–175. [CrossRef]

84. Almutairi, K.; Hosseini Dehshiri, S.J.; Hosseini Dehshiri, S.S.; Mostafaeipour, A.; Hoa, A.X.; Techato, K. Determination of optimal renewable energy growth strategies using SWOT analysis, hybrid MCDM methods, and game theory: A case study. *Int. J. Energy Res.* **2022**, *46*, 6766–6789. [CrossRef]
85. Chen, Y.; Liu, Y. Integrated Optimization of Planning and Operations for Shared Autonomous Electric Vehicle Systems. *Transp. Sci.* **2023**, *57*, 106–134. [CrossRef]
86. Zhou, P.; Wang, C.; Yang, Y. Design and Optimization of Solar-Powered Shared Electric Autonomous Vehicle System for Smart Cities. *IEEE Trans. Mob. Comput.* **2023**, *22*, 2053–2068. [CrossRef]
87. Othman, K. Impact of Autonomous Vehicles on the Physical Infrastructure: Changes and Challenges. *Designs* **2021**, *5*, 40. [CrossRef]
88. Rana, M.M.; Hossain, K. Connected and Autonomous Vehicles and Infrastructures: A Literature Review. *Int. J. Pavement Res. Technol.* **2023**, *16*, 264–284. [CrossRef]
89. Vo, H.V.; Chae, B.; Olson, D.L. Dynamic MCDM: The Case of Urban Infrastructure Decision Making. *Int. J. Inf. Technol. Decis. Mak.* **2002**, *1*, 269–292. [CrossRef]
90. Elassy, M.; Al-Hattab, M.; Takruri, M.; Badawi, S. Intelligent Transportation Systems for Sustainable Smart Cities. *Transp. Eng.* **2024**, *16*, 100252. [CrossRef]
91. Liu, J.; Zhao, H.; Ogata, Y. Game Based Multi-Agent Equilibrium in Intelligent City Transportation System Design. *DEStech Trans. Comput. Sci. Eng.* **2018**, 1–5. [CrossRef]
92. Dabiri, A.; Hegyi, A.; Hoogendoorn, S. Optimized Speed Trajectories for Cyclists, Based on Personal Preferences and Traffic Light Information-A Stochastic Dynamic Programming Approach. *IEEE Trans. Intell. Transp. Syst.* **2022**, *23*, 777–793. [CrossRef]
93. Guo, Z.; Sun, D.; Zhou, L. Game Algorithm of Intelligent Driving Vehicle Based on Left-Turn Scene of Crossroad Traffic Flow. *Comput. Intell. Neurosci.* **2022**, *2022*, 9318475. [CrossRef]
94. Schlossberg, M.; Riggs, W.; Adam Millard-Ball, L.A.; Shay, E. Rethinking the Street in an Era of Driverless Cars. 2018. Available online: www.urbanismnext.com (accessed on 30 September 2024).
95. Radakovic, D.; Singh, A.; Varde, A.S.; Lal, P. Enriching Smart Cities by Optimizing Electric Vehicle Ride-Sharing through Game Theory. In Proceedings of the 2022 IEEE 34th International Conference on Tools with Artificial Intelligence (ICTAI), Macao, China, 30 October–2 November 2022; IEEE: Piscataway, NJ, USA, 2022; pp. 755–759. [CrossRef]

smart cities MDPI

Review

Toward Sustainable Urban Mobility: A Systematic Review of Transit-Oriented Development for the Appraisal of Dubai Metro Stations

Oussama Yahia [1], Afaq Hyder Chohan [2,*], Mohammad Arar [2] and Jihad Awad [3]

1 Department of Architecture, College of Architecture, Art & Design (CAAD), Ajman Campus, Ajman P.O. Box 346, United Arab Emirates; archosama4@gmail.com
2 Department of Architecture, Ajman University, Ajman P.O. Box 346, United Arab Emirates; m.arar@ajman.ac.ae
3 Healthy and Sustainable Built Environment Research Center (HSBERC), Architecture Ajman University, Ajman P.O. Box 346, United Arab Emirates; j.awad@ajman.ac.ae
* Correspondence: a.chohan@ajman.ac.ae

Highlights:

- Implementing TOD principles in Dubai's metropolitan planning is crucial for advancing a connected, efficient, and sustainable urban future.
- The study highlights the advantages of adopting TOD strategies at Dubai metro stations such as Al Rashidiya, Al Qusais, and Mall of the Emirates, aiming to boost transit efficiency, support pedestrian-friendly areas, and curtail dependence on private vehicles, aligning with sustainable urban development goals.

What are the main findings?

- This study concludes that implementing Transit-Oriented Development (TOD) near the Mall of the Emirates Metro Station in Dubai can significantly reduce commuting distances and enhance connectivity. SVM with Radial kernels outperformed other models, capturing non-linear patterns effectively, while DNN models showed signs of overfitting.
- The TOD design initiative would encourage residents to rely on efficient public transit instead of private cars. This shift can alleviate traffic congestion and lower emissions, promoting a healthier urban environment.

What is the implication of the main finding?

- TOD strategies near the Mall of Emirates Metro station, could implicate the transforming urban areas by increasing property values and living costs near transit hubs, which may inadvertently lead to gentrification and affordability challenges, displacing lower-income residents.

Abstract: In Dubai's rapidly expanding urban landscape, addressing the adverse impacts of increasing automobile reliance is critical. Growing vehicle usage contributes to urban sprawl, prolonged commutes, infrastructure strain, and diminished green spaces. As a sustainable alternative, Transit-Oriented Development (TOD) promotes compact density, mixed-use environments, and transit-focused design, particularly suited for Dubai's evolving context. This study evaluates the applicability of Transit-Adjusted Development (TAD) and TOD appraisal models, specifically the 3D and 6D frameworks, to stations on both the Red and Green Lines of the Dubai Metro. By examining Dubai's complex urban form, the research identifies strategic interventions to enhance urban mobility and mitigate sprawl. Through an extensive literature review, key factors shaping sustainable urban transport such as accessibility, land-use diversity, density, design, distance to transit, and demand management are analyzed. This investigation highlights the suitability of implementing TOD principles at prominent metro stations, including Al Rashidiya, Al Qusais, and Mall of the Emirates. These stations hold significant potential for strengthening transit efficiency, fostering pedestrian-friendly neighborhoods, and reducing dependency on private vehicles. The findings underscore the importance of integrating TOD strategies into Dubai's metropolitan planning. By doing so, Dubai can move toward a more connected, efficient, and environmentally responsible urban future.

Keywords: transit-oriented development; compact and integrated development; 3D and 6D TOD models

1. Introduction

Transit-Oriented Development (TOD) has emerged as an innovative urban planning approach that bridges the divide between the charm of historical, walkable neighborhoods and the complexities of modern urban forms. By reinstating the principles of compact, mixed-use developments, TOD offers a sustainable alternative to the widespread trend of urban sprawl prevalent in many cities today. This methodology effectively revives the essence of pre-automobile urban design, where communities were structured to enable residents to live, work, shop, and engage in leisure activities within accessible distances. The core philosophy of TOD revolves around creating vibrant, self-sufficient neighborhoods that promote a harmonious blend of residential, commercial, and recreational spaces, thereby enhancing the quality of urban life and reducing dependence on motorized transportation [1,2].

In order to attain the envisioned benefits of TOD, numerous city municipalities, in collaboration with urban planners and researchers, have crafted implementation policies tailored to their distinct urban forms. TOD initiatives can involve the creation of entirely new developments or the revitalization of existing neighborhoods centered around public transit stations. By fully leveraging these transit hubs, the goal is to cultivate more "livable communities" that encourage sustainable living and reduced dependency on private vehicles. Rather than abolishing automobile use altogether, this strategy aims to diminish its dominance by seamlessly integrating cars with other modes of public transportation [3,4].

A number of researchers assert that urban challenges, especially in cities of developing nations, largely stem from rapid urbanization and the surge in motor vehicle usage. These factors have significantly contributed to pervasive urban sprawl, which is a major catalyst for traffic congestion, fragmented urban landscapes, social inequity, overextended infrastructure, and environmental degradation [5,6]. This pattern of urban sprawl, first observed in the United States, prompted urban planners and scholars to propose TOD as a solution to the ongoing dispersion of communities. TOD has demonstrated considerable success in enhancing urban quality, particularly within inner-city neighborhoods, and fostering overall metropolitan growth. These areas experienced urban decay and declining population densities due to extensive road construction and disjointed land-use patterns [7,8]. By implementing TOD principles such as mixed-use and medium- to high-density developments and improving the built environment surrounding transit hubs, targeted communities were revitalized. This strategy often resulted in enhanced livability and mobility, along with the more efficient utilization and management of urban spaces and infrastructure [9,10].

Although TOD is often promoted as a solution to various urban issues, its success hinges on conducting a thorough analysis of the specific area in question. This analysis should address current urban challenges, define clear objectives aligned with the area's characteristics, and anticipate future goals. Evidence from studies suggests that the active participation of diverse stakeholders, including government agencies, private and public sectors, investors, and the local community, is crucial. Guidance from experts such as urban planners, geographers, engineers, and economists ensures that the TOD initiative is comprehensive and multidisciplinary, thereby maximizing its potential benefits [11,12].

In this regard, critical metrics that need careful evaluation encompass land-use planning, transportation system analysis, property market trends, housing affordability, accessibility, the availability of open spaces, population density, walkability, and financial feasibility. By meticulously examining these factors, a TOD project can be tailored to effectively address the specific needs of the area, leading to more sustainable and livable urban environments [13,14].

The discussion above can be précised as follows: A collaborative effort with diverse stakeholders, including government entities, private sectors, investors, community members, and experts in urban planning, engineering, and economics, is essential to ensure that TOD projects are comprehensive and multidisciplinary. By critically evaluating factors such as land-use planning, transportation systems, property market trends, housing affordability, accessibility, the availability of open spaces, population density, walkability, and financial viability, it is possible to tailor TOD initiatives to specific urban contexts. By meticulously addressing these elements, TOD can effectively revitalize urban areas, reduce social inequities, alleviate traffic congestion, and foster environmental sustainability.

Evidence from various studies by Thomas, R., and Bertolini, L. [7]; Abdi, M. H. [4]; and Chen, H., et al. [13] demonstrates that TOD not only enhances the quality of inner-city neighborhoods but also contributes to overall metropolitan growth by creating vibrant, self-sufficient communities that balance residential, commercial, and recreational needs. Therefore, by embracing TOD, it is possible to develop more sustainable and livable urban environments.

It is noteworthy that the interpretation and application of TOD can significantly vary from one location to another, influenced by specific objectives and the urban typology of the area in question. This variation affects the methodologies employed in implementing TOD. Projects may range from individual buildings that incorporate TOD principles, such as optimal building orientation, enhanced accessibility, thoughtful façade design, and street frontages that respect and integrate with the surrounding environment, especially pivotal transit stops, to larger-scale developments encompassing entire neighborhoods, where the transit stop is a fundamental component of the urban structure [15–17].

Nevertheless, the benefits of TOD are extensive and can be categorized in multiple ways. Studies suggest that these benefits can be divided into primary and secondary classes and further distinguished based on whether they cater to private or public entities. Secondary benefits are indirect outcomes influenced by the implementation of TOD and may also result from other intervention programs or manifest in the long term. Examples of such benefits include a reduction in traffic congestion, decreased crime rates, improved air quality, and increased retail sales. Primary benefits, on the other hand, are direct results of applying TOD principles. These include the revitalization of communities and neighborhoods, increased transit ridership, enhanced property values, the provision of affordable housing, the promotion of social equity, and the stimulation of economic development [18].

Emerging cities in the GCC region, India, and China face rapid urbanization challenges, including traffic congestion, environmental degradation, and social inequity. TOD plays a crucial role in addressing these issues by integrating high-density, mixed-use developments with efficient public transportation. In the GCC, TOD can reduce reliance on private vehicles and promote sustainable growth. In India, it offers solutions to urban sprawl and improves accessibility in densely populated cities. China's implementation of TOD has enhanced connectivity and reduced carbon emissions. However, challenges such as adapting TOD principles to local contexts and ensuring affordability and effective stakeholder coordination remain. Critically, TOD holds significant potential but requires tailored strategies to meet the unique needs of each region [19–21].

As Dubai rapidly evolves into a leading economic, business, and residential hub within the GCC region and on the international stage, the necessity for efficient and sustainable urban planning strategies becomes paramount [22]. A significant challenge faced by such burgeoning cities is the integration of new transit systems into an already established urban fabric. Constructing transit infrastructure in these environments poses difficulties in ensuring that new developments seamlessly align with existing urban structures while meeting the needs of a growing population [23].

In this context, this study suggests that Transit-Oriented Development (TOD) emerges as a critical solution. TOD focuses on creating high-density, mixed-use communities centered around public transit hubs, promoting walkability, reducing reliance on private vehicles, and enhancing overall urban livability. For Dubai, implementing TOD is essential to effectively integrate new transit systems into its complex urban landscape. By adopting TOD principles, the city can ensure that its transit infrastructure not only efficiently connects various districts but also supports sustainable urban growth.

Moreover, as Dubai continues to attract international businesses and residents, the demand for accessible and efficient public transportation will inevitably increase. TOD offers a framework to meet this demand by developing interconnected communities that facilitate economic activity while minimizing environmental impact. Embracing TOD is not just a planning necessity but a strategic imperative for Dubai's future development. It enhances global competitiveness and improves the quality of life for residents, solidifying Dubai's position as a modern, sustainable metropolis.

Recognizing the critical need for Transit-Oriented Development (TOD) in Dubai, particularly for its metro stations, this study aims to thoroughly understand TOD principles, standards, key components, and their beneficial outcomes. By examining successful TOD practices that have transformed urban conditions around transit stops in Dubai, the research highlights the importance of integrating TOD into the city's urban planning strategies. Given Dubai's rapid growth as a leading economic, business, and residential hub in the GCC region and internationally, implementing TOD at metro stations is essential to enhance connectivity, reduce traffic congestion, and promote sustainable urban development. This study emphasizes how applying TOD principles can improve urban livability and support Dubai's ongoing expansion.

2. TOD: A Tool for Urban Mobility Design and Integrated Community

Transit-Oriented Development (TOD) originally emerged as a response to urban sprawl, significantly evolving over the past three decades. An early example is the Twin Cities Rapid Transit (TCRT) system in Minneapolis and St. Paul between 1900 and 1930. Championing 524 miles of streetcar tracks and serving over 200 million riders, Thomas Lowry aimed to stimulate commercial and real estate growth along transit corridors, fostering smart city expansion. This enabled affordable suburban living with convenient access to city centers. However, post-World War II automobile proliferation shifted focus away from transit, leading to urban sprawl and reduced transit services. Today, TOD seeks to revive transit corridors and mixed-use developments around rail lines to counter these effects [24–26].

In the aftermath of World War II, the widespread proliferation of automobiles prompted significant shifts in urban planning, elevating the importance of TOD. Cities restructured their layouts to accommodate the unprecedented influx of private cars, with highways becoming prominent features that encouraged urban sprawl and the expansion of suburban subdivisions. Public transit services were deprioritized in planning agendas. Policymakers focused on zoning regulations that segregated residential neighborhoods from hazardous industrial areas and distanced commercial activities from homes. These post-war urban strategies were prevalent worldwide, leading to adverse effects such as car-dependent land use, low-density suburban growth, inner-city decline, and increased traffic congestion stemming from transportation infrastructure and land-use policies [27,28].

The discussion above portrays TOD as a significant urban planning strategy, though its implementation varies across cities due to unique circumstances and planning issues. However, the definition of TOD has been subject to debate among researchers and planners because the concept has evolved over time. Nonetheless, a common understanding positions TOD as a tool or approach directly associated with promoting intelligent city growth, improving quality of life, and sustaining mobility.

According to Patnala et al. [29], TOD involves development projects of moderate to high density located within a comfortable walking distance of a major transit stop. These projects typically offer a blend of residential, employment, and shopping opportunities designed to be pedestrian-friendly while still accommodating automobiles. TOD can entail new construction or the redevelopment of existing buildings, with designs oriented to facilitate transit use.

Similarly, Calthorpe [30] describes TOD as a mixed-use community situated within an average walking distance of one-fourth of a mile from a transit stop and a central commercial area. The focus is on design, configuration, and a mix of uses that emphasize a pedestrian-oriented environment and strengthen the use of public transportation. These communities integrate residential, retail, office spaces, open areas, and public facilities within a convenient walking distance, making it easy for residents and employees to travel by transit, bicycle, foot, or car. Additionally, Steuteville and Langdon [31] view TOD as part of the "new urbanism" movement initiated by architects and planners in the 1990s. This approach aims to revive the urban fabric reminiscent of the pre-World War II era, relying heavily on transit principles. In its most fundamental form, TOD is considered a New Urbanist neighborhood, often referred to as a traditional neighborhood development, centered around a mass transit station.

Furthermore, Laaly et al. [32] assert that TOD is characterized by neighborhoods centered around a transit station, deliberately planned and designed to make transit use convenient and appealing for residents, workers, shoppers, and others. These neighborhoods feature a variety of housing types, including multi-family units, located within no more than a quarter mile of an existing or planned rail station. Additionally, the Federal Transit Administration (FTA) [33] describes TOD in two ways under the concept of livable communities. First, transit-oriented communities are distinguished by design and development patterns that encourage the use of transit, bicycling, and walking to access shopping, business centers, services, housing, and more. Second, transit-oriented development and community-sensitive transit are seen as means to reverse the trends of urban sprawl, which lead to increasingly longer trips, inadequate pedestrian access, traffic congestion, and negative environmental impacts.

From the various definitions of Transit-Oriented Development (TOD) discussed, several key characteristics and features consistently emerge as fundamental components of the concept. A primary feature of TOD is the strategic location of developments within easy walking distance, typically a quarter mile from major transit stops such as rail stations or bus terminals. This proximity ensures that residents and visitors have convenient access to public transportation, thereby encouraging its use and reducing reliance on private vehicles.

TOD emphasizes the integration of various land uses, including residential, commercial, retail, office, and public spaces. By combining these elements within a single community, TOD creates vibrant neighborhoods where people can live, work, shop, and engage in leisure activities without the need for long commutes.

A critical aspect of TOD is the focus on creating pedestrian-friendly environments. This includes designing accessible sidewalks, engaging street frontages, and providing amenities that encourage walking and cycling. The built environment prioritizes pedestrians and cyclists, promoting healthier lifestyles and reducing traffic congestion. Furthermore, TOD neighborhoods are characterized by moderate to high density, which supports efficient public transit services. A higher concentration of residents and businesses within a compact area creates a critical mass that sustains local amenities and enhances the viability of transit operations. Effective integration with public transportation systems is essential in TOD. The design and orientation of buildings and public spaces facilitate easy access to transit services while still accommodating automobile access when necessary. This seamless integration encourages more people to choose public transit over private cars.

TOD also serves as a strategic approach to counteract the spread of low-density, car-dependent developments. By focusing growth on transit hubs, it promotes smarter, more sustainable urban expansion, conserving land resources and reducing environmental impacts associated with sprawl. Redeveloping or enhancing existing neighborhoods around transit hubs is a significant benefit of TOD, as such developments improve the quality of life by stimulating economic vitality, attracting investments, and fostering a sense of community. Encouraging modes of transportation that reduce carbon emissions is a core objective of TOD. By promoting public transit, walking, and cycling, TOD contributes to lowering environmental footprints and combating climate change. Ensuring that essential services, employment opportunities, and recreational activities are easily accessible through various transportation modes enhances the overall functionality of urban areas. TOD aims to create well-connected communities that facilitate mobility and access for all residents.

It can be concluded that, despite variations in specific definitions, the overarching objective of transit-oriented development remains consistent: to create sustainable, livable communities centered around efficient public transit systems. By enhancing mobility options and reducing dependence on private vehicles, TOD contributes to more dynamic, resilient, and environmentally responsible urban environments. Based on the discussion in Section 2, Figure 1 was formulated to encircle the key features of TOD.

Despite several advantages, there are some downsides of the TOD framework that influences the neighborhood of a transit hub. Transit-Oriented Development (TOD) has been increasingly recognized as a catalyst for urban transformation, influencing property values, rental prices, and overall living costs. Research suggests that TOD projects enhance accessibility and connectivity, leading to increased demand for properties near transit hubs. According to Cervero and Murakami [34], properties within TOD areas witness a significant appreciation in value due to improved accessibility and infrastructure. Similarly, Duncan [35] found that proximity to transit stations increases real estate prices, as potential buyers and investors prioritize convenient access to public transport.

However, the economic benefits associated with TOD often lead to unintended consequences, such as rising housing costs and affordability challenges. Studies by Hess and Almeida [36] indicate that while TOD boosts property values, it can contribute to gentrification, displacing lower-income residents. Chatman and Noland [37] further argue that increased demand for housing in TOD zones results in higher rental costs, potentially outweighing the benefits of reduced transportation expenses.

Despite these challenges, TOD can contribute to a balanced urban economy by reducing transportation expenditures and promoting sustainable living. Belzer and Autler [38] emphasize that strategic policies, such as affordable housing requirements within TOD frameworks, can help mitigate rising costs and ensure social inclusivity. Additionally, Renne [39] argues that TOD reduces household transportation expenditures, potentially offsetting higher rents and enhancing overall affordability in the long term.

In conclusion, TOD has a profound impact on real estate prices and living costs. While it fosters economic growth and urban revitalization, careful planning and policy interventions are required to prevent socio-economic disparities and ensure equitable access to affordable housing in transit-rich areas.

Figure 1. Features of the TOD model.

3. Parallel Concepts of Urban Mobility Design

It is interesting to note that the concept of Transit-Adjacent Development (TAD) emerged alongside Transit-Oriented Development (TOD) as urban planners and researchers observed that mere proximity to transit stations did not guarantee the benefits associated with TOD. While TOD focuses on creating high-density, mixed-use communities that actively promote public transit use and reduce reliance on private vehicles, TAD refers to developments that are physically near transit facilities but lack the necessary design and planning to encourage transit ridership effectively [40].

The emergence of TAD highlights the critical distinction between physical adjacency and functional integration with transit systems. In many instances, developments are constructed near transit stations without considering the essential elements that make TOD successful, such as pedestrian-friendly infrastructure, mixed land uses, and seamless connectivity between different modes of transportation [41]. This oversight often results in underutilized transit services and continued dependence on automobiles. One reason for the rise of the TAD concept is the recognition that simply placing developments near transit stops does not automatically create an activity center or achieve high transit performance. Factors such as zoning policies favoring low-density or single-use developments, inadequate pedestrian pathways, and a lack of amenities can hinder the effectiveness of transit stations in attracting ridership [42]. Additionally, without intentional planning, these areas may fail to provide vibrant, accessible environments that encourage people to choose public transit over personal vehicles [43].

The identification of TAD underscores the need for comprehensive planning and design strategies that align with the principles of TOD. It serves as a cautionary concept, emphasizing that the benefits of transit investments can only be realized when developments are specifically designed to integrate with transit services. This includes creating walkable environments, implementing mixed-use developments, and ensuring connectivity to surrounding areas [44]. Understanding why TAD occurs is crucial for policymakers and planners aiming to promote sustainable urban development. By addressing the shortcomings that lead to TAD, such as revising zoning regulations and improving pedestrian infrastructure, cities can transition these areas into true TODs. This transformation maximizes the social, economic, and environmental advantages of transit systems, leading to enhanced urban vitality and reduced car dependency Cervero, R., Ferrell, C., & Murphy, S. 2002 [41].

The discussion above can be summarized as follows: Transit-Adjacent Development (TAD) is characterized by developments that are physically near transit stations but lack functional integration with the transit system, thereby failing to effectively encourage its use. These developments often exhibit a lack of transit-supportive design. the planning and architecture do not prioritize or facilitate public transit ridership. Typically, TADs are single-use or low-density developments, characterized by zoning that does not support the mixed-use environments found in successful Transit-Oriented Developments (TODs).

Furthermore, this study indicates that TADs suffer from insufficient pedestrian infrastructure, lacking accessible sidewalks, crosswalks, and pedestrian pathways, which makes it inconvenient for individuals to walk to transit stations. There is also poor connectivity between transportation modes, with inadequate integration of cycling paths and bus networks, leading to isolated transit facilities. Zoning policies may not be conducive to transit use, often favoring developments with expansive parking lots or automobile-centric designs that do not support high transit ridership. The absence of nearby amenities and services such as shops, restaurants, and recreational facilities fails to attract people to the area or encourage the use of public transit. As a result, residents and visitors continue to depend heavily on automobiles despite the proximity to transit stations due to the unaccommodating design of the area. TADs also fail to create vibrant activity centers, lacking the lively and accessible environments necessary to draw people to choose public transit over personal cars. Consequently, transit services adjacent to these developments are often underutilized, experiencing lower ridership because the surrounding area does not support or encourage transit use. This represents a missed opportunity for sustainable development, as TADs fail to leverage the potential social, economic, and environmental benefits that come from well-integrated transit and land-use planning. The design may prioritize parking facilities over easy access to transit, further encouraging automobile use.

By highlighting these features in Figure 2, it becomes clear that TAD represents a missed opportunity to fully capitalize on the advantages of proximity to transit. Understanding these shortcomings is crucial for urban planners and policymakers aiming to transition TADs into true TODs that promote sustainable urban development and reduce reliance on private vehicles.

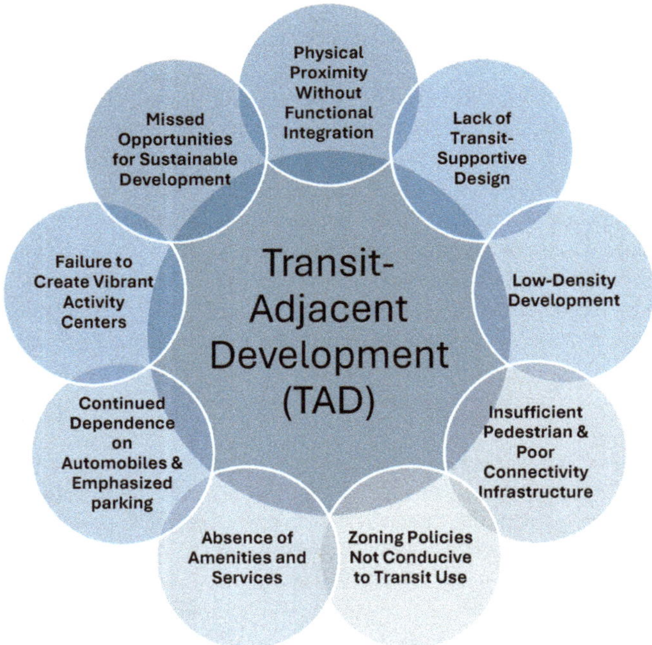

Figure 2. TAD model.

The 3D and 6D Concepts of Transit Mobility Developments

Transit-Oriented Development (TOD) is often linked to the "3D" framework—Density, Diversity, and Design—introduced by Cervero and Kockelman [45] to explain how urban form influences travel behavior. Density supports efficient transit by concentrating people and jobs. Diversity mixes land uses to enhance accessibility and foster vibrant communities. Design encourages active transportation through pedestrian-friendly environments.

The framework was expanded to the "6Ds" with the addition of Distance to Transit, Destination Accessibility, and Demand Management [46]. Distance to transit increases ridership by reducing the proximity to transit stations. Destination accessibility improves mobility by easing access to key locations via transit. Demand management promotes sustainable travel through policies like parking management. By integrating these dimensions, TOD offers a holistic approach to sustainable urban development, addressing sprawl, congestion, and environmental degradation [47].

Researchers have extensively adopted the "3D" framework to develop and quantify indicators of TOD, such as population density, employment density, land-use mix, and street network connectivity Cervero and Kockelman [45]. This framework serves as a foundational tool in understanding how TOD contributes to smarter and more sustainable urban development.

Peter Calthorpe's [30] seminal work, *The Next American Metropolis: Ecology, Community, and the American Dream*, positions TOD as a pivotal planning model to address fragmented spatial urban patterns. He argues that low-density, segregated land-use patterns are primary catalysts for the emergence of "car-dependent cities", highlighting the necessity for integrated urban planning to mitigate this issue. Fundamentally, most definitions of TOD emphasize core principles: the integration of mixed land uses, increased urban density, the enhancement of the built environment, and the promotion of compact development to encourage alternative transportation modes such as transit, walking, and cycling Dittmar, H., & Ohland, G. [44]. Initially conceptualized by Cervero and Kockelman [45], the

"3Ds"—Density, Diversity, and Design—were shown to significantly reduce Vehicle Kilometers Traveled (VKT) and increase public transit usage. As the concept evolved, additional dimensions were incorporated, expanding the framework to the "6Ds" by including Distance to Transit, Destination Accessibility, and Demand Management Ewing, R., & Cervero, R. 2010 [46]. This expansion reflects a broader understanding of TOD as a comprehensive planning instrument encompassing transportation, land use, urban design, policies, regulations, politics, finance, and economics Suzuki, H., Cervero, R., & Iuchi, K. 2013 [47].

The "D" framework is instrumental in planning, policymaking, and implementing TOD strategies [48]. Introducing the six "Ds" aims to promote TOD within local and regional planning, recognizing that successful implementation requires collaboration among diverse stakeholders. The primary objective is to overcome political, financial, and planning obstacles that hinder TOD by formulating policies that stimulate market opportunities for creating compact, high-density, and diverse land uses within walkable areas around transit stations [49].

However, several barriers impede the effective implementation of TOD. Challenges related to economics, finance, and politics include hesitant private developers wary of investing in TOD projects, governmental resistance to high-density developments, and deficiencies in collaboration between transit agencies, public land-use authorities, and private stakeholders [50–52].

The six "Ds" are widely regarded as critical variables in achieving TOD objectives, each contributing uniquely to the overall success of transit-oriented initiatives. **Diversity** refers to providing a mix of land uses within the area surrounding a transit station. A balanced integration of residential, commercial, employment, recreational, and civic spaces enhances the quality of life by promoting walkability, cycling, and public transit use for shorter trips, thereby reducing reliance on private vehicles Suzuki, H., Cervero, R., & Iuchi, K. 2013. [47]. This land-use diversity fosters social equity by allowing people of varying income levels to live near workplaces and access affordable transportation, mitigating socio-economic disparities and enhancing the vibrancy of public spaces [53].

Density is crucial for improving transit performance. High-density developments are advocated to minimize urban sprawl and decrease vehicle dependency. Studies indicate that densifying neighborhoods while considering land-use diversity reduces commuting trips and enhances transit connectivity, frequency, and availability [54]. Moreover, study insists that medium- to high-density mixed-use urban zones centered around transit stops create lively, active, and secure environments that encourage residents to live, work, and engage in recreational activities.

Empirical research supports the positive relationship between urban density and livability. Studies in New Zealand and Australia reveal that residents in high-density areas report increased housing satisfaction due to the proximity of amenities such as shops, schools, and public spaces accessible without cars [55]. These findings underscore the importance of density in enhancing the overall urban experience.

Design plays a pivotal role in making TOD areas attractive and functional. Thoughtful urban design prioritizes pedestrian and cycling infrastructure, street connectivity, and aesthetic appeal encourages active transportation modes and enhances the usability of public transit [56]. High-quality urban design contributes to creating environments where people prefer to spend time, increasing social interaction and community cohesion.

Distance to transit affects the likelihood of transit use. Shorter distances between residences or workplaces and transit stations increase the convenience of using public transit, thereby boosting ridership [57]. Planning that minimizes the distance to transit points is essential for maximizing the utility of the transit system.

Destination accessibility refers to the ease with which people can reach desired destinations using the transit system. Enhanced accessibility requires efficient transit networks that connect key urban areas, enabling seamless movement across the city [58]. Improving destination accessibility makes public transit a more competitive alternative to private vehicles.

Demand management involves strategies to influence travel behavior toward more sustainable modes. Policies such as parking management, congestion pricing, and incentives for using public transit or non-motorized transportation are integral to this dimension [59]. Effective demand management reduces traffic congestion and environmental impacts while promoting healthier lifestyles.

The discussion above can be concluded as follows: The foundational "3D" framework (Density, Diversity, and Design) serves as a critical tool in understanding how Transit-Oriented Development (TOD) fosters smarter and more sustainable urban environments. Density refers to the concentration of people, jobs, and housing units within a specific area. High density supports efficient public transit services by providing a critical mass of users, encouraging shorter commuting distances, reducing reliance on private vehicles, and minimizing urban sprawl. Diversity involves a mix of land use—residential, commercial, industrial, recreational, and civic spaces within the same area. This diverse land-use mix enhances accessibility to various services and amenities, promoting walkability and cycling, reducing the need for long commutes, and fostering vibrant, active communities. Design focuses on the physical layout, aesthetics, and functionality of the urban environment. Good design enhances the user experience and encourages the use of public spaces by prioritizing pedestrian-friendly infrastructure, street connectivity, and attractive public realms, thereby increasing public transit usability.

Furthermore, the discussion reveals that the concept of TOD evolved an additional dimension to address a broader range of factors influencing sustainable urban development, expanding the framework to the "6Ds". Alongside density, diversity, and design, the three new dimensions are distance to transit, destination accessibility, and demand management. Distance to transit measures the proximity of residences, workplaces, and amenities to public transit stations. Closer distances increase the convenience and likelihood of using public transit, enhancing ridership and maximizing the utility of the transit system. Destination accessibility refers to the ease with which people can reach key destinations using the transit network. Efficient connectivity between various urban areas makes public transit a competitive alternative to private vehicles, enhancing overall mobility, reducing travel times, and improving the appeal of public transportation. Demand management involves strategies and policies aimed at influencing travel behavior toward more sustainable modes of transportation. By addressing congestion and environmental concerns through measures such as parking management, congestion pricing, and incentives for non-motorized transport, demand management reduces traffic congestion and lowers environmental impacts.

Collectively, the "6D" framework provides a comprehensive approach to planning and implementing TOD strategies. By integrating these six dimensions, urban planners and policymakers can more effectively promote sustainable, livable, and equitable urban environments. Addressing factors from high-density development and mixed land use to strategic transit accessibility and demand management policies, the expanded framework ensures that TOD can meet the complex challenges of modern urban development, ultimately leading to cities that are more dynamic, resilient, and environmentally responsible. The following model in Figure 3 was developed to extract the dynamic and limitations of the 3D and 6D approaches in transit-oriented development.

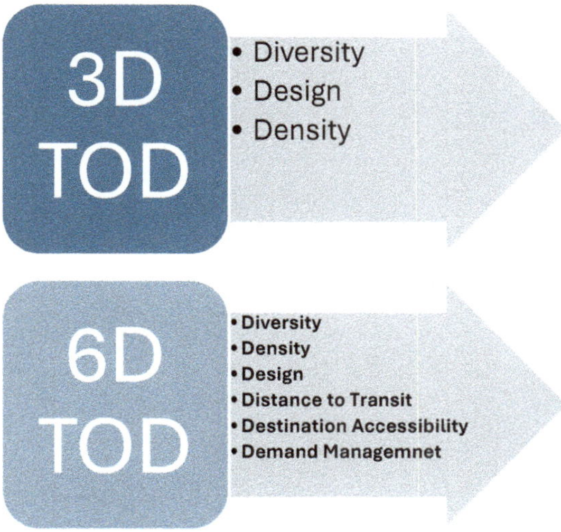

Figure 3. Model of 3D and 6D TOD.

4. Dubai Urban Sprawl and Needs Analysis for TOD

Dubai's urbanization involves significant economic and social transformations beyond simple population growth. The city's rapid expansion has led to environmental and societal challenges such as overcrowding, transportation inefficiencies, unrest among foreign workers, and social security issues. The surge in job opportunities has attracted a surplus labor force from other countries, providing cheap and plentiful workers for emerging economic activities. While industrial development is underway, the primary focus is on constructing tourist attractions, hotels, and resorts. This concentration of investments draws numerous migrants seeking employment and housing. This situation appeals to both the Dubai government and foreign construction companies involved in the city's development [60].

According to Pacione [61], the accelerated pace of development in Dubai has strained its infrastructure and resources. The city's population has surged due to both natural growth and a significant influx of expatriates, leading to increased demand for housing, transportation, and public services. This rapid expansion often outpaces the city's ability to sustainably manage growth, resulting in urban sprawl and infrastructural bottlenecks.

Another study suggests that Dubai's rapid expansion over the past two decades has led to significant urban sprawl, characterized by the fragmentation and segregation of urban land activities. The emergence of satellite towns such as Internet City, Media City, Motor City, and Knowledge Village on the city's periphery has resulted in disconnected urban developments. This lack of integration has led to long commuting distances, increased traffic accidents due to expansive highways, higher energy consumption, and the depletion of land resources [62].

Based on the discussion above in Section 4, this study considers that the problems associated with the rapid urban growth of Dubai include socio-economic problems, environmental threats, the fragmentation and segregation of urban land activities, increased commute travel distance and time, and the damaging effects on social equity and the physical quality of the built environment, as portrayed in Figure 4.

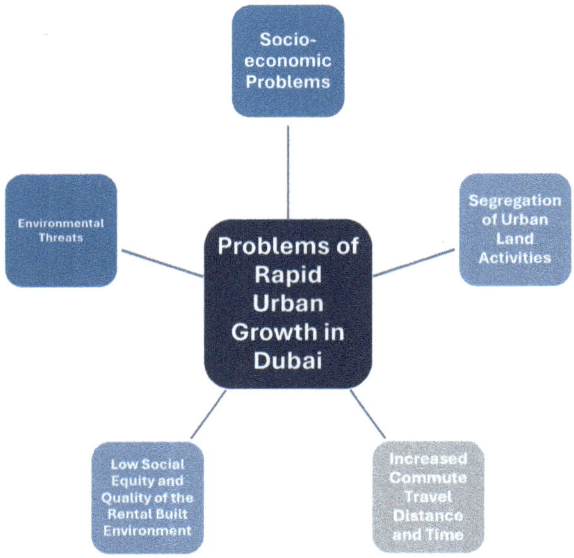

Figure 4. Urban issues in Dubai.

4.1. Issues of Rapid Urban Growth and Urban Mobility in Dubai

Regarding the segregation of urban land activities, Scholz [63] noted that the improper relationship between Dubai's urban developments has led to activity and zoning segregation. This segregation contributes to increased commuting times and distances, as residents must travel longer to reach workplaces, schools, and amenities. The reliance on vast highways exacerbates traffic congestion, increases the likelihood of accidents, and elevates carbon emissions due to higher energy consumption. Sprawling urban forms consume more land resources, leading to environmental degradation and the loss of natural habitats [64]. This pattern of development results in the physical disconnection of communities, weakening social ties and community cohesion. The separation of residential and commercial zones prevents the formation of vibrant, mixed-use neighborhoods essential for sustainable urban living. Furthermore, the lack of integration hampers the effectiveness of public transportation systems, making it difficult to implement efficient transit solutions Cervero, R., Guerra, E., Al, S. [27].

Additionally, urban planning in Dubai often emphasizes specialized zones such as business districts, residential enclaves, and leisure areas, leading to spatial fragmentation [65]. This segregation creates disconnected urban spaces that hinder the integration of different land uses. The lack of mixed-use developments limits opportunities for daily social interactions among diverse population groups, reinforcing social divisions [66]. Such fragmentation challenges the creation of cohesive communities and affects the overall urban experience.

Dubai is also experiencing the spatial separation of residential, commercial, and industrial zones, which has led to increased reliance on private vehicles for commuting [67]. Despite investments in public transportation infrastructure like the Dubai Metro, many residents find it more convenient to use cars due to insufficient coverage and connectivity. This reliance contributes to traffic congestion, longer commute times, and elevated greenhouse gas emissions. The lack of pedestrian-friendly infrastructure further discourages the use of alternative transportation modes such as walking or cycling [68].

Increased commuter travel distances and times are another significant issue in urban Dubai. Worku [69] also noted that low-density, sprawling developments significantly increase commute times and distances. This issue is evident in the daily movement of residents from suburban housing developments to central urban areas for work. The situation is exacerbated for those commuting from neighboring emirates like Sharjah and Ajman, where a 2007 survey indicated that 80% of Ajman's population worked in Sharjah or Dubai, averaging a 20–45 km daily commute. This heavy reliance on private vehicles stems from inadequate public transportation options, leading to increased traffic congestion and environmental pollution. Moreover, extended commute times contribute to reduced productivity, higher stress levels among commuters, and a negative impact on overall quality of life [69,70]. The inefficiency of the current transportation infrastructure highlights the need for TOD, which emphasizes accessibility and connectivity, reducing the necessity for long commutes.

Many studies consider that low social equity and the quality of rental housing are consequences of urban sprawl in Dubai. This phenomenon adversely affects social equity and the physical quality of the built environment. Peripheral dormitory suburbs, lacking diverse activities, force residents to travel considerable distances for social interaction, recreation, and essential services. This separation diminishes social vitality and creates monotonous living environments devoid of cultural and community engagement [71]. Furthermore, Aljoufie, M., et al. [65] mentioned that spatial segregation contributes to socio-economic disparities, as lower-income residents may be marginalized to areas with fewer amenities and opportunities. The lack of affordable housing options within the city center exacerbates this divide, leading to inequality in access to resources and services. Physical fragmentation also undermines the potential for sustainable urban design, resulting in environments that are not conducive to walking or cycling, thereby increasing dependency on cars [72].

In addition, the study of Davidson, C. M. [71] declares that a prominent socio-economic issue in Dubai is the disparity between the affluent and the working-class populations. The city heavily relies on migrant workers, primarily from South Asia, who contribute to its construction and service sectors under challenging conditions. These workers often face low wages, inadequate living standards, and limited labor rights, leading to social stratification and potential unrest [73]. Furthermore, the lack of good affordable housing and the rising cost of living further exacerbate these disparities, affecting social cohesion. Dubai's development boom has significant environmental implications. Extensive construction activities have led to habitat destruction and increased carbon emissions. The city's high consumption of water and energy resources driven by factors like air conditioning demands and water-intensive landscaping contributes to resource depletion. The reliance on desalination for water supply poses ecological risks due to the discharge of brine into the Arabian Gulf, affecting marine ecosystems. Additionally, the urban heat island effect intensifies due to vast concrete surfaces and the lack of green spaces, exacerbating climatic challenges [74,75].

4.2. The Need for Transit-Oriented Development in Dubai

Implementing Transit-Oriented Development (TOD) in Dubai presents a strategic solution to counteract the negative effects of urban sprawl that have resulted from rapid urbanization and car-dependent planning. By focusing on high-density, mixed-use developments centered around efficient public transit systems, TOD can significantly reduce commuting distances, enhance connectivity, and improve social equity [30,66]. This approach aligns with Dubai's vision for sustainable growth and its efforts to become a global city with a high quality of life for its residents [76].

Integrating residential, commercial, and recreational spaces encourages walkability and cycling, fostering healthier lifestyles and stronger community interactions [44,75].

In Dubai, where car ownership and usage are high, promoting alternative modes of transportation can alleviate traffic congestion and reduce environmental pollution. TOD supports the creation of vibrant, livable neighborhoods where people can live, work, and play without relying heavily on private vehicles. By introducing sustainable urban mobility options, TOD addresses environmental concerns by reducing reliance on private vehicles, thereby lowering carbon emissions and mitigating traffic congestion [77]. Given Dubai's commitment to sustainability and reducing its carbon footprint, integrating TOD principles can contribute to meeting these environmental objectives. By promoting public transit, walking, and cycling, TOD reduces air pollution and energy consumption associated with transportation [78].

In the context of economics, TOD supports growth by improving access to employment opportunities and stimulating local businesses through increased foot traffic in well-connected urban areas [79]. These ideas are duly endorsed by the Dubai Urban Plan 2040 [76], which mentions that enhanced accessibility can attract investment, boost property values, and promote economic diversification, which are key goals in Dubai's strategic plans. TOD can also reduce infrastructure costs by optimizing land use and decreasing the need for extensive road networks [80]. Furthermore, TOD can improve social equity by providing affordable housing options near transit hubs, enabling low- and middle-income residents to access employment centers and services more easily [81]. This inclusivity helps to bridge social divides and foster a more cohesive community. In Dubai, where socio-economic disparities exist, TOD can play a role in creating more balanced and equitable urban development Elsheshtawy, Y. [70].

In terms of land use, in the GCC region, particularly in Dubai, urban planning has often been characterized by single-use zoning and sprawling developments. Adopting TOD models can transform the urban landscape into a more sustainable and efficient form [82]. The development of the Dubai Metro and the expansion of public transit infrastructure provides an opportunity to implement TOD around transit stations [83]. Strategically planning developments around these transit nodes can optimize land use, reduce urban sprawl, and enhance the overall urban experience. Encouraging active transportation modes like walking and cycling within TOD promotes healthier lifestyles [84]. The design of pedestrian-friendly environments with accessible public spaces can improve the physical and mental well-being of residents. In Dubai, integrating green spaces and recreational areas within TOD can enhance the quality of life and make urban living more attractive [85].

However, implementing TOD requires supportive policies, incentives, and collaboration among government agencies, private developers, and the community [86]. Whereas the studies of Acuto, M. [65] and Worku, G. B. [68] suggest that Dubai's regulatory framework and planning policies need to facilitate higher densities, mixed-use developments, and pedestrian-oriented design to realize the benefits of TOD. Addressing potential challenges such as land ownership issues, financing, and cultural preferences for car usage is essential for successful implementation.

The discussion at end of this section can be summarized as follows: Dubai's urban sprawl poses significant challenges to sustainable development, social equity, and the quality of the built environment. The fragmentation and segregation of land uses, extended commute times, and social inequities highlight the urgent need for a paradigm shift in urban planning. Transit-oriented development offers a viable framework to create integrated, accessible, and vibrant communities. By adopting TOD principles, Dubai can enhance its urban infrastructure, promote sustainable growth, and improve the quality of life for its residents. Adopting TOD models in Dubai addresses multiple urban challenges, including environmental sustainability, social equity, economic development, and quality of life. By leveraging the existing and planned public transit infrastructure, Dubai can create a more connected, sustainable, and livable city that meets the needs of its diverse population. Integrating TOD principles aligns with global best practices and supports Dubai's ambition to be a leading city in sustainable urban development. Table 1 portrays the key information and secondary data discussed in Section 4. Table 1 will help readers to understand the characteristics of transportation and needs of TOD in Dubai.

Table 1. Needs of TOD. Source: [30–79].

Key Issue	Description	Implications	Source
Urban Segregation	Dubai's urban planning has led to the segregation of residential, commercial, and industrial zones.	Increased commuting times, reliance on highways, traffic congestion, environmental degradation, social fragmentation.	[60–63]
Spatial Fragmentation	Specialized zoning for business, residential, and leisure areas results in disconnected spaces.	Limited social interactions, weakened community cohesion, inefficient public transport integration.	[63,64]
Reliance on Private Vehicles	Public transportation coverage is insufficient, leading residents to prefer private cars.	Traffic congestion, longer commutes, higher carbon emissions.	[27,65]
Long Commutes	Many residents commute long distances, particularly from Ajman and Sharjah to Dubai.	Increased stress, lower productivity, reduced quality of life.	[66,67]
Urban Sprawl Effects	Peripheral dormitory suburbs with limited amenities force residents to travel long distances.	Monotonous environments, diminished social interaction, socio-economic disparities.	[64,68]
Socio-economic Disparities	Migrant workers face poor living conditions and low wages.	Social stratification, potential unrest, lack of social cohesion.	[68,70]
Environmental Impact	High water and energy consumption, habitat destruction, urban heat island effect.	Resource depletion, marine ecological risks, climate challenges.	[71,72]
Need for TOD	TOD promotes mixed-use, high-density developments centered around public transit.	Reduced commuting distances, enhanced connectivity, improved social equity.	[30,64]
Health and Well-being	Encouraging walkability and cycling through TOD.	Healthier lifestyles, stronger community interactions.	[38,72]
Economic Growth	TOD improves accessibility to jobs, boosts local businesses, and attracts investment.	Increased property values, economic diversification.	[73,77]
Infrastructure Efficiency	Optimizing land use with TOD reduces the need for extensive road networks.	Lower infrastructure costs, sustainable growth.	[78]
Social Equity	Affordable housing near transit hubs to support low- and middle-income residents.	Improved access to employment and services, reduced social divides.	[67,79]
Car Ownership	69% of residents own a private car.	High reliance on private cars increases traffic congestion and emissions.	[65]
Common Transport Modes	68% use private cars, 9% use public buses, 2.6% use the metro.	Limited use of public transport indicates a preference for convenience over sustainability.	[65]
Family Transportation Preferences	78.6% of families primarily use private cars.	Increased household expenses and dependency on fuel consumption.	[65]
Public Transport Frequency	On average, residents use public buses once per day.	Low public bus usage suggests inefficiencies and dissatisfaction with service.	[65]
Travel Time (Round Trip)	▪ Public bus: 1 h 15 min ▪ Private car/taxi: 55 min ▪ Metro: 42 min	Longer travel times on buses impact productivity and leisure time.	[65]
Transportation Cost Perception	▪ 41% rate it as expensive/very expensive ▪ 50% rate it as fair ▪ 9% rate it as cheap/very cheap	High transportation costs affect disposable income and the affordability of other essentials.	[65]

Table 1. *Cont.*

Key Issue	Description	Implications	Source
Average Monthly Transportation Cost	AED 858 per household	Significant financial burden on families, especially low-income groups.	[65]
Public Transport Service Quality	▪ 25% rate it as very good or excellent ▪ 6% rate it as poor	Most residents bear the cost themselves, reducing savings potential.	[65]
Metro Service Rating	Rated superior to public buses in all service indicators.	Service inefficiencies discourage potential users, increasing car dependency.	[65]
Concerns About Public Bus Use	61.2% unwilling to use public buses even if there are nearby stations.	The metro provides a viable alternative, but its coverage may be limited.	[65]

5. Problem Statement

Dubai's rapid urbanization and increasing reliance on private vehicles have resulted in urban sprawl, prolonged commute times, infrastructure strain, and a decline in green spaces. Although metro lines are available, the absence of well-planned Transit-Oriented Development (TOD) around stations has limited their effectiveness in promoting sustainable urban mobility. This research aims to systematically assess the application of Transit-Oriented Development (TOD) and Transit-Adjusted Development (TAD) frameworks to identify targeted, station-level interventions. By analyzing critical factors such as density, land-use diversity, urban design, accessibility, proximity to transit, and demand management, the study seeks to provide strategic insights for enhancing connectivity, efficiency, and environmental sustainability in Dubai's urban growth.

6. Objectives

1. **Identification of Key Factors:** To explore the TAD and TOD models and determine which urban form and mobility variables such as accessibility, land-use mix, density, and design most significantly influence the potential for TOD implementation in Dubai's metro station neighborhoods.
2. **Framework Validation:** To assess the effectiveness of TAD and TOD appraisal models (3D and 6D frameworks) in evaluating the current conditions of Dubai's metro stations on the Red and Green Lines.
3. **Strategic Recommendations:** To propose targeted interventions and design guidelines for selected appraised metro stations that enhance transit efficiency, improve pedestrian friendliness, and reduce reliance on private vehicles.

7. Methodology

This research employed a mixed-methods approach, integrating quantitative spatial analysis with qualitative assessments to address the three stated objectives. First, secondary data was compiled through an extensive literature review to identify the key urban form and mobility factors shaping TOD potential. Academic sources and international best practices on TOD and TAD frameworks [87] were reviewed to establish a robust theoretical foundation. This review also helped to refine the assessment criteria for accessibility, land-use mix, density, and urban design elements.

Second, primary data collection and spatial analysis were undertaken to validate the frameworks. Primary data was collected through multiple case studies, and site visits of T1, T2, and T3 types of Dubai Metro stations were conducted. Geographic Position Systems (GPSs) and multi-criteria evaluations based on the 3D and 6D frameworks (Density, Diversity, Design, Destination Accessibility, Distance to Transit, and Demand Management) were applied to selected Dubai Metro stations along the Red and Green Lines. Indicators such as pedestrian catchment areas, land-use distribution, and built-form density were quantified. Third, qualitative interviews and stakeholder consultations involving urban planners, transportation experts, and representatives from Dubai's Roads and Transport Authority further validated the findings and ensured contextual relevance, refer to Appendix A. These consultations guided the translation of the assessment results into practical design guidelines and targeted interventions [87].

This mixed-methods approach was justified, as it integrated robust theoretical insights, spatial–quantitative assessments, and stakeholder perspectives to produce holistic, context-sensitive recommendations. By triangulating data sources, this study ensured reliability and validity, enhancing the applicability of proposed TOD strategies for improving accessibility and pedestrian friendliness while reducing car dependency in Dubai's evolving urban landscape.

8. Case Study: Dubai Metro Stations and the Implementation of 3D and 6D TOD Models

Dubai has consistently pursued innovation and sustainability, setting benchmarks in urban development to position itself among the world's leading cities. The Dubai Metro stands as a prime example of this ambition. As the longest driverless metro system globally and the first of its kind in the GCC region, it reflects the city's commitment to creating a sustainable, "people-based" urban environment. This ambitious project, envisioned by Sheikh Mohammed bin Rashid, symbolizes Dubai's journey toward becoming a global leader in sustainable infrastructure and efficient urban transit systems. Sheikh Mohammed (Prime Minster of UAE) described the Dubai Metro as a "worthy addition" to the city, aimed at easing lives through safe and sustainable transportation. Despite initial skepticism from detractors, the project has proven instrumental in addressing traffic congestion, supporting economic growth, and enhancing the overall quality of life for Dubai's residents.

The Dubai Metro's roots trace back to 1992, driven by alarming urban growth and an annual increase of 17% in car numbers, far exceeding the global average of 4%. Under the leadership of the Roads and Transport Authority (RTA), extensive research and planning culminated in the metro's inauguration in 2009. The government's primary objective was to create a robust public transit system that would alleviate traffic congestion by reducing private car reliance and cutting city traffic by 30%. Dubai's urban landscape was divided into seven zones to design an efficient metro system. These zones, categorized by density, land-use diversity, and growth opportunities, guided the alignment of metro routes. The zoning strategy also facilitated seamless integration with other transportation modes such as buses and ferries, ensuring a cohesive and multi-modal commuting experience [88].

8.1. Metro Lines and Routes

The Dubai Metro currently operates two main lines, the Red Line and the Green Line, spanning a total of 74.25 km. Stretching 52.1 km, the Red Line was inaugurated in 2009 with 10 stations and expanded to 29 stations by 2010. It connects Al Rashidiya to the UAE Exchange station in Jebel Ali, covering key commercial and business districts. Major stops include Dubai Mall, Deira City Centre, Burjuman Mall, and Mall of the Emirates. Running along Sheikh Zayed Road, the Red Line supports current businesses and anticipates future developments, such as the Al Satwa redevelopment project. The Green Line, introduced in 2010, spans 22.15 km with 20 stations and serves central Dubai and older districts around Dubai Creek. Starting from Etisalat Station and ending at Creek Station, it connects dense urban areas and community hubs like Al Rigga, Al Ras, and Al Fahidi. It also facilitates access for residents commuting from Sharjah through Al Nahda station. The metro's expansion includes the Expo 2020 Route, a 14.5 km extension planned in 2017, featuring seven new stations. This route connects Nakheel Harbour and Towers Station to the Expo

2020 site, passing through key communities such as Discovery Gardens, Al Furjan, and Dubai Investment Park [89].

8.2. Station Architecture and Types

Dubai Metro's 49 stations are designed with architectural themes symbolizing earth, water, air, and fire, reflecting the city's identity and cultural heritage. Stations such as Al Ghubaiba and Al Ras on the Green Line incorporate elements of old Dubai, while others feature shell-like structures inspired by the UAE's pearl-diving history. These designs not only enhance aesthetic appeal but also serve functional purposes, such as noise isolation and improved transit efficiency. Stations are categorized into three main types based on their structure:

(a) At-Grade Stations: Ground-level stations like the UAE Exchange station.
(b) Elevated Stations: Divided into three subtypes, these stations accommodate elevated platforms for smooth traffic and pedestrian movement. Examples include Type 1 stations along the Red Line and Type 2 stations elevated 5 m above the ground in dense areas.
(c) Underground Stations: These include regular and transfer stations like Union and Burjuman, designed for seamless connectivity and efficient commuter flow.

In this context, approximately 45% of Red Line stations are elevated, while 55% of Green Line stations are elevated to address the compact urban fabric of older areas. The average distance between stations is 1.5 km, ensuring optimal coverage of key districts [90,91].

8.3. Facilities and Accessibility

Dubai Metro stations are built to world-class standards, prioritizing safety, accessibility, and commuter comfort. Key facilities include pedestrian bridges, clear signage, moving walkways, and lifts. Accessibility features cater to individuals with special needs, aligning with RTA's goals of creating safe and inclusive transit environments. The metro also offers "Park-N-Ride" services to commuters from neighboring emirates like Sharjah and Ajman. Three main parking facilities accommodate approximately 8000 vehicles: Al Rashidiya with 2700 parking spaces, Nakheel Harbour and Towers with 3000 parking spaces, and Etisalat Station with 2300 parking spaces RTA 2024 [88]. Table 2 was developed to summarize the findings from Section 8.

Table 2. Key data and facts about the Dubai Metro. Source: [89–91].

Important Aspect	Details
Inception and Vision	The Dubai Metro is the largest investment in the city's transit infrastructure, initiated under Sheikh Mohammed bin Rashid's vision to enhance sustainability and urban connectivity, making Dubai a global leader in public transit.
Significance	First driverless metro in the GCC and the longest globally.
Planning Timeline	Studies began in 1992; construction launched in 2004; the Red Line was inaugurated in 2009; the Green Line was inaugurated in 2010.
Urban Challenges Addressed	Reducing car dependency, combating traffic congestion, minimizing urban sprawl, and integrating multi-modal transportation.
Zone System	Dubai is divided into seven zones based on urban density, land-use diversity, and growth opportunities; zoning also influences fare differentiation.
Metro Lines	- **Red Line**: 52.1 km with 29 stations. Covers major business districts and shopping areas along Sheikh Zayed Road. - **Green Line**: 22.15 km with 20 stations. Covers dense areas in old Dubai and communities around Dubai Creek.

Table 2. *Cont.*

Important Aspect	Details
Future Plans	Expo 2020 Route: Planned 14.5 km extension with seven stations. Future Blue Line: Planned to connect Al Jaddaf station on the Green Line to Emirates Road, running parallel to the Red Line.
Infrastructure	Elevated tracks dominate, except 5.7 km (Red Line) and 8.3 km (Green Line) of underground tunnels.
Station Themes	49 stations feature architectural themes of earth, water, air, and fire; station designs incorporate local heritage and context, such as Al Ghubaiba and Al Ras on the Green Line.
Station Types	**Type 1—At-Ground Stations**: Ground-level stations (e.g., UAE Exchange on the Red Line). **Type 2—Elevated Stations**: 5 M elevated stations for operational needs. **Type 2—Underground Stations**: Regular and transfer stations (e.g., Union, Burjuman). **Type 3—End Stations and Extendable:** Al Rashidya and Al Qusias.
Integration	Designed to integrate multiple transit modes (metro, buses, ferries) for seamless connectivity; "Park-N-Ride" facilities available at key stations with a total capacity of 8000 parking spaces.
Facilities	Stations with features such as pedestrian crossing bridges, clear signage, moving walkways, lifts, noise-isolated shells, and specialized amenities for accessibility.
Park-N-Ride	- Al Rashidiya (Red Line): 2700 spaces. - Nakheel Harbour and Towers (Red Line): 3000 spaces. - Etisalat (Green Line): 2300 spaces.
Average Station Distance	1.5 km between stations, ensuring optimal coverage of key districts.
Travel Time	Estimated one hour for end-to-end travel on either the red or green line.

9. Projections for Implementing TOD Models at Dubai Metro Stations

At this stage of the research, this study has comprehensively examined Transit-Oriented Development (TOD) as a strategic framework for urban mobility design in Section 2. Building on this foundation, Section 3 explored additional concepts and methodologies related to urban mobility design, providing a broader contextual understanding. Subsequently, Section 4 offered an in-depth analysis of urban sprawl and its significant implications within the context of Dubai, highlighting the critical challenges posed by rapid urban expansion.

In Section 8, the focus shifted to an extensive review of the Dubai Metro system, encompassing its operational framework, technical specifications, and infrastructural characteristics. This examination not only highlighted the metro's pivotal role in addressing Dubai's urban mobility challenges but also set the stage for integrating TOD principles with existing transit infrastructure.

Section 9 synthesizes the insights gained from previous sections by presenting a consolidated data profile and a structured checklist. These tools serve as a foundation for determining the most suitable TOD models tailored to the diverse transit hubs and stations located along the Dubai Metro's Green and Red Lines. The goal is to establish an adaptable and efficient TOD framework that aligns with the specific characteristics, requirements, and future development potential of each transit hub. Table 3 was developed as a testing matrix to check whether various types of Dubai Metro stations conform to the urban mobility criteria. Data was collected through a matrix after visiting various types of metro stations in Dubai and performing unstructured interviews with users.

Table 3. Testing matrix of urban mobility models and the Dubai Metro.

Station Typology			Models	Factors of Urban Mobility Design	Frequency			Projections for TOD Stations			
T1	**T2**	**T3**			**T1**	**T2**	**T3**		**T1**	**T2**	**T3**
11 Stations Only on the Red line	9 Stations on the Red Line and 11 on the Green Line	2 Stations, Al Rashidya on the Red Line and Al Qusais on the Green Line	Features of TAD Model	Lack of Transit-Supportive Design	M	M	X	Projections for Expansion as TOD Stations	UAE Exchange; Oud Mehta; Mall of Emirates; Ibn Battuta Mall	Nill—All Stations Are Elevated and Have Space Limitations	Al Rashidya on the Red Line, Al Qusais on the Green Line.
				Inadequate Pedestrian Infrastructure	O	O	O				
				Poor Connectivity Between Transportation Modes	O	O	O				
				Favoring Automobile Use	O	O	O				
				Absence of Amenities and Services	X	O	O				
				Failure to Create Vibrant Activity Centers	O	O	O				
				Missed Opportunities for Sustainable Development	O	O	O				
			Features of TOD (3D Model)	Diversity	X	O	X				
				Design	G	EX	G				
				Density	MD	CD	LD				
			Features of TOD (6D Model)	Diversity	X	O	X				
				Density	MD	CD	LD				
				Design	G	EX	G				
				Distance to Transit	F	F	F				
				Destination Accessibility	M	P	P				
				Demand Management	X	X	X				

X—Not Available/No; O—Available/Yes; CD—Commercial Density; MD—Mixed Density; N—Near; F—Far; M—Moderate; G—Good; P—Poor; EX—Excellent; LD—Low Density.

10. Analysis

Table 3 provides a detailed assessment of key urban mobility and form-related factors that influence the potential for successful Transit-Oriented Development (TOD) in select Dubai Metro stations. Using a classification system that evaluates each factor's presence and quality ranging from "X" (Not Available) to "EX" (Excellent), the table reveals distinct patterns across three station typologies (T1, T2, and T3).

T1 comprises 11 stations located exclusively on the Red Line; these sites predominantly exhibit automobile-centric planning and limited TOD-aligned features. Their characteristic indicators such as "X" for transit-supportive design and inadequate pedestrian infrastructure highlight a systemic reliance on private vehicles and a corresponding shortage of walkable, mixed-use environments. Destination accessibility, often rated as "X" or "P", confirms that the integration of surrounding land uses and transit networks remains weak in T1 stations.

The T2 typology of stations includes 20 stations (9 on the Red Line and 11 on the Green Line) that demonstrate more moderate progress toward TOD ideals. Although certain elements, like diversity (MD) and density (M), show improvement compared to T1 stations, these ratings remain mixed. For example, some T2 stations feature better pedestrian connections ("G" for walkability) or moderately improved land-use patterns (MD for Mixed Density), reflecting incremental yet uneven advancement. While the overall environment is still evolving, T2 stations illustrate the positive influence of integrating multiple transit lines and encouraging a partial shift away from car dependency.

In contrast, T3 stations, consisting of Al Rashidiya (Red Line) and Al Qusais (Green Line), represent the highest potential for TOD implementation. Here, many criteria move from "M" or "G" to "EX" in terms of diversity, indicating that these areas are already primed for enhanced walkability and mixed-use development. Still, certain factors—like demand management or destination accessibility—may receive only Moderate ("M") ratings, indicating opportunities for targeted improvements. The strategic geographic positioning of these stations offers a strong foundation for comprehensive TOD strategies, including reinforcing pedestrian priority, optimizing land-use mixes, and reducing reliance on private vehicles.

The analysis suggests a clear progression in TOD maturity across the typologies. T1 stations, largely auto oriented, require foundational investments in pedestrian infrastructure and integrated planning. T2 stations are on the cusp of more pronounced TOD alignment, benefiting from partial enhancements yet still facing gaps in connectivity and mixed-use intensity. T3 stations already showcase advanced TOD potential, with conditions ripe for shaping them into exemplary transit hubs that seamlessly blend land use, transit accessibility, and pedestrian comfort. Furthermore, the analysis underscores those certain high-density urban areas (e.g., Burj Khalifa–Dubai Mall Station vicinity) can serve as models for the broader network. By applying tailored interventions such as improving walkable networks, refining land-use mixes, and enhancing intermodal connections, older, less integrated stations can be retrofitted to align more closely with TOD principles. Ultimately, the table's data emphasizes that targeted planning and policy interventions, informed by nuanced assessments of station-level conditions, can propel Dubai's transit infrastructure toward a more sustainable, accessible, and resilient future.

11. Conclusions

The findings underscore a logical, incremental progression in TOD maturity across the three station typologies, reflecting the interplay of urban design, land-use integration, and transit accessibility. The T1 category, predominantly representing older or less integrated Red Line stations, demonstrates fundamental deficiencies. These shortcomings, such as weak pedestrian infrastructure, minimal mixed-use density, and inadequate connections to other transportation modes, logically follow from their initial design, which likely prioritized automobile-oriented patterns over integrated transit solutions. Without a foundation of pedestrian comfort and intermodal coordination, these stations cannot effectively attract ridership or foster dynamic, transit-supportive neighborhoods.

In contrast, T2 stations show that even partial enhancements in infrastructure and urban design strategies lead to measurable improvements. The move toward mixed land uses, marginally better pedestrian networks, and emerging intermodal linkages logically improves station-area vitality. Although some elements, like demand management or robust accessibility frameworks, remain underdeveloped, the incremental nature of these improvements signifies that pragmatic modifications to existing networks and urban layouts can create momentum toward full TOD realization. These moderate gains illustrate how gradual, targeted upgrades can set the stage for more comprehensive interventions in the future.

At the apex, T3 stations exemplify the culmination of strategic planning and contextual suitability. They benefit not only from their geographic positioning, intersecting major lines or anchoring vibrant urban districts, but also from deliberate urban form considerations. Their comparatively stronger performance in areas like diversity and density stems from integrated land-use planning that aligns mobility options with local economic activities and community amenities. Nonetheless, the gaps that persist, especially regarding demand management and destination accessibility, remind us that no single factor suffices. Success emerges when all elements—land use, infrastructure, services, and policy—work in concert.

Ultimately, the logical trajectory from T1 to T3 demonstrates that TOD success is neither accidental nor immediate. It requires continuous improvement, evidence-based interventions, and policy adjustments informed by station-level data and broader urban contexts. Emphasizing strategic locations like high-density nodes (e.g., Burj Khalifa–Dubai Mall) or retrofitting older stations along the Red Line is not merely an ad hoc strategy;

it aligns with fundamental TOD principles, ensuring that each incremental step moves toward a cohesive, sustainable, and people-centric urban fabric. In essence, the conclusion points to a rational, iterative process: applying insights gleaned from station assessments, refining urban design, and interweaving land use and transit to create truly transit-oriented communities.

This study concludes that implementing Transit-Oriented Development (TOD) near the Mall of the Emirates Metro Station in Dubai can significantly reduce commuting distances and enhance connectivity, encouraging residents to rely on efficient public transit instead of private cars. This shift can alleviate traffic congestion and lower emissions, promoting a healthier urban environment. TOD fosters vibrant, walkable neighborhoods, improving social equity and fostering stronger community interactions. Additionally, it attracts investment, boosts property values, and supports economic diversification, contributing to Dubai's sustainable urban growth. Table 4 is presented to summarize the results and to explicitly explain the appraisal of each station under the TOD protocol.

Table 4. Appraisal of Dubai Metro stations under the TOD framework.

Station Type	TOD Factors					Proposed Interventions
	Urban Design	Land Use Integration	Transit Accessibility	Demand Management	Destination Accessibility	
T1 (Older/Red Line Stations)	Weak pedestrian infrastructure, automobile-oriented layout	Minimal mixed-use density, limited land use diversity	Inadequate connections to other transportation modes	Nonexistent demand management strategies	Limited accessibility to key destinations	Pedestrian-friendly upgrades, intermodal connections
T2 (Moderately Improved Stations)	Moderate improvements in pedestrian networks and urban layout	Emerging mixed-use developments with some integration	Improved intermodal linkages, but still underdeveloped	Marginal efforts towards demand management	Improved but not fully optimized	Targeted improvements in urban layout and policies
T3 (Strategically Planned Stations)	Strategic urban form, integrated with economic activities	Comprehensive land-use planning, aligning with local economy	Strong connectivity with multiple transit options	Structured strategies but still areas for improvement	Well-connected with major urban districts	Holistic integration of infrastructure, services, and policy

This study considers that future research on Transit-Oriented Development (TOD) in Dubai should address station-specific challenges, including land-use constraints and pedestrian accessibility, to enhance integration at individual stations. The impact of existing zoning policies and urban planning strategies on TOD outcomes must also be evaluated, alongside potential policy adjustments, particularly for T1 stations. Longitudinal studies are needed to assess TOD's long-term effects on ridership, economic growth, and sustainability. Research should explore integrating smart technologies, such as real-time transit data, to improve intermodal connectivity while examining TOD's role in promoting social equity, affordable housing, and accessible services. Comparative studies with international examples can identify adaptable best practices for Dubai. Sustainability metrics are critical for evaluating TOD's environmental benefits, while demand management strategies like congestion pricing should be analyzed to reduce car dependency. Finally, dynamic land-use modeling and collaboration between stakeholders are essential for optimizing TOD planning and governance in Dubai's high-density urban zones.

Author Contributions: Methodology, O.Y.; Investigation, O.Y.; Resources, O.Y.; Writing original draft, A.H.C.; Writing review and editing, J.A. and M.A. All authors have read and agreed to the published version of the manuscript.

Funding: This research received no external funding.

Data Availability Statement: Data supporting reported results can be requested from the author: a.chohan@ajman.ac.ae.

Acknowledgments: The authors would like to thank Ajman University and the Healthy and Sustainable Built Environment Research Center (HSBERC) for their support and for providing the research facilities used in this study.

Conflicts of Interest: There are no conflicts of interest related to this study.

Appendix A

TOD Survey Questionnaire (Professional Stake Holders)

Ajman University, College of Architecture, Art, and Design (CAAD)

Title of Research: The Assessment of Mall of The Emirates Station Area According to Transit-Oriented Development Strategy (TOD).

This research assesses one of the station areas in Dubai, which is the Mall of The Emirates (MOE) station area according to the TOD strategy. It is based on the Institute of Transportation and Development Policy report (ITDP). The research is conducted because the Dubai metro has emerged in an existing urban setting with no previous urban spatial assessment in these areas. Additionally, the increasing number of suburban communities and cars justifies the need for this assessment. The spatial assessment of the MOE station area will relate the existing urban setting to the areas with the TOD approach and will act as a sample and reference for future transit developments in Dubai.

Principal Researcher: Oussama Yahia
Research Supervisor: Dr. Afaq Chohan

Part A - Professional Profile

1. **Name:** ________________________________
2. **Specialization:** ________________________
3. **Designation and workplace:** ________________

Part B - Questionnaire

1. **Are there any standards and regulations in RTA Dubai that refer/follow to design the following urban elements, especially within metro transit station areas?** *(You can mention the standard if known)*
    - Walkways and crosswalks: **Yes** ☐ **No** ☐
    - Ground floor activities and retail frontages: **Yes** ☐ **No** ☐
    - Creating shaded areas (trees, shading devices, porches): **Yes** ☐ **No** ☐
    - Cycling paths and parking: **Yes** ☐ **No** ☐
    - Street parking management: **Yes** ☐ **No** ☐
    - Roadways area: **Yes** ☐ **No** ☐
2. **What standards does RTA Dubai follow to monitor the urban planning/transport process?**
    - American Standards ☐
    - British Standards ☐
    - Other (Please specify): ____________
3. **What is the standard length of the urban block in Dubai communities if available?**
    - 110m ☐
    - 130m ☐
    - 150m ☐
    - 170m ☐
    - 190m ☐
    - No Standard ☐

4. **What is the standard radius of the catchment area (open safety zones) around Dubai metro stations?**
    - 500m - 800m ☐
    - 800m - 1000m ☐
    - More than 1000m ☐
    - No Standard ☐
5. **What is the standard percentage of mixing residential to non-residential land uses in Dubai communities if available?**
    - 50% to 60% ☐
    - 51% to 70% ☐
    - 71% to 80% ☐
    - More than 80% ☐
    - No Standard ☐
6. **Are there any policies in Dubai city to transform vacant unused lands into parks, plazas, or playgrounds?**
    - **Yes ☐ No ☐**
7. **Is there any policy or regulation RTA refers to control the building heights within the communities nearby metro stations?**
    - **Yes ☐ No ☐**
8. **Are there any policies or regulations to build new metro stations in mixed communities of various income groups?**
    - **Yes ☐ No ☐**
9. **Are there any plans to build more stations and extend metro lines in Dubai city to connect more communities?**
    - **Yes ☐ No ☐** *(If yes, please mention possible areas: ______________)*
10. **According to RTA data, what is the most popular mode of public transportation in Dubai?**

- Metro ☐
- Bus ☐
- Ferry ☐
- Taxi ☐

If you have any other comments or additional information, please write them below:

Thank you for your time in answering the questionnaire.

References

1. Ma, Y. Compact Urban Development and Vitality: A New Vision for Guangzhou Central Area. Ph.D. Thesis, Politecnico di Torino, Turin, Italy, 2022.
2. Cruceru, A. Urban form, transit space, and the public realm. Ph.D. Thesis, University of Guelph, Guelph, ON, Canada, 2011.
3. Ndebele, R. *Leveraging Transit Oriented Development (TOD) Through Public Private Partnerships (PPPs): South Africa*; University of Johannesburg: Johannesburg, South Africa, 2018.
4. Abdi, M.H. Towards Transit-Oriented Development in Iran Understanding Policy, Planning and Urban Design Prerequisites. Ph.D. Thesis, Universidad Politécnica de Madrid, Madrid, Spain, 2021.
5. Pucher, J.; Peng, Z.R.; Mittal, N.; Zhu, Y.; Korattyswaroopam, N. Urban transport trends and policies in China and India: Impacts of rapid economic growth. *Transp. Rev.* **2007**, *27*, 379–410. [CrossRef]
6. Pojani, D.; Stead, D. Sustainable urban transport in the developing world: Beyond megacities. *Sustainability* **2015**, *7*, 7784–7805. [CrossRef]
7. Thomas, R.; Bertolini, L. *Transit-Oriented Development*; Palgrave Pivot: Cham, Switzerland, 2020. [CrossRef]

8. Jacobson, J.; Forsyth, A. Seven American TODs: Good practices for urban design in transit-oriented development projects. *J. Transp. Land Use* **2008**, *1*, 51–88. [CrossRef]
9. Shamskooshki, H. *Toward Mixed-Use Communities by Transit-Oriented Development (TOD) in the United States*; Real Corp.: Omaha, NE, USA, 2012.
10. Renne, J.L.; Listokin, D. The opportunities and tensions of historic preservation and transit oriented development (TOD). *Cities* **2019**, *90*, 249–262. [CrossRef]
11. Cervero, R. *Transit-Oriented Development in the United States: Experiences, Challenges, and Prospects*; National Academies Press: Washington, DC, USA, 2004.
12. Noland, R.B.; Weiner, M.D.; DiPetrillo, S.; Kay, A.I. Attitudes towards transit-oriented development: Resident experiences and professional perspectives. *J. Transp. Geogr.* **2017**, *60*, 130–140. [CrossRef]
13. Chen, H.; Zhao, K.; Zhang, Z.; Zhang, H.; Lu, L. Methods for the Performance Evaluation and Design Optimization of Metro Transit-Oriented Development Sites Based on Urban Big Data. *Land* **2024**, *13*, 1233. [CrossRef]
14. Wan, T.; Lu, W.; Sun, P. Equity impacts of the built environment in urban rail transit station areas from a transit-oriented development perspective: A systematic review. *Environ. Res. Commun.* **2023**, *5*, 9. [CrossRef]
15. Su, S.; Zhang, H.; Wang, M.; Weng, M.; Kang, M. Transit-oriented development (TOD) typologies around metro station areas in urban China: A comparative analysis of five typical megacities for planning implications. *J. Transp. Geogr.* **2021**, *90*, 102939. [CrossRef]
16. Ibraeva, A.; de Almeida Correia, G.H.; Silva, C.; Antunes, A.P. Transit-oriented development: A review of research achievements and challenges. *Transp. Res. Part A Policy Pract.* **2020**, *132*, 110–130. [CrossRef]
17. Noland, R.B.; Ozbay, K.; DiPetrillo, S.; Iyer, S.; Center, A.M.V.T. *Measuring Benefits of Transit Oriented Development*; (No. CA-MNTRC-14-1142); Mineta Transportation Institute: San Jose, CA, USA, 2014.
18. Simonson, K.K.A. Advancing TOD in Boston's Suburbs: Advantages and Obstacles in the Entitlement Process. Ph.D. Thesis, Massachusetts Institute of Technology, Cambridge, MA, USA, 2011.
19. Jumah, G.M. Enhancing Urban Connectivity and Regeneration of Historic Areas: Case Study of Heart of Sharjah. Master's Thesis, The British University in Dubai, Dubai, United Arab Emirates, 2020.
20. El Adli, K.Z.; Abd El Aziz, N.A. Greening the urban environment: An integrated approach to planning sustainable cities the case of greater Cairo. In *The Importance of Greenery in Sustainable Buildings*; Springer: Berlin/Heidelberg, Germany, 2022; pp. 47–72.
21. Kumar, P.P.; Parida, M.; Sekhar, C.R. Developing context sensitive planning criteria for transit oriented development (TOD): A fuzzy-group decision approach. *Transp. Res. Procedia* **2020**, *48*, 2421–2434. [CrossRef]
22. Chohan, A.H.; Awad, J. Shaping the Architects of Tomorrow, Interplay of Teaching Philosophies and Practice Requirements: An Empirical Taxonomy of Professional Architectural Practice in the UAE. *Buildings* **2023**, *13*, 1231. [CrossRef]
23. Mathew, N.M. Understanding Space, Politics and History in the Making of Dubai, a Global City. Ph.D. Thesis, University of the Witwatersrand, Johannesburg, South Africa, 2014.
24. Jariwala, P. Transit Oriented Development (TOD), A Tool for Effective Urban Growth. 2019. Available online: https://www.academia.edu/38506568/Transit_Oriented_Development_TOD_A_Tool_for_Effective_Urban_Growth (accessed on 29 December 2024).
25. Knowles, R.D.; Ferbrache, F.; Nikitas, A. Transport's historical, contemporary and future role in shaping urban development: Re-evaluating transit oriented development. *Cities* **2020**, *99*, 102607. [CrossRef]
26. Carlton, I. *Histories of Transit-Oriented Development: Perspectives on the Development of the TOD Concept*; Institute of Urban and Regional Development: Berkley, CA, USA, 2009.
27. Cervero, R.; Guerra, E.; Al, S. *Beyond Mobility: Planning Cities for People and Places*; Island Press: Washington, DC, USA, 2017.
28. Parker, T.; McKeever, M.; Arrington, G.B.; Smith-Heimer, J.; Brinckerhoff, P. *Statewide Transit-Oriented Development Study: Factors for Success in California*; California Department of Transportation: Sacramento, CA, USA, 2002.
29. Patnala, P.K.; Parida, M.; Chalumuri, R.S. A decision framework for defining Transit-Oriented Development in an Indian city. *Asian Transp. Stud.* **2020**, *6*, 100021. [CrossRef]
30. Calthorpe, P. *The Next American Metropolis: Ecology, Community, and the American Dream*; Princeton Architectural Press: New York, NY, USA, 1994.
31. Steuteville, R.; Langdon, P. *New Urbanism: Best Practices Guide*; New Urban News Publications: Ithaca, NY, USA, 2009.
32. Laaly, S.; Jeihani, M.; Lee, Y.J. A multiscale, transit-oriented development definition based on context-sensitive paradigm. *Transp. Res. Rec.* **2017**, *2671*, 31–39. [CrossRef]
33. FTA. Federal Transit Administration. (2024). TOD Changes Lives. Transit Oriented Development. U.S. Department of Transportation. Federal Transit Administration. 1200 New Jersey Avenue, SE Washington, DC 20590. Available online: https://www.transit.dot.gov/TOD (accessed on 2 December 2024).
34. Cervero, R.; Murakami, J. Rail and property development in Hong Kong: Experiences and extensions. *Urban Stud.* **2009**, *46*, 10. [CrossRef]

35. Duncan, M. The impact of transit-oriented development on housing prices in San Diego, CA. *Urban Stud.* **2011**, *48*, 1. [CrossRef] [PubMed]
36. Hess, D.B.; Almeida, T.M. Impact of proximity to light rail rapid transit on station-area property values in Buffalo, New York. *Urban Stud.* **2007**, *44*, 1041–1068. [CrossRef]
37. Chatman, D.G.; Noland, R.B. Transit service, physical agglomeration and productivity in US metropolitan areas. *Urban Stud.* **2014**, *51*, 917–937. [CrossRef]
38. Belzer, D.; Autler, G. *Transit-Oriented Development: Moving from Rhetoric to Reality*; The Brookings Institution: Washington, DC, USA, 2002.
39. Renne, J.L. Transit-oriented development: Developing a strategy to measure success. *Transp. Res. Rec.* **2005**, *1927*, 151–157.
40. Kamruzzaman, M.; Baker, D.; Washington, S.; Turrell, G. Advance transit oriented development typology: Case study in Brisbane, Australia. *J. Transp. Geogr.* **2014**, *34*, 54–70. [CrossRef]
41. Cervero, R.; Ferrell, C.; Murphy, S. *Transit-Oriented Development and Joint Development in the United States: A Literature Review*; TCRP Research Results Digest 52; Transportation Research Board: Washington, DC, USA, 2002; Available online: https://trid.trb.org/View/726711 (accessed on 2 December 2024).
42. Belzer, D.; Autler, G. *Transit Oriented Development: Moving from Rhetoric to Reality*; Brookings Institution Center on Urban and Metropolitan Policy: Washington, DC, USA, 2002; pp. 6–15.
43. Cervero, R.; Landis, J. Twenty years of the Bay Area Rapid Transit system: Land use and development impacts. *Transp. Res. Part A Policy Pract.* **1997**, *31*, 309–333. [CrossRef]
44. Dittmar, H.; Ohland, G. *The New Transit Town: Best Practices in Transit-Oriented Development*; Island Press: Washington, DC, USA, 2012.
45. Cervero, R.; Kockelman, K. Travel Demand and the 3Ds: Density, Diversity, and Design. *Transp. Res. Part D Transp. Environ.* **1997**, 2, 199–219. [CrossRef]
46. Ewing, R.; Cervero, R. Travel and the Built Environment: A Meta-Analysis. *J. Am. Plan. Assoc.* **2010**, *76*, 265–294. [CrossRef]
47. Suzuki, H.; Cervero, R.; Iuchi, K. *Transforming Cities with Transit: Transit and Land-Use Integration for Sustainable Urban Development*; World Bank Publications: Washington, DC, USA, 2013.
48. Tumlin, J.; Millard-Ball, A. How to Make Transit-Oriented Development Work. *Planning* **2003**, *69*, 14–19.
49. Dunphy, R.; Myerson, D.; Pawlukiewicz, M. *Ten Principles for Successful Development Around Transit*; Urban Land Institute: Washington, DC, USA, 2004.
50. Boarnet, M.G.; Crane, R. Public Finance and Transit-Oriented Planning: New Evidence from Southern California. *J. Plan. Educ. Res.* **1998**, *17*, 206–219. [CrossRef]
51. Mathur, S.; Gatdula, A. Review of planning, land use, and zoning barriers to the construction of Transit-oriented developments in the United States. *Case Stud. Transp. Policy* **2023**, *12*, 100988. [CrossRef]
52. Duffhues, J.; Mayer, I.S.; Nefs, M.; Van Der Vliet, M. Breaking barriers to transit-oriented development: Insights from the serious game SPRINTCITY. *Environ. Plan. B Plan. Des.* **2014**, *41*, 770–791. [CrossRef]
53. Fainstein, S.S. Cities and Diversity: Should We Want It? Can We Plan for It? *Urban Aff. Rev.* **2005**, *41*, 3–19. [CrossRef]
54. Bramley, G.; Dempsey, N.; Power, S.; Brown, C. What is 'social sustainability', and how do our existing urban forms perform in nurturing it. In *Sustainable Communities and Green Futures' Conference*; Bartlett School of Planning, University College London: London, UK, 2006; pp. 1–40.
55. McCrea, R.; Walters, P. Impacts of Urban Consolidation on Urban Livability: Comparing an Inner and Outer Suburb in Brisbane, Australia. *Hous. Theory Soc.* **2012**, *29*, 190–206. [CrossRef]
56. Ewing, R.; Handy, S. Measuring the Unmeasurable: Urban Design Qualities Related to Walkability. *J. Urban Des.* **2009**, *14*, 65–84. [CrossRef]
57. O'Sullivan, S.; Morrall, J. Walking Distances to and from Light-Rail Transit Stations. *Transp. Res. Rec.* **1996**, *1538*, 19–26. [CrossRef]
58. Geurs, K.T.; van Wee, B. Accessibility Evaluation of Land-Use and Transport Strategies: Review and Research Directions. *J. Transp. Geogr.* **2004**, *12*, 127–140. [CrossRef]
59. Litman, T. *Transportation Demand Management: Strategies, Planning and Evaluation*; Victoria Transport Policy Institute: Victoria, BC, Canada, 2013.
60. Fazli, R.F.; Faridi, R.A. Urbanization in Dubai: Process, Problems and Challenges. *J. West Asian Stud.* **2008**, *22*. Available online: http://www.publishingindia.com/GetBrochure.aspx?query=UERGQnJvY2h1cmVzfC8xMjMucGRmfC8xMjMucGRm (accessed on 2 December 2024).
61. Pacione, M. City Profile: Dubai. *Cities* **2005**, *22*, 255–265. [CrossRef]
62. Alawadi, K.; Benkraouda, O. The debate over neighborhood density in Dubai: Between theory and practicality. *J. Plan. Educ. Res.* **2019**, *39*, 18–34. [CrossRef]
63. Scholz, W. The nexus of urban development and infrastructure. In *Transport Planning and Mobility in Urban East Africa*; Routledge: New York, NY, USA, 2020; pp. 7–30.

64. Bertolini, L. Integrating mobility and urban development agendas: A manifesto. *Disp. Plan. Rev.* **2012**, *48*, 16–26. [CrossRef]
65. Acuto, M. High-rise Dubai: Urban Entrepreneurialism and the Technology of Symbolic Power. *Cities* **2010**, *27*, 272–284. [CrossRef]
66. Aljoufie, M.; Zuidgeest, M.; Brussel, M.; van Maarseveen, M. Spatial–temporal analysis of urban growth and transportation in Jeddah City, Saudi Arabia. *Cities* **2013**, *31*, 57–68. [CrossRef]
67. Bagaeen, S. Brand Dubai: The instant city; or the instantly recognizable city. *Int. Plan. Stud.* **2007**, *12*, 173–197. [CrossRef]
68. Worku, G.B. Demand for improved public transport services in the UAE: A contingent valuation study in Dubai. *Int. J. Bus. Manag.* **2013**, *8*, 108. [CrossRef]
69. Tranter, P.J. Speed kills: The complex links between transport, lack of time and urban health. *J. Urban Health* **2010**, *87*, 155–166. [CrossRef] [PubMed]
70. Elsheshtawy, Y. *Dubai: Behind an Urban Spectacle*; Routledge: New York, NY, USA, 2009.
71. Davidson, C.M. *Dubai: The Vulnerability of Success*; Columbia University Press: New York, NY, USA, 2008.
72. Newman, P.G.; Kenworthy, R. *Cities and Automobile Dependence*; Gower Publishing Company: Aldershot UK, 1998.
73. Elsheshtawy, Y. Redrawing boundaries: Dubai, an emerging global city. In *Planning Middle Eastern Cities*; Routledge: New York, NY, USA, 2004; pp. 183–213.
74. Arabi, R. Assessment of Vegetation Cooling Effect Through the Application of Green Roofs in Tropical Campus Environment. Ph.D. Thesis, Universiti Putra Malaysia, Seri Kembangan, Malaysia, February 2018.
75. Gehl, J. *Cities for People*; Island Press: Washington, DC, USA, 2013.
76. Dubai Urban Plan 2040. Urban Master Plan. Government of Dubai: 2024. Available online: https://www.dm.gov.ae/municipality-business/planning-and-construction/dubai-2040-urban-master-plan/ (accessed on 6 December 2024).
77. Banister, D. The Sustainable Mobility Paradigm. *Transp. Policy* **2008**, *15*, 73–80. [CrossRef]
78. Ministry of Energy and Infrastructure (MOEI). Dubai Clean Energy Strategy 2050. *Government of Dubai*. 2017. Available online: https://u.ae/en/about-the-uae/strategies-initiatives-and-awards/strategies-plans-and-visions/environment-and-energy/uae-energy-strategy-2050 (accessed on 5 December 2024).
79. Suzuki, H.; Dastur, A.; Moffatt, S.; Yabuki, N.; Maruyama, H. *Eco2 Cities: Ecological Cities as Economic Cities*; World Bank: Washington, DC, USA, 2010.
80. Litman, T. *Evaluating Public Transit Benefits and Costs: Best Practices Guidebook*; Victoria Transport Policy Institute: Victoria, BC, Canada, 2011; Volume 65.
81. Thomas, R.; Bertolini, L. Beyond the Case Study Dilemma in Urban Planning: Using a Meta-Matrix to Distil Critical Success Factors in Transit-Oriented Development. *Urban Policy Res.* **2014**, *32*, 219–237. [CrossRef]
82. Salama, A.M.; Wiedmann, F. *Demystifying Doha: On Architecture and Urbanism in an Emerging City*; Routledge: New York, NY, USA, 2016.
83. Rode, P.; Hoffmann, C.; Kandt, J.; Smith, D.; Graff, A. Towards New Urban Mobility: The Case of London and Berlin. *Cities* **2015**, *60*, 1–14.
84. Frank, L.D.; Sallis, J.F.; Conway, T.L.; Chapman, J.E.; Saelens, B.E.; Bachman, W. Many Pathways from Land Use to Health: Associations between Neighborhood Walkability and Active Transportation, Body Mass Index, and Air Quality. *J. Am. Plan. Assoc.* **2006**, *72*, 75–87. [CrossRef]
85. Beatley, T. *Biophilic Cities: Integrating Nature into Urban Design and Planning*; Island Press: Washington, DC, USA, 2011.
86. Cervero, R.; Sullivan, C. Green TODs: Marrying Transit-Oriented Development and Green Urbanism. *Int. J. Sustain. Dev. World Ecol.* **2011**, *18*, 210–218. [CrossRef]
87. Curtis, C.; Renne, J.L.; Bertolini, L. *Transit Oriented Development: Making It Happen*; Ashgate: Farnham, UK, 2009.
88. Roads and Transport Authority RTA. 2024. Available online: https://www.rta.ae/wps/portal/rta/ae/home?lang=en (accessed on 2 December 2024).
89. Dubai Metro Tram Stations and Location Map. Available online: https://www.rta.ae/wps/portal/rta/ae/public-transport/metro-stations-map (accessed on 2 December 2024).
90. Smith, D.A.; Hewson, N.R.; Hendy, C.R. Design of the Dubai Metro light rail viaducts—Superstructure. *Proc. Inst. Civ. Eng. Bridge Eng.* **2009**, *162*, 55–62. [CrossRef]
91. Smith, D.; Smith, T.; Chiarello, M.; Ho, A. Dubai Metro footbridge design. *Proc. Inst. Civ. Eng. Bridge Eng* **2011**, *164*, 23–39. [CrossRef]

smart cities

MDPI

Article

Assaying Traffic Settings with Connected and Automated Mobility Channeled into Road Intersection Design

Maria Luisa Tumminello [1,*], Nazanin Zare [1], Elżbieta Macioszek [2,*] and Anna Granà [1]

[1] Department of Engineering, University of Palermo, 90133 Palermo, Italy; nazanin.zare@unipa.it (N.Z.); anna.grana@unipa.it (A.G.)

[2] Department of Transport Systms, Traffic Engineering and Logistics, Faculty of Transport and Aviation Engineering, Silesian University of Technology, 44-100 Katowice, Poland

[*] Correspondence: marialuisa.tumminello01@unipa.it (M.L.T.); elzbieta.macioszek@polsl.pl (E.M.)

Highlights

What are the main findings?

- Integrating road intersection geometric design with smart vehicle technologies enhances the benefits of well-designed road infrastructure, optimizing traffic efficiency as the prevalence of smart vehicles continues to rise.
- Traffic settings involving connected and automated vehicles (CAVs) can inform the design of road intersections within the context of smart cities, using microsimulation to analyze performance in mixed urban traffic environments.

What is the implication of the main finding?

- Support the development of expertise in the comparative analysis of the performance of alternative intersection designs that incorporate smart vehicle technologies.
- The microsimulation-driven framework underlines the importance of effective calibration to promote the development of evaluation methodologies for intersection designs and smart mobility integration in city planning.

Abstract: This paper presents a microsimulation-driven framework to analyze the performance of connected and automated vehicles (CAVs) alongside vehicles with human drivers (VHDs), channeled towards assessing project alternatives in road intersection design. The transition to fully automated mobility is driving the development of new intersection geometries and traffic configurations, influenced by increasing market entry rates (MERs) for CAVs (CAV-MERs), which were analyzed in a microsimulation environment. A suburban signalized intersection from the Polish road network was selected as a representative case study. Two alternative design hypotheses regarding the intersection's geometric configurations were proposed. The Aimsun micro-simulator was used to hone the driving model parameters by calibrating the simulated data with reference capacity functions (RCFs) based on CAV factors derived from the Highway Capacity Manual 2022. Cross-referencing the conceptualized geometric design solutions, including a two-lane roundabout and an innovative knee-turbo roundabout, allowed the experimental results to demonstrate that CAV operation is influenced by the intersection layout and CAV-MERs. The research provides an overview of potential future traffic settings featuring CAVs and VHDs operating within various intersection designs. Additionally, the findings can support project proposals for the geometric and functional design of intersections by highlighting the potential benefits expected from smart driving.

Keywords: sustainable road design; roundabouts; signalized intersections; knee-turbo roundabout; smart mobility; connected and automated vehicles; traffic microsimulation

1. Introduction

As urban populations grow, the demand for smart mobility is increasing, creating a need for reliable, efficient, and environmentally friendly transportation systems. Road engineering professionals face the challenge of developing innovative solutions that can accommodate rising mobility demands while integrating advanced technologies into the design of smart roads [1]. Recent advancements in Intelligent Transport Systems (ITS), particularly those involving connected and automated vehicles (CAVs), offer new opportunities to enhance road performance and promote sustainability from both research and policy perspectives [2]. CAVs use communication technologies to exchange information with each other, road infrastructure, and other entities for improved traffic management [3]. As automation levels increase, these advancements decrease driver workload and human unpredictability, resulting in a more efficient driving experience for all road users [4]. The deployment of CAVs is expected to provide numerous benefits, including fewer crashes, increased road capacity, optimized mobility, reduced emissions, improved road quality, and enhanced overall well-being for citizens [5,6]. However, several challenges related to the practical implementation and impact of CAVs remain unresolved, as high levels of automation are not yet commercially widespread. As CAVs gradually replace vehicles with human drivers (VHDs), their interactions become increasingly important, particularly at intersections where different traffic streams converge [6,7]. CAVs are equipped with communication and control devices like cooperative adaptive cruise control (CACC), which optimizes traffic efficiency by adjusting speed and distance relative to the preceding vehicle [8]. Although low CAV penetration rates present challenges, CAV platoons can travel closely to minimize delays safely, but CAVs may drive cautiously when interacting with VHDs lacking communication technology [9].

Interactions between CAVs and VHDs can vary significantly, particularly at unsignalized intersections, where the cautious and hesitant driving behavior of human-operated vehicles can complicate yield negotiations [10]. Conversely, signalized intersections facilitate decision-making for gap acceptance with clear traffic signals [1,7]. Although roundabouts are typically less expensive to install and maintain than signalized intersections, CAVs face challenges when navigating curved trajectories in multilane scenarios, especially regarding gap acceptance and lane-changing [11–13]. The transition from two-lane to turbo roundabouts introduces lane dividers or curbs to separate traffic streams, requiring the pre-selection of entry lanes and thus facilitating navigation for vehicles [14,15].

Research on CAVs increasingly utilizes micro-simulation models to assess performance in mixed traffic environments [2,16,17]. These simulations aim to analyze the dynamic CAV–infrastructure interactions and entry negotiation processes between CAVs or between a CAV and a VHD at intersections, while evaluating the effectiveness of algorithms in real-world scenarios [18,19]. Findings indicate that smart vehicles outperform VHDs in capacity and saturation headways due to their connectivity and automation, even at varying levels of communication [20,21]. Ongoing research also investigates trajectory design for CAVs and VHDs to improve operations at signalized intersections, employing deep reinforcement learning for optimal efficiency [22,23]. The optimization of CAV trajectories at roundabouts also shows significant performance improvements as penetration rates rise, leading to reduced travel times [24]. To establish effective navigation rules for roundabouts, optimal vehicle path management is essential; however, employing machine learning methods or more advanced algorithms can be beneficial but they often incur significant computational costs and prolonged processing times due to their iterative nature. Furthermore, while micro-simulation results offer valuable insights, they are typically specific to individual studies, necessitating diverse evaluations of case scenarios for broader generalizability across different contexts [13]. The study in [25] similarly highlights variability in CAV behaviors and the importance of mixed traffic with VHDs. Although focused on an expert system for a single-lane roundabout, it emphasizes broader applicability across intersection types, addressing the need for adaptable, inclusive traffic management solutions for emerging CAV technologies.

In the context of smart cities, although advancements in CAV technology and their testing on road infrastructures can inform the design of smart intersections and roundabouts [1,2,25], the broader challenges of managing and optimizing CAV operations at intersections remain unresolved and require further research [25,26]. Specifically, there is a gap in the literature regarding published studies that compare various intersection geometric design proposals in terms of CAV performance in mixed traffic environments. In this broader context of the interaction between CAVs and road infrastructure, this paper aims to address that gap by evaluating different intersection design options and analyzing their impact on the operational efficiency of CAVs. In this perspective, it aligns with the evolving needs of professionals in the field of road design.

Based on the above, a microsimulation-driven framework for analyzing the performance of CAVs alongside VHDs is proposed. The aim of this paper is to investigate how traffic scenarios involving connected and automated mobility can be channeled into the design of road intersections, thereby promoting the sustainable integration of smart vehicles into current and future road infrastructure. The primary objective is to explore the role of microsimulation, particularly using Aimsun (Version 20) [17], to compare the performance of CAVs in mixed traffic environments and various intersection layouts, focusing on the following key questions:

- To what extent can the calibration process of Aimsun's parameters replicate simulated data that align with the reference capacity functions (RCFs) for a given road entity?
- What is the impact of the geometric and functional design of intersections, along with traffic control modes, on the expected operational performance of CAVs as their market entry rates (MERs) vary?
- Is it feasible to develop a performance criterion that enables the comparison of alternative geometric designs of intersections based on the analogy of entry mechanisms?

A suburban, four-legged signalized intersection in a Polish city was selected as a case study. Various traffic scenarios with different MERs for CAVs (CAV-MERs) were modeled in Aimsun, including two proposed roundabout solutions: a two-lane roundabout and a knee-turbo counterpart, which served as testbeds in the validation-driven approach based on the analogy in entry mechanisms. Inspired by the turbo roundabout design detailed in [14,27], the knee-turbo geometry was developed as an alternative layout designed to accommodate travel demands within the examined intersection area and its surrounding built environment. Traffic and geometric data were configured in Aimsun to match the conceptual scenarios, and sensitive driving model parameters were adjusted to align the simulated data with the RCFs, considering the increasing CAV-MERs and correction factors sourced from [13].

Cross-referencing various geometric design solutions for the signalized intersection, including the two-lane roundabout and the knee-turbo scheme, revealed that CAV operations are affected by intersection layouts and the CAV-MERs. In this perspective, the study proposes a methodological framework to evaluate future intersection designs by integrating CAV technology with road engineering. Additionally, the framework can serve as a complementary tool in geometric and functional design, facilitating the assessment of how modifications in intersection schemes influence CAV performance. Thus, the paper offers a helpful methodology for assessments throughout the design process, aiding decision-making on alternative project proposals that leverage the benefits of smart driving—an approach commonly used in feasibility studies before detailed project development. However, additional case studies of intersections with various control types are needed to ensure the broader applicability of the proposed method across different road entities and urban environments and to enable the effective generalization of the results. In addition to its scientific contribution in assessing future traffic scenarios with CAVs and VHDs at intersections, the paper fosters a culture of ITS and promotes smart mobility, aiming to raise awareness among citizens and road users about the advantages of a safer, more appealing, and less polluted road environment.

The paper is structured as follows: It begins with an introduction, followed by the research materials and methods in Section 2, outlining the case study description, design alternatives, and Aimsun network modeling. Section 3 presents the results, while Section 4 discusses them. Finally, Section 5 concludes the paper.

2. Materials and Methods

The microsimulation-driven framework proposed in this paper for analyzing the performance of CAVs alongside VHDs at various intersection designs follows the steps shown in Figure 1.

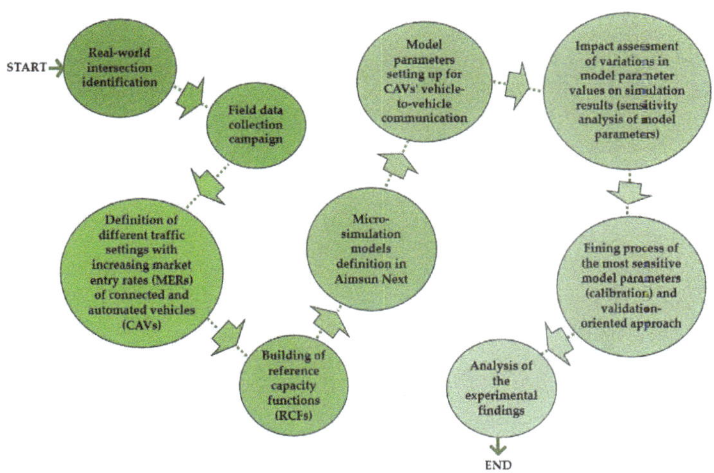

Figure 1. Bubble chart to map the research methodology.

The initial phase of the study focused on selecting a representative signalized intersection within a suburban urban setting, taking into account its spatial layout, traffic patterns, and operational features. Field data collection was carried out to gather detailed information on traffic flows, vehicle types, and signal control mechanisms. Using the existing intersection design as a foundation, a new proposal was developed to replace the signalized intersection with a two-lane roundabout, with the goal of converting it into a turbo roundabout. This approach aims to improve traffic throughput, safety, and efficiency by optimizing the intersection's geometric and operational characteristics.

The next step involved establishing traffic settings with CAV-MERs, varying from 0% to 100% in 20% intervals. Subsequently, for each intersection, capacity functions were developed corresponding to these CAV-MER levels. These functions incorporated correction factors specific to CAVs from [13], reflecting their impact on intersection performance as their MER increases. This analysis helps in understanding how different CAV-MERs can influence intersection capacity and operational efficiency.

In Aimsun, vehicle-to-vehicle (V2V) communication for CAV fleets was implemented to simulate traffic conditions for the proposed intersection layouts. Subsequently, a sensitivity analysis of the model parameters was conducted before calibrating them. The calibration process involved comparing the simulated output data—based on the most sensitive model parameters—with the RCFs at roundabouts. The calibration accuracy for the signalized intersection model was assessed by varying the green time values within the traffic light cycle at different CAV-MER levels.

The knee-turbo roundabout served as a counterpart for validating the simulated two-lane roundabout model, based on the analogy in their entry mechanisms by entry approach. In the subject approach selected for the analysis, the left lane experienced conflicts from two circulating traffic flows, while the right lane encountered a single circulating flow. Thus, the right lane of the two-lane roundabout was considered a de facto right-turn lane, conflicting with the one-lane circulatory roadway, to facilitate a comparative operational analysis between the two conceptualized roundabouts. This validation-driven approach offered an additional tool for assessing the calibration effectiveness of the intersections under study. Ultimately, selected key performance metrics were employed to evaluate and compare CAV performance across the different intersection layouts and traffic settings.

2.1. Case Study Description

To analyze the performance of a mixed fleet of CAVs and VHDs at various intersection types and their associated traffic regulation modes, we selected a suburban, isolated four-legged signalized intersection within the road network of Opole City, Poland (50°39′17″ N; 17°57′16″ E), which was well-suited to effectively facilitate the development of diverse design proposals. The selected signalized intersection consists of a main road, "Jerzego i Ryszarda Kowalczyków", which has two entry lanes, each 3.25 m wide, and one exit lane measuring 3.50 m in width. The main road directs traffic towards the center of Opole City in the north-west direction, while in the south-eastern direction it transitions into the DK 94 highway, which leads to the hamlet of Strzelce Opolskie. The minor road consists of "Mieszka I" street, extending in the north-easterly direction, featuring two entry lanes, each 3.25 m wide, and one exit lane measuring 3.50 m wide. In the south-westerly direction, "Jagiellonów" street includes three entry lanes that are 3.25 m wide and one exit lane that is 3.50 m wide. The signalized intersection comprises four approaches, each featuring one shared right-turn and through lane (R&T) plus an exclusive left-turn lane (L), except for the south-west (S-W) approach, which has three entry lanes for each movement. Figure 2 illustrates key details of the intersection under study, while Figure 3 shows the main results of the field data collection that took place during morning and afternoon peak hours in September 2024.

Data on the signal timing plan, traffic movements, and vehicular composition were recorded using a camera positioned along the west-north (W-N) approach. Post-processing of this field data revealed an average traffic volume of 2263 vehicles per hour, primarily consisting of cars, with a small percentage of heavy vehicles, as shown in Figure 3a. Additionally, minimal percentages of pedestrians and cyclists were observed, likely due to the suburban context of the intersection's location. The traffic light timing was analyzed without including pedestrians and cyclists, but in urban areas with higher pedestrian volumes, their phase would be incorporated into and extend the cycle duration.

Due to the low presence of pedestrians and cyclists in the suburban setting, these users were excluded from the subsequent analysis of the two-lane roundabout and its knee-turbo counterpart. Typically, pedestrians and bicyclists navigate roundabouts via separated shared-use paths, with crossings set back from the circulatory roadway and protected by splitter islands for safe two-stage crossing [12,28]. Bikers should feel as safe as or safer than approaching drivers, but their safety must not compromise pedestrian safety. Multi-lane roundabouts often incorporate active traffic control devices, requiring pedestrians to activate signals [12]. Pedestrian accommodations in turbo roundabouts are like those found in modern roundabouts [27].

(**a**) (**b**)

Figure 2. The signalized intersection under study: (**a**) geometric scheme sketch; (**b**) photo of the E-S entry approach. Note: the acronyms are provided in the legend.

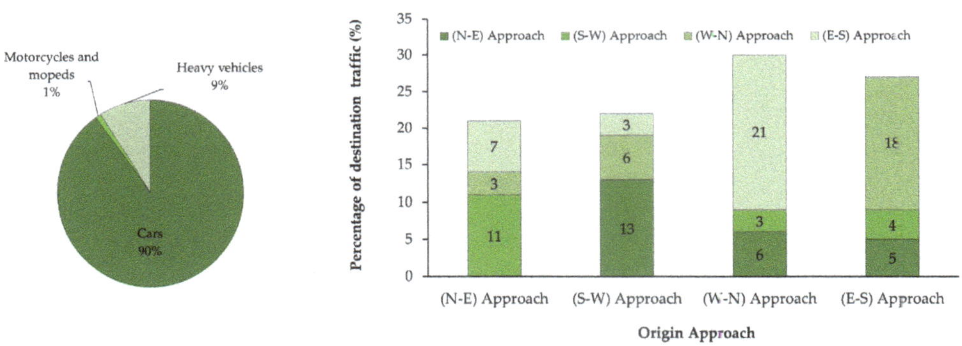

(**a**) (**b**)

Figure 3. Traffic features: (**a**) pie graph on the vehicular composition; (**b**) histogram chart of origin–destination traffic by approach.

The distribution of maneuvers among various approaches at the signalized intersection under study is as follows: 39% of the total traffic on the main road engaged in crossing maneuvers, while 10% of users made left turns and 8% made right turns. Figure 3b illustrates the percentages of destination traffic for each origin approach. Crossing traffic on the minor road constituted 24%, which included 11% of traffic flowing from the north-eastern (N-E) origin to the south-western (S-W) destination, while the remaining 13% pertained to the opposite direction. Additionally, the percentage of left turns from the S-W approach to the main road was approximately 6%, with right turns at around 3%. Conversely, from the N-E approach to the main road, left turns were about 7%, while right turns were around 3%.

The traffic light operated on a four-phase timing plan in the fixed mode, with a total cycle length of 120 s, as illustrated in Figure 4.

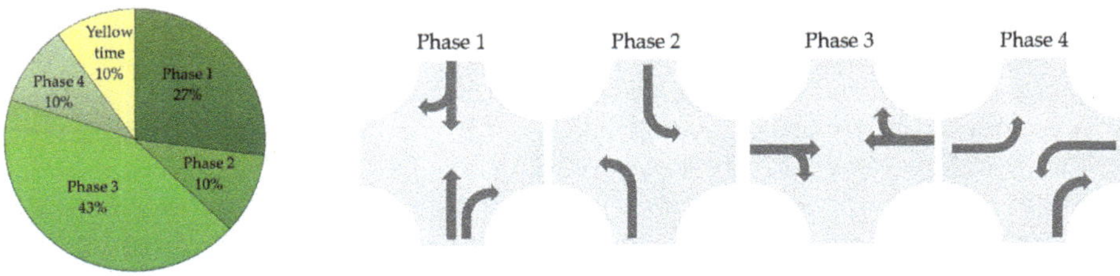

(a) (b)

Figure 4. The timing plan: (**a**) pie graph on the green time percentage for each phase of the traffic signal; (**b**) the four-phase timing plan diagram with the allowable movements.

Figure 4a displays the green time percentage for each phase, while Figure 4b shows the four-phase timing plan diagram with the allowable movements during each phase at the signalized intersection. In Phase 1, right turns and through traffic on the north-eastern (N-E) and south-western (S-W) minor roads receive a green time of 27% of the total cycle length. During Phase 2, left-turning traffic from the minor road to the main road is allocated a green time of 10% of the cycle. Phase 3 grants 43% of the cycle to right turns and through movements on the main road. Finally, Phase 4 permits left-turning traffic from the west-north (W-N) and east-south (E-S) main roads, as well as right-turning traffic from the south-west (S-W) minor road, to operate for 10% of the cycle.

Consistent with [13], lane groups were established at the signalized intersection: exclusive left turns were classified as an individual lane group, while right turns and through movements shared a single lane group. This arrangement facilitated the conceptualization and formulation of various design proposals, including the potential for roundabouts.

Considering the available space for a circular layout, the suburban context of the installation, and the traffic composition, an initial design for a two-lane roundabout was developed to transition the existing signalized intersection into a knee-turbo roundabout [14,27]. Figure 5 presents the geometric sketches of the designed two-lane roundabout (see Figure 5a) and the knee-turbo roundabout (see Figure 5b). Additionally, Figure 5c,d illustrates the conflict points for the two-lane roundabout and the knee-turbo roundabout, respectively. Changes in conflict points and vehicle path designs can aid in the comparative analysis of the roundabouts, offering better insights into how different geometric layouts influence CAV performance.

The designed four-legged roundabout featured two entry lanes and one exit lane for each approach except for the S-W approach, as shown in Figure 5a. It included a non-traversable central island and an outer diameter of 26 m, along with two circulatory lanes, each measuring 3.75 m wide. The geometric design also integrated raised splitter islands and deflection angles exceeding 45 degrees.

A knee-turbo roundabout was subsequently conceptualized and designed (see Figure 5b). Inspired by the turbo roundabout design detailed in [14,27], this geometry was developed as an alternative tailored to meet the travel demands within the intersection area and its surrounding built environment. The knee-turbo roundabout is a specialized variation of the basic turbo roundabout, distinguished by its distinctive geometric features [14,27].

Unlike the basic version, the knee-turbo features a more pronounced angular shape—resembling a knee—often involving tighter curves and strategic lane arrangements to enhance the integration of multiple traffic streams. It typically incorporates dedicated lanes for different movement types, such as turning and straight-through flows, which further reduce weaving conflicts common in modern roundabouts [12,14,27]. One key advantage of the knee-turbo design is its improved ability to effectively slow vehicles before they enter the circulating roadway, thereby reducing crash risks and enhancing adaptability in complex traffic environments, all while maintaining efficient vehicle circulation. Although it provides similar traffic throughput to the basic turbo roundabout, the knee-turbo roundabout offers greater flexibility in configuring entry approaches, allowing for various lane combinations in the entry, exit, and circulatory roadway along the opposing approaches [14,27]. While safety parameters for turbo roundabouts are generally well established—particularly when comparing the basic turbo to two-lane roundabouts [12,14,27]—considering these characteristics during the design process is essential to mitigate the risks of over- or under-dimensioning in the evaluated schemes.

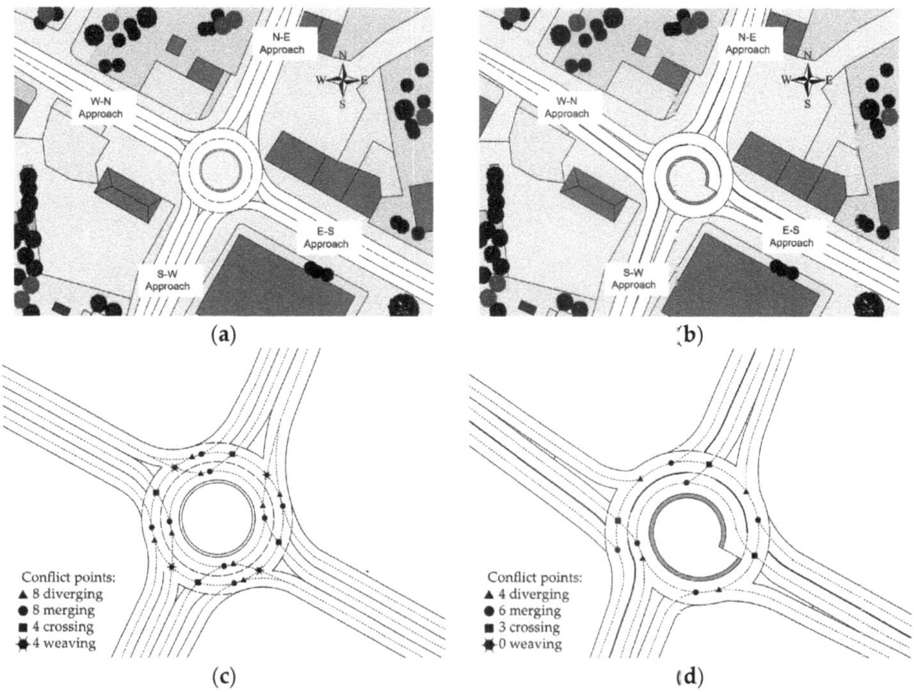

Figure 5. Geometric sketch of (**a**) the designed two-lane roundabout; (**b**) the knee-turbo roundabout; (**c**) the conflict points at the two-lane roundabout; (**d**) the conflict points at the knee-turbo roundabout. Note: N-E stands for north-east, E-S stands for east-south, S-W stands for south-west, and W-N stands for west-north. Note: The acronyms are provided in the legend.

The footprint of the knee-turbo roundabout was adapted to accommodate the two-lane layout, optimizing the available installation space (see Figure 5b). The main road features two entry lanes and one exit lane in the west-north (W-N) and east-south (E-S) directions. The minor road includes two entry lanes and one exit lane in the north-east (N-E) direction, while in the south-west (S-W) direction, it has one entry lane and two exit lanes. Each entry lane is 3.75 m wide, each exit lane is 3.50 m wide, and each circulating lane measures 4.00 m in width. The splitter islands comply with Polish guidelines, ensuring effective traffic management and safety within the roundabout design [29].

The knee-turbo roundabout is introduced as a new conceptual variant in geometric design, supporting the proposed validation-driven approach based on the analogy in entry mechanisms. It was used as a reference to validate the simulated two-lane roundabout model due to the similarities in vehicle approach and navigation. Entry mechanisms in roundabouts or turbo roundabout approaches include the following [12,13,27,29]: single-lane entry conflicts with a single circulatory lane, as is typical in the approach to a single-lane roundabout and a turbo roundabout with one circulatory lane (e.g., S-W approach at the knee-turbo roundabout in Figure 5b); single-lane entry with two circulating lanes, where one entry lane conflicts with both circulating lanes (the driver enters the roundabout and selects the lane based on the desired exit) [12,14]; two-lane entry with a single circulatory lane is used in two-lane roundabouts or turbo roundabouts, where an inside circulatory lane is added at one approach (e.g., the E-S approach for the knee turbo roundabout in Figure 5b) or at approaches aligned in the same direction (e.g., basic or egg turbo roundabouts) [14,27]; two-lane entry with two circulatory lanes where drivers preselect the lane before entering and then maneuver towards the desired exit in a two-lane roundabout (e.g., Figure 5a) or turbo roundabouts (e.g., N-E and W-N approaches at knee-turbo roundabout in Figure 5b).

Operationally, turbo roundabouts' spiral geometry and lane dividers require drivers to choose the correct lane prior to entry to ensure proper exiting. At the two-lane exits, drivers in the inside lane turn to exit similarly to those in two-lane roundabouts. However, turbo roundabouts eliminate the need for inside-lane drivers to cross the outside lane by design, physically preventing this crossing and requiring drivers in the outside lane to exit directly [28,29]. However, the circular design of roundabouts and turbo roundabouts helps control vehicle speeds, enhancing safety and traffic flow during all movement stages [12,27]. Such design characteristics facilitate predictable traffic patterns and similar operational dynamics. In both the knee-turbo roundabout and the two-lane roundabout shown in Figure 5, drivers follow comparable entry rules, such as yielding to circulating traffic and selecting appropriate lanes based on their intended direction. This consistency enhanced the effectiveness of validating the simulated models, as vehicle behavior can be expected to align at both roundabouts due to their similar geometries and traffic operations.

In the selected approach for analysis (i.e., the W-N approach in Figure 5a,b), the left lane was conflicted by two circulating traffic flows, while the right lane intersected with only one flow. Consequently, drivers entering from the right lane used the outer circulating lane, while those entering from the left lane traveled in the inner circulating lane at both roundabouts. As a result, the right lane of the two-lane roundabout effectively functioned as a de facto right-turn lane, conflicting with one circulating lane. This setup enabled a comparative operational analysis between the two designs. Overall, this validation approach provided an additional method for assessing the accuracy of the intersection calibration.

The passage highlights opportunities for CAVs and V2V communication at roundabouts [1,13]. It notes that for a VHD waiting to enter, routing decisions of conflicting vehicles remain uncertain until the vehicle commits to entering, circulating, or exiting, which can lead to delays and reduced capacity. When both yielding and incoming vehicles are CAVs, V2V communication enables them to share routing information, potentially enhancing flow and efficiency, as a CAV approaching a roundabout can detect incoming vehicles through line-of-sight sensors or V2V communication. The entering CAV prioritizes the closest approaching vehicles, assesses for route conflicts, and can enter before the incoming vehicle signals its intention to circulate or exit [13,27]. If the first vehicle in the approaching stream is not a CAV or if route conflicts exist, the CAV's behavior is governed by gap acceptance criteria [12]. In this context, path guidance strategies, including advanced pavement markings, dynamic signage, lane delineation, and enhanced lighting, improve the safe navigation of CAVs on curved layouts by reducing conflicts and congestion [1,12].

2.2. *Intersection Operation Analysis*

Beginning with a baseline scenario featuring only VHDs, five traffic settings were simulated in Aimsun, using increments of 20% CAVs. The baseline traffic setting contained 100% VHDs and 0% CAVs, while the subsequent traffic settings included (1) 80% VHDs and 20% CAVs, (2) 60% VHDs and 40% CAVs, (3) 40% VHDs and 60% CAVs, (4) 20% VHDs and 80% CAVs, and (5) 0% VHDs and 100% CAVs. The research began by defining micro-simulation models in Aimsun, focusing on the conceptualized intersection layouts and traffic settings. The calibration process concentrated on the signalized intersection and the roundabout models by leveraging, in this latter case, the analogy of yield negotiation at the entries of both roundabouts shown in Figure 5. Validation used the knee-turbo roundabout as a control scheme to assess the calibration accuracy. Due to the absence of observed data for vehicles with high automation levels, RCFs were developed for both the signalized intersection and the two-lane roundabout to serve as target data. The equations in Table 1 were used to create RCFs that reflect varying CAV-MERs at the intersection level [13].

Table 1. Equations for signalized intersections and roundabout capacity calculation.

Intersection Type	Capacity Formula	Parameters Description	
Signalized intersection	$C_{CAVs} = \frac{G \cdot N_l \cdot S_{fr,\,CAVs}}{c}$	C_{CAVs}: entry capacity (veh/h); G: the effective green time (s); N_l is the number of lanes in a lane group; $S_{fr,CAVs}$: the saturation flow rate (veh/h/ln); and c: the total cycle length (s).	(1)
	$S_{fr,CAVs} = S_b \cdot \prod_{i=1}^{n} f_i$	$S_{fr,CAVs}$: the saturation flow rate (veh/h/ln); S_b: the base saturation flow rate (veh/h/ln); f_i: the correction factors.	(2)
Roundabout	$C_{e,CAVs} = f_{(a)} \cdot a \cdot e^{-f_{(b)} \cdot b \cdot Q_c}$	$C_{e,CAVs}$: CAVs' capacity (pc/h); Q_c: circulating flow rate (pc/h); a: intercept parameter (equal to 1380 pc/h for the right entry lane; 1350 pc/h for the left entry lane); b: slope parameter (equal to 0.00102 and 0.00092 for the right and left entry lanes, respectively); and $f_{(a)}$ and $f_{(b)}$: correction factors for the parameters a and b, respectively [13].	(3)

Model parameters were iteratively refined based on the alignment of simulated outputs with the RCFs during the calibration process. The green time and the traffic light cycle length in Table 1 (see Equation (1)) were collected in the field at the signalized intersection, while the saturation flow rate was calculated according to [13]. The base value for the saturation flow was initially set at 1900 pc/h, but this was corrected by factors (f_i) that consider specific characteristics of the subject intersection, such as lane width, turning movements, heavy vehicles, and CAVs in traffic (see Equation (2) in Table 1).

As the CAV-MERs changed, the base saturation flow rate for through movements also varied, while the saturation flow rate for left turns was adjusted with an additional correction factor based on the CAV-MERs. Modifications in CAV capacity for right turns were consistent with [13]. The surfaces in Figure 6 illustrate the RCFs with varying CAV-MERs at the roundabout and the signalized intersection. Specifically, Figure 6a displays the RCF of the left entry lane (LEL) at the roundabout, while Figure 6b shows the RCF of the right entry lane (REL) at the roundabout. Meanwhile, Figure 6c shows the RCF of the exclusive left-turn lane (L) at the signalized intersection as the green time increased, while Figure 6d shows the RCF of the shared right-turn and through lane (R&T) at the signalized intersection as the green time increased, highlighting how effective green time varied in response to different CAV-MERs while keeping the total cycle length constant.

2.3. Modeling Intersection Case Studies in AIMSUN to Evaluate the Validation-Driven Approach

Aimsun micro-simulator (Version 20) was used to assess the impact of different intersection layouts on the operational effectiveness of CAVs mixed with VHDs [17]. Given that lane divisors (e.g., curbs or road markings) are used to separate each lane at the knee-turbo roundabout, the functionality with curbs was simulated to ensure entry lane preselection and prevent lane changes. Micro-simulation models were defined based on the geometrical features of the conceptualized intersections, simulating traffic data and settings using origin–destination (O—D) matrices.

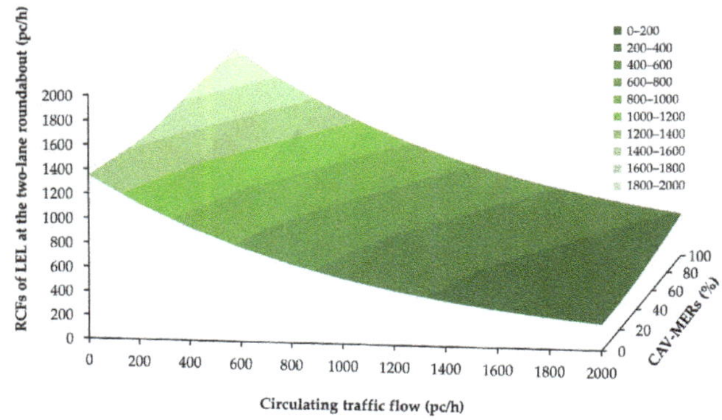

(a)

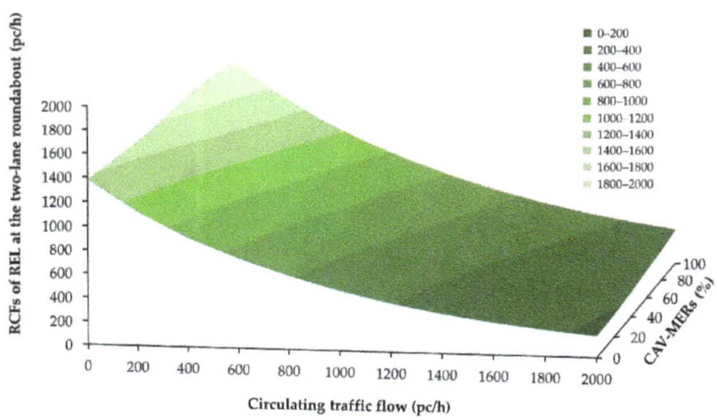

(b)

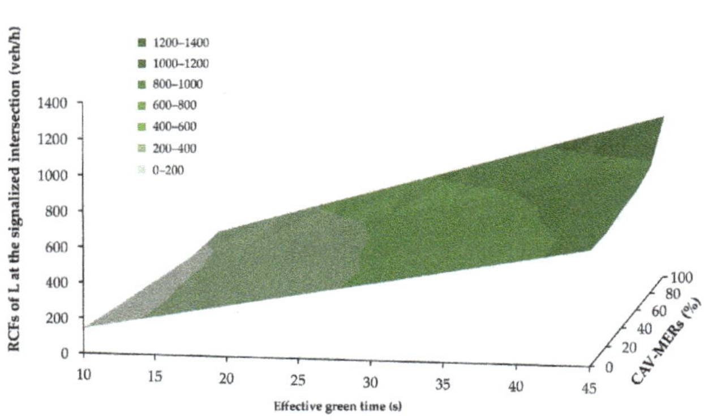

(c)

Figure 6. *Cont.*

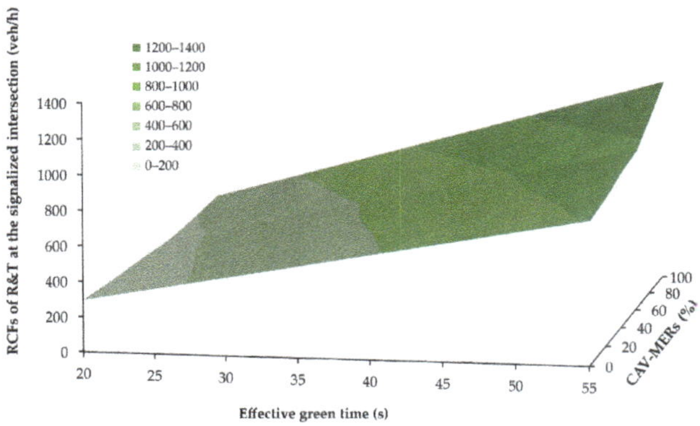

(d)

Figure 6. RCFs with varying CAV-MERs for (**a**) the reference left entry lane (LEL) capacity at the two-lane roundabout; (**b**) the reference right entry lane (REL) capacity at the two-lane roundabout; (**c**) the reference exclusive left-turn lane (L) group capacity at the signalized intersection as the green time increased; and (**d**) the reference shared right-turn and through lane (R&T) group capacity at the signalized intersection as the green time increased. Note: the acronyms are provided in the legend.

Based on the mentioned traffic settings, the total traffic matrix was divided into two distinct matrices: one matrix represented a specific percentage (x) of CAVs, while the other matrix contained the remaining percentage (1 − x) of VHDs. For every intersection, an appropriate type of traffic control was systematically assigned: give-way signals were allocated to each entry lane at roundabouts, while a time control plan was developed at the signalized intersection based on comprehensive field observations and data analysis. Detectors were installed on specific road sections to continuously monitor changes in entry capacity. Ten one-hour simulation trials with only VHDs were executed to assess Aimsun's accuracy in replicating the observed traffic flows.

The simulation results emphasized the need to calibrate the model parameters, revealing a percentage difference of approximately 12% between the simulated and field data. Assuming high communication reliability, V2V communication for CAVs was enabled through the CACC module, allowing for information exchange to minimize the distance to the vehicle ahead and improve overall traffic efficiency and safety [17,30]. The Gipps' car-following model analyzed vehicle behavior by measuring the spacing and timing between cars [31]. In turn, the lane-changing model assessed the available gaps for safe lane changes, while gap-acceptance models evaluated driver decision-making at intersections based on kinematic parameters, determining the feasibility of maneuvers in varying traffic conditions.

Before booting the calibration process, a sensitivity analysis was conducted to evaluate how changes in model parameters for each intersection affected the simulation results. Regardless of the intersection model considered, the sensitivity analysis primarily involved global parameters affecting all vehicles in the network and specific vehicle class parameters targeting distinct road user categories. Figure 7 summarizes the sensitivity analysis results for the case study, illustrating the variation of entering traffic within the range of the most sensitive parameter values.

The most sensitive parameters were then calibrated by iteratively adjusting their values to closely align the simulated capacity data with the RCFs, ensuring a more accurate representation of traffic behavior in the model. Notably, among the global modeling parameters, the reaction time of moving vehicles was identified as the most sensitive across all defined models.

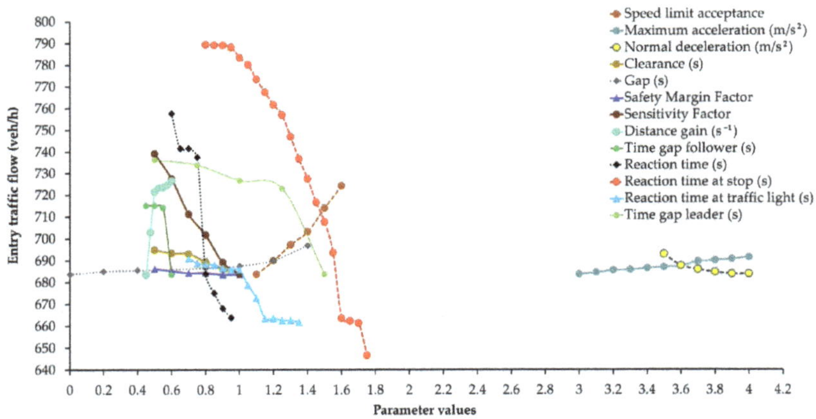

Figure 7. The graphic development of the sensitivity analysis of the model parameters.

This sensitivity also applied to reaction times at stops and traffic lights, impacting vehicles in stationary conditions significantly. The reaction time, measured in seconds, is a parameter in car-following models that reflects the time taken by a vehicle to respond to speed variations of the vehicle ahead [17]. If fixed, this parameter aligns with the simulation step for all users' classes; however, it can also be defined as a variable based on a probability function tailored to the specific characteristics of each user class, enhancing model accuracy.

The reaction time at a stop refers to the duration a queuing vehicle takes to initiate movement once the preceding vehicle restarts. Similarly, the reaction time at traffic lights occurs at signalized intersections and represents the duration taken by the first vehicle at the stop line to accelerate in response to the green signal.

Table 2 shows the calibrated values of the model parameters for the signalized intersection and the two-lane roundabout.

Table 2. Calibrated values of the model parameters for the signalized intersection and the two-lane roundabout.

Model Parameters	Default Values	Tuned-up Model Parameter Values							
		Signalized Intersection				Two-Lane Roundabout			
		L		R&T		LEL		REL	
		VHDs	CAVs	VHDs	CAVs	VHDs	CAVs	VHDs	CAVs
Speed limit acceptance	1.10	1.30	1.60	1.30	1.60	0.97	1.10	1.00	1.10
Maximum acceleration (m/s^2)	3.00	3.20	3.40	3.20	3.40	3.00	4.00	3.00	4.00
Normal deceleration (m/s^2)	4.00	4.00	3.50	4.00	3.50	4.00	4.00	4.00	4.00
Clearance (s)	1.00	1.00	0.50	1.00	0.50	1.00	1.00	1.00	1.00
Gap (s)	0.00	0.00	0.00	0.00	0.00	1.33	0.00	1.58	0.00
Safety Margin Factor	1.00	1.00	1.00	1.00	1.00	1.00	0.50	1.00	0.50
Sensitivity Factor	1.00	1.00	1.00	1.00	1.00	1.00	0.50	1.00	1.00
Distance gain (s^{-1})	0.45	0.45	0.60	0.45	0.60	0.45	0.45	0.45	0.45
Time gap leader (s)	1.50	1.50	0.50	1.50	0.50	1.50	1.50	1.50	1.50
Time gap follower (s)	0.60	0.60	0.50	0.60	0.50	0.60	0.60	0.60	0.60
Reaction time (s)	0.80	0.85 [1]	0.62 [2]	0.80 [1]	0.60 [2]	0.95 [1]	0.67 [2]	0.86 [1]	0.63 [2]
Reaction time at stop (s)	1.20	1.75 [1]	1.06 [2]	1.60 [1]	0.82 [2]	1.20	1.20	1.20	1.20
Reaction time at traffic light (s)	1.60	1.36 [1]	1.02 [2]	1.20 [1]	0.73 [2]	1.60	1.60	1.60	1.60

Note: The reaction time is the weighted mean of the corresponding values based on CAV-MER, with each MER used as a weight: (1) 0% of CAVs, (2) 100% of CAVs. Based on my own research using the default data presented in [17].

The parameters for calibrating both VHDs and CAVs included speed limit acceptance, which refers to the driver's compliance with speed limits, and maximum acceleration (m/s^2), representing the highest acceleration that a vehicle can achieve within each network model, consistent with the geometry or traffic control mode of the intersection type. The speed acceptance values greater than one mean that vehicles can assume maximum speeds exceeding the speed limit, otherwise below the speed limit, thus denoting a more cautious driving behavior. Consistent with the traffic control mode, VHDs and CAVs at signalized intersection assumed a decisive driving behavior, since both the speed limit acceptance and maximum acceleration parameters resulted in higher values than the default values of 1.10 and 3.00 m/s^2, respectively (see Table 2). On the contrary, the speed limit acceptance was set below the default threshold for the VHDs at the two-lane roundabout, while the maximum acceleration was maintained as the default value. Differently from [11], although the curved geometry forced CAVs to adhere to the speed limit at the examined two-lane roundabout, they adopted an assertive driving behavior, assuming maximum acceleration values higher than the default one, thanks to the V2V communication. Further parameters that were highly sensitive in the context of CAVs operating at the signalized intersection included normal deceleration (m/s^2) and clearance (s). The former denotes the maximum deceleration employed by any vehicle in typical driving situations, while the latter indicates the distance between two stationary vehicles. These two parameters were found to have less effect on the calibration of both vehicle classes in roundabouts, where the dominant factor influencing driving behavior is the effect induced by the curvilinear geometry of the layout [11]. A parameter typically tuned for CAVs operating at roundabouts is the safety margin factor, which determines vehicle priorities at intersections [11,17]. It was set to 0.50 for both roundabout entry lanes, replacing the default value of 1.00, while the default value remains in effect for the cautious driving behavior of VHDs at roundabout entry lanes [11]. The adjusted value of the safety margin employed here is consistent with the recommendations by Aimsun [17], where this parameter can be adjusted for a specific maneuver to reflect the road geometry under examination. The set of modeling parameters for the traffic roundabout scenarios with CAVs included the sensitivity factor, which relates to the distance between vehicles based on the deceleration assessments of the following vehicle to estimate the leader's deceleration, collectively modeling CAVs' driving behavior at the roundabout. Aimsun [17] allows you to adjust the vehicle headway distribution to indicate cautious driving or assertive driving. Therefore, a sensitivity factor greater than 1.00 indicates cautious driving, while a value below 1.00 reflects assertive driving (see Table 2). The value tuned for CAVs was a compromise to simulate changes in driving behavior or interactions among different vehicles on curved paths to evaluate CAV driving skills in mixed traffic conditions [17]. This parameter was set to 0.50 for the LEL, replacing the default value of 1.00, while the default value remains in effect for the cautious driving behavior of VHDs at roundabout entry lanes. In turn, the gap parameter, which measures the distance in seconds between the front bumpers of consecutive vehicles, was calibrated for VHDs as representative of their behavior at roundabouts, while the default value remains in effect for CAVs.

In CAV modeling, distance gain (s^{-1}), time gap leader (s), and time gap follower (s) were used to represent the spatial and temporal distances between vehicles equipped with the CACC module while queuing. Similar considerations were found in [2,32] to approximate the driving behavior of CACC-CAVs during cruising. Aimsun allows the analyst to set the distribution of VHDs and the percentage of CACC-CAVs in the respective tabs [17]. Consequently, the aforementioned parameters—characteristic of the CAV guide—have been kept at their default values for VHDs. Additionally, these parameters were found to have a greater influence on signalized intersections than on roundabouts, where the dominant factor affecting driving behavior is primarily governed by the curvilinear geometry of the layout and its smoothing effect.

Finally, the calibrated reaction times are shown in Table 2. In CAV modeling, the reaction time is set to 0.62 s for the L group and 0.60 s for the R&T group, while it is set to 0.67 s for the LEL and 0.63 s for the REL, replacing the default value of 0.80 s across all lanes. Additionally, the reaction times at stop and traffic lights commonly tuned for signalized intersections, were maintained at their respective default values for both CAVs and VHDs in both entry lanes at roundabouts [17].

The calibration results in Table 2 highlight the assertiveness of CAVs and VHDs at the signalized intersection, where speed acceptance and maximum acceleration values exceeded their respective default thresholds [16]. Conversely, for VHDs at the roundabout, the speed acceptance value fell below the default value, while the reaction time surpassed the predefined threshold, indicating slightly risk-averse behavior. CAVs demonstrated a more decisive driving attitude at the examined roundabout, reacting more readily to the leader's actions with higher acceleration than default values and slower reaction times and safety margins below the thresholds. The tuned CACC-CAV parameters enabled these vehicles to use intra-vehicle communication more effectively, allowing them to drive decisively at the signalized intersection compared to the roundabout, which typically accommodates smoother traffic flow.

The Geoffrey index (GEH) measured the deviation between simulated data and RCFs, aiming for discrepancies of less than 5 in at least 85% of cases [16]. The percentage difference between the two datasets indicated calibration success and confirmed the validity of the calibration process. The GEH index was used in the calibration of the two-lane roundabout by comparing simulated data with the RCFs as CAV-MERs increased, achieving over 90% accuracy for the LEL and REL. Additionally, the accuracy of the signalized intersection calibration was determined by a 5% percentage difference between simulated capacity and RCFs.

Table 3 presents statistics for the lanes' groups (i.e., L and R&T) at the signalized intersection. The t-statistic indicated no significant difference between the means of simulated data and RCFs, while the F-statistic confirmed equal sample variances. The root mean squared normalized error (RMSNE) demonstrated close alignment between simulated data and capacity values, corroborating the calibration process and the models' effectiveness in replicating the examined phenomenon. Similar promising results were observed for both lanes (i.e., LEL and REL) of the roundabout, as shown in Table 4.

Table 3. Summary statistics comparing the simulated capacity values and RCFs for different CAV-MERs at the signalized intersection.

Statistics	CAV-MERs by Lane Group (%)											
	0		20		40		60		80		100	
	L	R&T	L	R&T	L	R&T	L	R&T	L	R&T	L	R&T
Mean $_1$	362.22	566.67	371.00	594.22	395.33	648.22	417.44	690.22	449.33	760.11	551.78	858.33
s.e. $_1$	62.04	61.86	62.78	64.70	64.66	69.84	70.77	72.19	74.85	79.13	91.94	89.39
Mean $_2$	367.54	574.80	371.21	605.05	393.27	650.43	407.97	680.68	444.72	771.44	573.36	877.32
s.e. $_2$	60.66	60.91	61.27	64.12	64.91	68.93	67.33	72.13	73.40	81.75	94.63	92.97
95% c.i. *	(−189.3, 178.6)	(−192.2, 175.9)	(−186.2, 185.7)	(−203.9, 182.4)	(192.2, 196.3)	(−210.4, 205.7)	(−197.6, 216.6)	(−206.7, 226.0)	(−217.6, 226.9)	(−252.6, 229.9)	(−301.3, 258.1)	(−292.4, 254.4)
t-value	−0.06	−0.09	−0.002	−0.12	0.02	−0.02	0.10	0.09	0.04	−0.10	−0.16	−0.15
t-critical	2.12	2.12	2.12	2.12	2.12	2.12	2.12	2.12	2.12	2.12	2.12	2.12

Table 3. *Cont.*

Statistics	CAV-MERs by Lane Group (%)											
	0		20		40		60		80		100	
	L	R&T	L	R&T	L	R&T	L	R&T	L	R&T	L	R&T
$p(\alpha)$value	0.95	0.93	0.98	0.91	0.98	0.98	0.92	0.93	0.96	0.93	0.87	0.88
F-statistic	1.05	1.03	1.05	1.02	1.01	1.03	1.10	1.01	1.04	1.07	1.06	1.08
F-critical	3.44	3.44	3.44	3.44	3.44	3.44	3.44	3.44	3.44	3.44	3.44	3.44
F-prob	0.95	0.97	0.95	0.98	0.99	0.97	0.89	0.98	0.96	0.93	0.94	0.91
RMSNE	0.04	0.02	0.02	0.03	0.02	0.01	0.03	0.02	0.02	0.02	0.04	0.02

* c.i. stands for confidence interval. Note: Mean $_{1,2}$ (s.e. $_{1,2}$) are referred to each mean value and standard error of the two groups of data (namely the simulated data and RCFs).

Table 4. Summary statistics comparing the simulated capacity values and RCFs for different CAV-MERs at the two-lane roundabout.

Statistics	CAV-MERs by Entry Lane (%)											
	0		20		40		60		80		100	
	LEL	REL	LEL	REL	LEL	REL	LEL	REL	LEL	REL	LEL	REL
Mean $_1$	774.50	821.69	745.93	876.45	786.75	943.93	881.15	1046.77	980.67	1128.53	1067.83	1211.58
s.e. $_1$	57.99	61.22	56.60	63.20	58.35	65.86	62.70	71.86	67.64	72.67	71.40	72.09
Mean $_2$	731.11	810.41	738.00	883.28	799.77	954.29	887.45	1019.71	1001.66	1060.85	1107.11	1115.14
s.e. $_2$	67.97	72.37	66.28	75.90	68.27	77.33	72.47	79.67	78.52	79.79	85.92	77.71
95% c.i. *	(−135.3; 222.0)	(−178.8; 201.4)	(−165.9; 181.8)	(−204.8; 191.2)	(−192.2; 166.1)	(−214.0; 193.3)	(−197.4; 184.8)	(−188.1; 242.2)	(−227.7; 185.7)	(−148.7; 284.1	(−262.1; 183.5)	(−116.1; 309.0)
t-value	0.51	0.12	0.13	0.07	0.15	0.10	0.11	0.25	0.20	0.63	0.35	0.91
t-critical	2.000	2.006	1.995	2.007	1.995	2.006	1.995	2.006	1.995	2.005	1.995	2.005
$p(\alpha)$value	0.63	0.91	0.93	0.95	0.88	0.92	0.95	0.80	0.84	0.53	0.73	0.37
F-statistic	1.37	1.40	1.37	1.44	1.37	1.38	1.34	1.23	1.35	1.21	1.45	1.16
F-critical	1.76	1.91	1.76	1.91	1.76	1.91	1.76	1.91	1.76	1.91	1.76	1.91
F-prob	0.38	0.40	0.35	0.35	0.36	0.41	0.40	0.60	0.38	0.63	0.28	0.70
RMSNE	0.13	0.16	0.09	0.14	0.07	0.10	0.07	0.08	0.06	0.11	0.07	0.13

* c.i. stands for confidence interval. Note: Mean $_{1,2}$ (s.e. $_{1,2}$) are referred to each mean value and standard error of the two groups of data (namely the simulated data and RCFs).

The scatter plots in Figure 8 drive the validation of the analogy in entry mechanisms for the roundabout and knee-turbo roundabout (W-N approach). By way of example, Figure 8a displays capacity data for the LEL, and Figure 8b shows the REL with 40% CAVs and 60% VHDs. The narrow prediction intervals and high R^2 values observed across the different cases indicated a strong positive correlation between the datasets for both roundabout types, thereby supporting the extension of the two-lane roundabout calibration to the knee-turbo roundabout model.

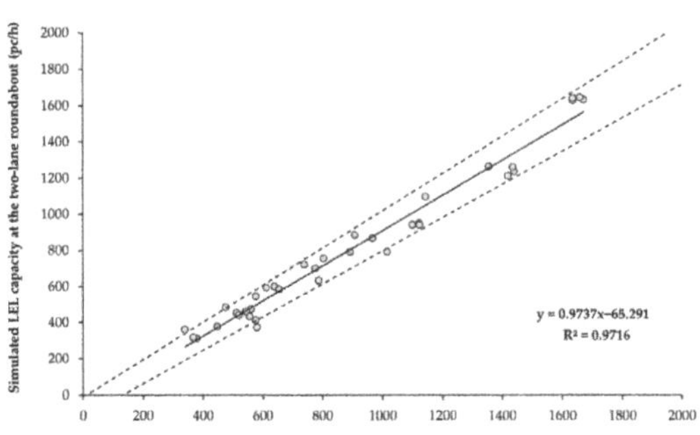

(a)

Figure 8. *Cont.*

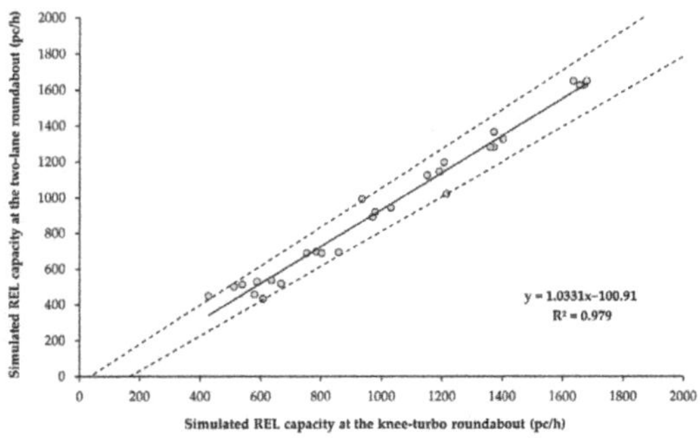

(**b**)

Figure 8. Scatter plots at 40% CAVs and 60% VHDs comparing the capacities of the simulated two-lane and the knee-turbo roundabouts for (**a**) the left entry lane (LEL); (**b**) the right entry lane (REL). Note: the acronyms are provided in the legend.

3. Results of the Simulation Experiments

A microsimulation approach was deemed to assay the efficiency of CAVs and VHDs across different intersection layouts and traffic scenarios as part of the move towards automated smart mobility. CAVs were assumed to operate in either realistic traffic scenarios with perfect communication conditions [30]. Entry capacity (pc/h), delay (s), and travel time (s) were selected as the operational performance metrics for simulation trials across six traffic scenarios with varying CAV-MERs. Capacity represents the maximum number of vehicles entering a road section, while delay indicates the difference between the expected travel time through the intersection and the actual time experienced by users without the intersection, averaged across all vehicles. Additionally, travel time reflects the average duration spent by all vehicles moving through the network [13,17].

Focusing on the W-N approach in Figures 2 and 5, Figure 9 compares the knee-turbo roundabout, two-lane roundabout, and signalized intersection regarding the percentage differences in entry capacity by entry lane or lane group (i.e., LEL and L capacities in Figure 9a and REL and R&T capacities in Figure 9b) between the baseline setting with 0% CAVs and traffic settings with progressively increasing CAV-MERs.

Overall, the simulation results showed that operational conditions, from low to high conflicting flows approaching entry capacity, improved with increasing CAV-MERs across all examined intersections. The bar graph in Figure 9a concerning LEL and L capacities illustrates higher percentage increases in the LEL capacity at roundabouts for CAV-MERs ranging from 20% to 60%, compared to traffic settings with CAV-MERs of 80% and above. In turn, with CAV-MERs of 80% the signalized intersection exhibited a rapid increase in percentage capacity of approximately 34% for the considered lane group, followed by the corresponding entry lanes of the knee-turbo and two-lane roundabouts, with capacity growths of 27% and 26%, respectively. When only CAVs travelled on the network (see Figure 9a), the considered lane group at the signalized intersection showed a percentage capacity growth of 58% compared to the baseline scenario with only VHDs, followed by the knee-turbo roundabout with an entry lane capacity growth of 41%. Meanwhile, the LEL capacity at the two-lane roundabout maintained a similar growth trend as observed in the previous cases.

In turn, Figure 9b concerning REL and R&T capacities displays the percentage increase compared to the baseline scenario at the three examined intersections. The bar graph in Figure 9b indicates that the two-lane roundabout experienced a higher percentage REL capacity growth compared to the R&T capacity at the signalized intersection for CAV-MERs ranging from 20% to 40%. Conversely, for CAV-MERs of 60% to 100%, the signalized intersection exhibited a more pronounced increase in capacity compared to the roundabouts, where operational performance generally depends on traffic flow matrices [12,33]. Moreover, the L group at the signalized intersection, designed for a specific traffic movement, generated higher capacity increments compared to the R&T, particularly at elevated CAV-MERs (Figure 9a,b). The presence of curbs in the knee-turbo roundabout may enhance LEL capacity, particularly at high CAV-MER values, facilitating improved traffic flow and movement efficiency; however, these curbs can also hinder the capacity increase for the REL case as CAV-MERs rise by preventing traffic from spreading across two lanes of the circulatory roadway. However, the presence of traffic lights, which eliminate the yielding negotiation process, allows for higher capacity compared to the baseline scenario, particularly when CAV-MERs reach 100%.

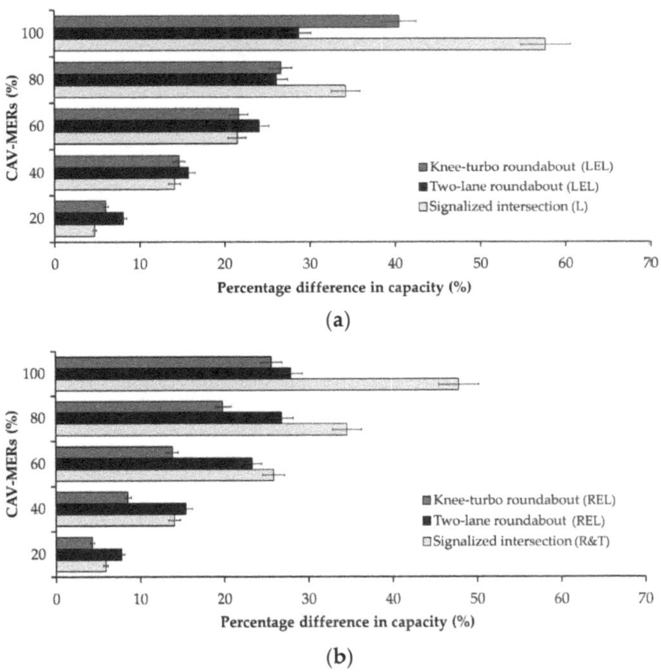

Figure 9. Comparison among the knee-turbo roundabout, two-lane roundabout, and signalized intersection in terms of percentage differences in capacity (pc/h) across various CAV-MERs for (**a**) the left-entry lanes (LELs) of two-lane and knee-turbo roundabouts and exclusive left-turn lane (L) of the signalized intersection; (**b**) the right-entry lanes (RELs) of two-lane and knee-turbo roundabouts and the shared right-turn and through lane (R&T) of the signalized intersection. Note: the acronyms are provided in the legend.

Based on traffic data collected in the field for the signalized intersection, an unbalanced traffic matrix was also used to simulate undersaturated conditions at the conceptualized intersections. Figure 10 displays a comparison among the three intersections (i.e., LEL and L delays in Figure 10a and REL and R&T delays in Figure 10b), highlighting the percentage differences in delay (s) for the CAV-MERs. The figure illustrates a larger percentage reduction in delay for both entry lanes at the knee-turbo roundabout compared to other intersections. The LEL delay at the knee-turbo roundabout in Figure 10a decreases by 40% to 65%, with CAV-MERs ranging from 40% to 100%. Similarly, the REL delay in Figure 10b decreases by 9% to 46%, with CAV-MERs ranging from 20% to 100%.

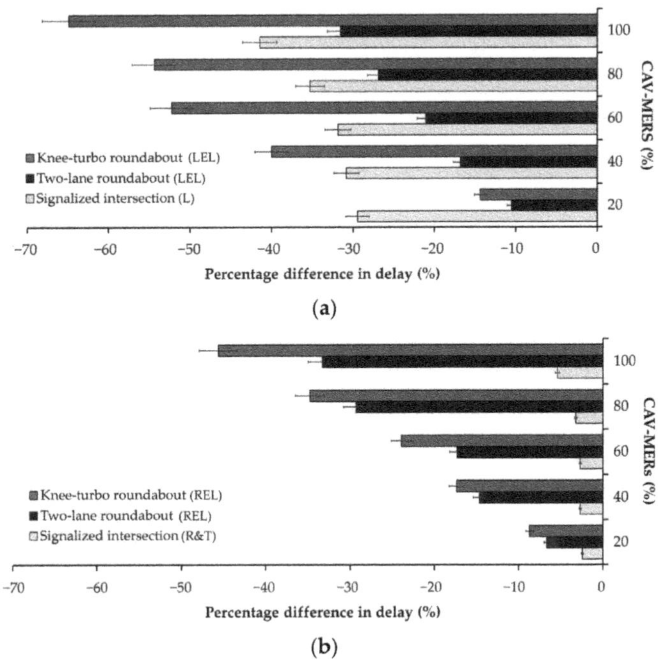

Figure 10. Comparison among the knee-turbo roundabout, two-lane roundabout, and signalized intersection in terms of percentage differences in delay (s) across various CAV-MERs for (**a**) the left-entry lanes (LELs) of the two-lane and knee-turbo roundabouts and the exclusive left-turn lane (L) of the signalized intersection; (**b**) the right-entry lanes (RELs) of the two-lane and knee-turbo roundabouts and the shared right-turn and through lane (R&T) of the signalized intersection.

The travel time trends for the entry lanes are illustrated in Figure 11 (i.e., LEL and L travel times in Figure 11a and REL and R&T travel times in Figure 11b). Overall, as CAV-MERs increase, the knee-turbo roundabout demonstrated a significant percentage reduction in travel time for both entry lanes compared to the other intersections across nearly all traffic settings. Figure 11a indicates that, for a CAV-MER of 20%, the signalized intersection experienced a percentage decrease in L travel time of 22%, compared to 10% for the corresponding entry lane in the knee-turbo roundabout. Percentage decreases in travel time of approximately 23% were observed for the considered lane (LEL) at the knee-turbo roundabout and lane group (L) at the signalized intersection for a CAV-MER of 40% (see Figure 11a). Additionally, the percentage reduction in travel time compared to the baseline scenario of VHDs is more pronounced for the REL at the knee-turbo roundabout, fluctuating from 7% to 31% for CAV-MERs starting from 20% onwards, followed by the REL at the two-lane roundabout and the R&T signalized intersection (Figure 11b).

Despite the similar decreases in delay and travel times across the three intersections, the knee-turbo roundabout geometry had a greater impact on the LEL performance than on the REL one; similarly, at the signalized intersection, the same effect was observed but differed by entry lane or lane group (i.e., L compared to R&T) due to lane group planning and the traffic control mode. In contrast, lane-changing behavior balanced the decrease in delay for both lanes of the two-lane roundabout.

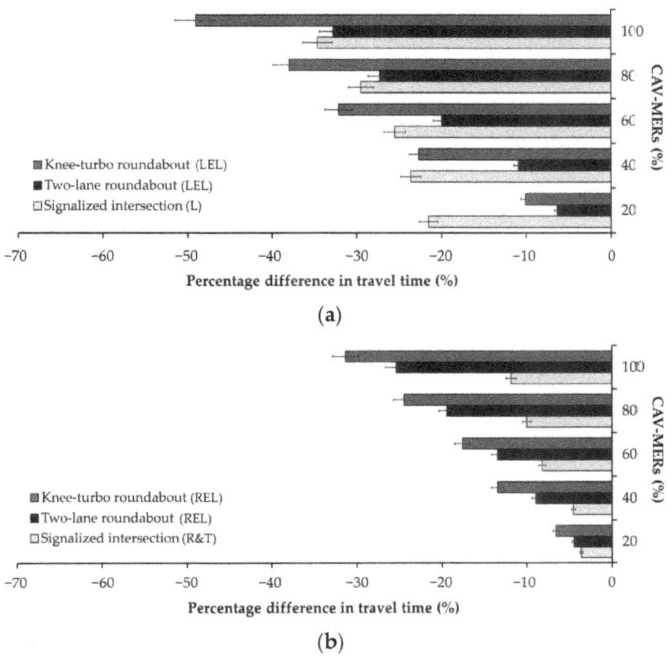

Figure 11. Comparison among the knee-turbo roundabout, two-lane roundabout, and signalized intersection in terms of percentage differences in travel times (s) across various CAV-MERs for (**a**) the left-entry lanes (LELs) of the two-lane and knee-turbo roundabouts and the exclusive left-turn lane (L) of the signalized intersection; (**b**) the right-entry lanes (RELs) of the two-lane and knee-turbo roundabouts and the shared right-turn and through lane (R&T) of the signalized intersection. Note: the acronyms are provided in the legend.

4. Discussion

Despite extensive discussions on CAV–infrastructure interaction, it is not always possible to directly transfer results obtained from other road infrastructures, such as road segments, to intersections [1,2,6,34]. A similar issue may arise concerning the specific objectives or methods proposed in the research—particularly related to the nodes of the road network—which can often hinder immediate comparisons, especially within the framework of smart cities [5,25,26]. Although the literature provides valuable insights, this is often because the design characteristics and vehicle dynamics of intersections and roundabouts are more complex, differing significantly from linear road sections due to conflict points or being influenced by the installation context. These aspects necessitate tailored approaches to design, management, and optimization [13]. Consequently, there is a proliferation of different case studies (e.g., [11,19,20,24,25]) which, while not always producing generalizable results, provide methodological pathways that can serve as valuable references for future research.

In this context, the study aimed to investigate how traffic scenarios involving connected and automated mobility can be channeled into the design of road intersections, thereby supporting the sustainable integration of smart vehicles into current and future road infrastructure.

Given that the current rate of replacing older vehicles with newer ones has not yet kept pace with advancements in the automotive industry, this study aims to support the real-world deployment of CAV technologies by emphasizing the importance of pilot studies to tackle practical and technical challenges. This approach aligns with the evolying needs of road engineering and underscores the critical role of ITS in facilitating this transition [1,2,34].

The results indicated significant traffic improvements with CAVs at the examined intersections (see Figures 9–11). As the transition to fully automated smart mobility advances, various traffic scenarios with increasing CAV-MERs were modeled in a microsimulation environment [17]. According to [13], due to the limited availability of smart vehicles for empirical analysis, RCFs for increasing CAV-MERs were developed to meet calibration requirements. To achieve optimal alignment between the simulated capacity outputs and RCFs, more sensitive driving behavior parameters were fine-tuned based on the sensitivity analysis for the signalized intersection and the two-lane roundabout, as illustrated in Figure 7. The calibration results in Table 2 indicated that CAVs and VHDs at the signalized intersection exhibited assertive behaviors, with speed acceptance and maximum acceleration exceeding thresholds. At the roundabout, VHDs demonstrated cautious behavior, with lower speed acceptance and reaction times above the thresholds. Conversely, CAVs at the roundabout displayed more decisive driving, reacting quickly with higher acceleration and lower reaction times and safety margins, facilitated by effective intra-vehicle communication through calibrated CACC-CAV parameters. Although the roundabout generally facilitated smoother traffic flow, V2V-enabled CAVs navigated the signalized intersection assertively. Statistics subsequently confirmed the accuracy of the calibration process (see Tables 3 and 4). The results were promising in matching the two datasets for each entry lane or lane group at the signalized intersection and two-lane roundabout, thereby addressing key question 1 in Section 1. As a result, the Aimsun models were considered carefully calibrated for replicating the studied phenomenon. A validation-driven approach was then proposed to assess the potential applicability of the two-lane roundabout calibration to the proposed knee-turbo counterpart, as both roundabouts featured similar entry mechanisms by approach. The proposed methodology ensured the reliability of the calibration results across two different roundabout configurations, as demonstrated by the scattergram analysis of the simulated capacity data for the calibrated roundabout models in Figure 8. This highlights the potential to extend the calibration of the two-lane roundabout to other intersection designs with similar entry mechanisms.

Three performance metrics were chosen to evaluate how intersection patterns affect the operational efficiency of CAVs across different traffic settings. Cross-referencing the conceptualized intersection solutions, the results underscored that CAV operation depends on both the intersection layout and the CAV-MERs, thereby enabling us to address key question 2 in Section 1. CAVs are more likely to accept smaller traffic gaps than VHDs, leading to increased entry capacity and improved delays and travel times as CAV-MERs rise.

Specifically, simulation results demonstrated greater percentage increases in capacity for the signalized intersection compared to roundabouts at high CAV-MERs, across both entry lanes and lane groups. This suggests that at CAV-MER levels of 60% or higher, signalized intersections can accommodate more vehicles more effectively than roundabouts, whose operational performance typically depends on traffic matrices [12,13,33] (see Figure 9). The results in Figure 9 align with those of [20], where a signalized intersection was simulated in Vissim [35] as a testbed, using traffic scenarios based on varying market penetration rates of smart vehicles mixed with VHDs. Specifically, increasing the proportion of CAVs—expected to receive information about upcoming traffic light states to adjust their speeds accordingly—was shown to enhance intersection capacity. Additionally, the study revealed that the greatest increase in lane group capacity occurred in a traffic stream composed entirely of CAVs [20]. However, potential lane changes at the two-lane roundabout—where vehicles can switch lanes based on available gaps—facilitated increased REL capacity compared to the knee-turbo scheme (see Figure 9b).

Conversely, analyzing the simulation outcomes of the intersections operating in undersaturated traffic conditions, the signalized intersection experienced a smaller reduction in delay and travel times than roundabouts as the number of CAVs in traffic increased [7,12] (see Figures 10b and 11b). This was more evident for the R&T than for the other lane group at the signalized intersection, due to the traffic setup in lane groups and the control mode. The results on percentage differences in travel time across various CAV-MERs are consistent with the findings of Mohebifard and Hajbabaie [24] at a single-lane roundabout in the United States. Although their study focused on trajectory control of CAVs, the authors [24] found that increasing the CAV-MERs from 20% to 100% reduced the travel times by 2.8% to 35.8%. Similar trends in vehicle delay times, compared to the case with 0% CAVs, were reported by [11,24] at single-lane and two-lane roundabouts, which were used as case studies to illustrate the methodologies proposed in each of these studies.

In this context, the research showed that CAV operations are affected by both the intersection layout and CAV-MERs. The signalized intersection outperformed roundabouts in capacity at high CAV penetration levels, while roundabouts offered better reductions in delay and travel times under undersaturated conditions (Figures 9–11). Although the journey time savings and delay time improvements observed in the simulation trials of the roundabout models are promising, they should be interpreted in the context of the traffic arrival distribution and the underlying hypotheses of the study. The simulation results also demonstrated that the intersection scheme affects connected and automated driving performance, with certain layouts showing a tendency to enhance the operational efficiency of smart mobility during the transition to full CAV adoption. However, due to the current absence of high-automation-level CAVs in traffic—limiting the availability of empirical data—the findings are inevitably influenced by the assumptions underlying the study. This is also consistent with the simulation results reported by [19], which indicated that, among intelligent vehicles, CACC-CAVs can outperform other intelligent vehicle systems in mixed traffic flows with VHDs at fixed and actuated signalized intersections. Additionally, simulations demonstrated reductions in average delay from 20% MPR, with performance improvements measured at actuated signal-controlled intersections under high traffic demand [19].

Although the impact of CAVs on traffic safety and efficiency in scenarios involving imperfect communication warrants further investigation [30], improvements in throughput and delays were documented by [25] in a single-lane roundabout case study used to demonstrate the capabilities of an expert system for managing conflicting vehicles. Fully CAV traffic increased the roundabout's capacity by 58–73%, and an optimized coordination strategy significantly reduced delays compared to VHDs. However, the authors emphasized the necessity of accounting for greater variability in CAV behaviors across different road scenarios to ensure that the results are broadly applicable.

Therefore, in response to the third question in Section 1, the authors emphasize that while the proposed validation-driven approach aided the comparison of different intersection design solutions based on the analogy of entry mechanisms by approach, a broader array of case studies should be further investigated to establish a comprehensive general performance criterion for effective comparison. Additionally, analyzing a larger series of intersection schemes and control modes would enhance the identification of context-specific factors, which could be incorporated into a robust performance criterion based on similar entry mechanisms. This also emphasizes the importance of developing novel assessment approaches for integration into practical roadway design workflows.

Based on the above, this research enhances understanding of how connected and automated mobility interacts with different intersection designs, emphasizing the importance of layout, operational parameters, and accurate behavioral modeling in traffic system analysis.

These findings help establish a theoretical framework linking CAV-MERs, intersection design features, and operational efficiency, offering a reference for future studies in ITS and mobility modeling under evolving traffic conditions.

Practically, the study offers valuable insights for transportation planners and road engineers, illustrating how various intersection configurations respond to increasing CAV market shares. The results suggest that signalized intersections may be better suited to accommodate higher levels of CAV penetration—from free-flowing conditions to capacity—aligning infrastructure investments and policy strategies aimed at optimizing traffic efficiency. Furthermore, the calibration methodology and validation framework outlined in the research serve as a replicable approach for evaluating different intersection solutions, supporting data-driven decision-making, and guiding the development of adaptable, future-proof road infrastructure aligned with smart mobility objectives. Acknowledging current limitations due to the scarcity of empirical CAV data, the study highlights the necessity of ongoing research, pilot projects, and adaptive infrastructure strategies to enable efficient traffic management in an era increasingly dominated by autonomous mobility.

5. Conclusions

This paper presents a validation-driven approach within a microsimulation environment to compare various intersection solutions and analyze how traffic settings involving connected and automated mobility can be channeled into road intersection design, thereby promoting the sustainable integration of smart vehicles into current and future road infrastructure. Focusing on comparing a signalized intersection with a two-lane roundabout and an innovative knee-turbo roundabout design under varying CAV-MERs and guided by the key questions posed in Section 1, this approach addresses practical challenges in the transition to smart mobility. An additional contribution is the proposal of a turbo roundabout design variant, tailored to accommodate the specific traffic and surrounding features of the case study context. Thus, the research bridges emerging CAV technologies and road engineering, contributing to demonstrating how intersection design can be optimized for mixed traffic environments towards fully autonomous mobility. It establishes a detailed microsimulation setup, sensitivity analysis, and calibration against RCFs, supported by field data from a signalized intersection at Opole, Poland. The proposed framework for calibration and cross-comparison aims to provide a methodological contribution, especially for practitioners assessing the feasibility of future intersection models before detailed design work begins, which requires further technical elaboration.

Results reveal that CAV operations can be influenced by intersection layout and CAV-MERs: signalized intersections tend to outperform roundabouts in capacity at high CAV levels, whereas roundabouts more effectively reduce delays under undersaturated traffic conditions. These insights offer valuable guidance for planners and road engineers, assisting in design decisions and providing a prospective outlook on scenarios leading to fully smart navigation. This aligns with evolving road engineering needs and underscores the role of ITS in this transition [1,2,34].

While the findings lay the groundwork for future studies across different scales and urban contexts, they are limited by the hypotheses, assumptions, and core concepts underpinning the simulation environment. Nonetheless, the literature within the smart cities context supports results aligned with this study's objectives (e.g., [11,19,20,24,25]). Given the scarcity of empirical data on high communication-level smart vehicles, assumptions about CAV behavior remain vital for evaluating operational potential [1,20]. Despite current data limitations, the study emphasizes the importance of ongoing research, pilot projects, and adaptive infrastructure strategies to support efficient traffic management in the era of autonomous mobility. However, this research sets a precedent for systematically assessing a broader range of design alternatives, including advanced technological solutions like different traffic control strategies and adaptive signaling, to evaluate sustainability across design options. The approach can facilitate the assessment of CAV efficiency relative to road geometry and driving behavior shifts, as well as develop expertise in comparing diverse intersection proposals involving CAVs. Ultimately, the findings offer valuable insights into the operational benefits of smart mobility in relation to road geometry, while the methodology provides transportation planners and engineers with alternative tools to evaluate intersection designs aligned with the gains expected from smart mobility.

To generalize results and accurately model transition scenarios, future research should extend investigations of CAV driving behavior on a wider scale. Collecting real-world behavioral data will enhance robustness, applicability, and precision. Additionally, validation should go beyond the entrance mechanism analogy for isolated intersections, encompassing other control modes and traffic demands across diverse urban contexts. This would support broader generalization across road networks and traffic flow structures.

Further, future studies should consider incorporating greater behavioral variability of CAVs—not only at fixed signalized intersections and roundabouts but also at actuated intersections—to reflect operational realities more accurately. Achieving full integration of CAVs into road design assessment practices will require collaborative efforts among stakeholders involved in smart mobility development. Future directions include expanding empirical research on CAV behavior and creating frameworks for implementing innovative intersection designs that incorporate smart mobility advancements throughout their lifecycle, ultimately fostering sustainable transportation practices.

Author Contributions: Conceptualization, M.L.T., N.Z., E.M. and A.G.; methodology, M.L.T., N.Z., E.M. and A.G.; software, M.L.T.; validation, M.L.T.; formal analysis, M.L.T. and E.M.; investigation, M.L.T., E.M. and A.G.; resources, M.L.T., E.M. and A.G.; data curation, M.L.T., N.Z., E.M. and A.G.; writing—original draft preparation, E.M. and A.G.; writing—review and editing M.L.T., N.Z., E.M. and A.G.; visualization, M.L.T., E.M. and A.G.; supervision, E.M. and A.G.; project administration, M.L.T., E.M. and A.G.; funding acquisition M.L.T., E.M. and A.G. All authors have read and agreed to the published version of the manuscript.

Funding: This research received no external funding.

Data Availability Statement: The data are made available by the corresponding authors upon kind request.

Acknowledgments: This publication is supported by the Rector's Professor Grant, Silesian University of Technology, Poland, grant number 12/040/SDU/10-07-01.

Conflicts of Interest: The authors declare no conflicts of interest.

Abbreviations

The following abbreviations are used in this manuscript:

ITS	Intelligent Transport Systems
CAVs	Connected and Automated Vehicles
VHDs	Vehicles with Human Drivers
MERs	Market Entry Rates
CAV-MERs	Market Entry Rates for CAVs
RCFs	Reference Capacity Functions
CACC	Cooperative Adaptive Cruise Control
V2V	vehicle-to-vehicle
L	exclusive left-turn lane
R&T	shared right-turn and through lane
LEL	left entry lane

REL	right entry lane
N-E	north-east
E-S	east-south
S-W	south-west
W-N	west-north
CACC-CAVs	Connected and Automated Vehicles with the Cooperative Adaptive Cruise Control system
GEH	Geoffrey index
s.e.	standard error
RMSNE	root mean squared normalized error

References

1. Pompigna, A.; Mauro, R. Smart Roads: A State of the Art of Highways Innovations in the Smart Age. *Eng. Sci. Technol. Int. J* **2022**, *25*, 100986. [CrossRef]
2. Makridis, M.; Mattas, K.; Mogno, C.; Ciuffo, B.; Fontaras, G. The impact of automation and connectivity on traffic flow and CO_2 emissions: A detailed microsimulation study. *Atmos. Environ.* **2020**, *226*, 117399. [CrossRef]
3. Matin, A.; Dia, H. Impacts of connected and automated vehicles on road safety and efficiency: A systematic literature review *IEEE Trans. Intell. Transp. Syst.* **2023**, *24*, 2705–2736. [CrossRef]
4. SAE International. SAE J3016 Automated-Driving Graphic. Available online: https://www.sae.org/news/2019/01/sae-updates-j3016-automated-driving-graphic (accessed on 30 March 2025).
5. Malik, S.; Khan, M.A.; El-Sayed, H.; Khan, M.J. Should Autonomous Vehicles Collaborate in a Complex Urban Environment or Not? *Smart Cities* **2023**, *6*, 2447–2483. [CrossRef]
6. Sadaf, M.; Iqbal, Z.; Javed, A.R.; Saba, I.; Krichen, M.; Majeed, S.; Raza, A. Connected and Automated Vehicles: Infrastructure Applications, Security, Critical Challenges, and Future Aspects. *Technologies* **2023**, *11*, 117. [CrossRef]
7. Ni, D. *Signalized Intersections*, 1st ed.; Springer: Cham, Swizerland, 2020; p. XV, 335. [CrossRef]
8. Tahiri, M.A.; Rachid, A.; Boudmane, B.; Mortabit, I.; Laaroussi, S. Toward Cooperative Adaptive Cruise Control: A Mini-review In Proceedings of the International Conference on Circuit, Systems and Communication (ICCSC), Fes, Morocco, 28–29 June 2024 pp. 1–6. [CrossRef]
9. Xu, L.; Ma, J.; Zhang, S.; Wang, Y. Car following models for alleviating the degeneration of CACC function of CAVs in weak platoon intensity. *Transp. Lett.* **2023**, *16*, 599–611. [CrossRef]
10. Karbasi, A.; O'Hern, S. Investigating the Impact of Connected and Automated Vehicles on Signalized and Unsignalized Intersections Safety in Mixed Traffic. *Future Transp.* **2022**, *2*, 24–40. [CrossRef]
11. Tumminello, M.L.; Macioszek, E.; Granà, A.; Giuffrè, T. Simulation-Based Analysis of "What-If" Scenarios with Connected and Automated Vehicles Navigating Roundabouts. *Sensors* **2022**, *22*, 6670. [CrossRef]
12. National Academies of Sciences, Engineering, and Medicine. *Guide for Roundabouts;* The National Academies Press: Washington DC, USA, 2023. [CrossRef]
13. National Academies of Sciences, Engineering, and Medicine. *Highway Capacity Manual: A Guide for Multimodal Mobility Analysis* 7th ed.; The National Academies Press: Washington, DC, USA, 2022. [CrossRef]
14. Fortuijn, L.G. Turbo roundabouts: Design principles and safety performance. *Transp. Res. Rec.* **2009**, *2096*, 16–24. [CrossRef]
15. Tollazzi, T. Alternative Types of Roundabouts at Development Phases. In *Alternative Types of Roundabouts. Springer Tracts on Transportation and Traffic*, 1st ed.; Springer: Cham, Swizerland, 2015; Volume 6, pp. 157–169. [CrossRef]
16. Barceló, J. *Fundamentals of Traffic Simulation*, 1st ed.; Springer: New York, NY, USA, 2010; p. 442. [CrossRef]
17. Aimsun Next. *Version 20 Dynamic Simulator User Manual*; TSS-Transport Simulation Systems: Barcelona, Spain, 2020.
18. Ahmed, H.U.; Ahmad, S.; Yang, X.; Lu, P.; Huang, Y. Safety and Mobility Evaluation of Cumulative-Anticipative Car-Following Model for Connected Autonomous Vehicles. *Smart Cities* **2024**, *7*, 518–540. [CrossRef]
19. Song, L.; Fan, W.; Liu, P. Exploring the effects of connected and automated vehicles at fixed and actuated signalized intersections with different market penetration rates. *Transp. Plan. Technol.* **2021**, *44*, 577–593. [CrossRef]
20. Hajbabaie, A.; Tajalli, M.; Bardaka, E. Effects of connectivity and automation on saturation headway and capacity at signalized intersections. *Transp. Res. Rec.* **2023**, *2678*, 31–46. [CrossRef]
21. Jiang, Y.; Cong, H.; Chen, H.; Yao, Z. Safety evaluation for mixed traffic flow of CAVs with different automation and connection levels. *Expert. Syst. Appl.* **2025**, *261*, 125561. [CrossRef]
22. Tong, H.; Xu, C.; Ai, Q.; Ren, W.; Wang, C.; Peng, C.; Jiao, Y. Developing a jam-absorption strategy for mixed traffic flow at signalized intersections using deep reinforcement learning. *Transp. Lett.* **2024**, 1–12. [CrossRef]

23. Li, D.; Zhu, F.; Wu, J.; Wong, Y.D.; Chen, T. Managing mixed traffic at signalized intersections: An adaptive signal control and CAV coordination system based on deep reinforcement learning. *Expert. Syst. Appl.* **2024**, *238 Pt C*, 121959. [CrossRef]
24. Mohebifard, R.; Hajbabaie, A. Trajectory control in roundabouts with a mixed fleet of automated and human-driven vehicles. *Comput.-Aided Civ. Infrastruct. Eng.* **2022**, *37*, 1959–1977. [CrossRef]
25. Martin-Gasulla, M.; Elefteriadou, L. Traffic management with autonomous and connected vehicles at single-lane roundabouts. *Transp. Res. C Emerg. Technol.* **2021**, *125*, 102964. [CrossRef]
26. Wu, Y.; Zhu, F. Junction management for connected and automated vehicles: Intersection or roundabout? *Sustainability* **2021**, *13*, 9482. [CrossRef]
27. Turbo Roundabouts Informational Primer. FHWA Safety Program. In *Federal Highway Administration*; U.S. Department of Transportation: Washington, DC, USA, 2022. Available online: https://highways.dot.gov/sites/fhwa.dot.gov/files/2022-06/fhwasa20019_0.pdf (accessed on 20 March 2025).
28. Gkyrtis, K.; Kokkalis, A. An overview of the efficiency of roundabouts: Design aspects and contribution toward safer vehicle movement. *Vehicles* **2024**, *6*, 433–449. [CrossRef]
29. Ministry of Infrastructure. Guidelines for the design of road intersections. In *Part 3: Roundabouts*; WR-D-31-3; Ministry of Infrastructure: Warsaw, Poland, 2022.
30. Garg, M.; Bouroche, M. Can Connected Autonomous Vehicles Improve Mixed Traffic Safety Without Compromising Efficiency in Realistic Scenarios? *IEEE Trans. Intell. Transp. Syst.* **2023**, *24*, 6674–6689. [CrossRef]
31. Gipps, P.G. A Behavioural Car-Following Model for Computer Simulation. *Transp. Res. Part B* **1981**, *15*, 105–111. [CrossRef]
32. Sadid, H.; Antoniou, C. Modelling and simulation of (connected) autonomous vehicles longitudinal driving behavior: A state-of-the-art. *IET Intell. Transp. Syst.* **2023**, *17*, 1050–1070. [CrossRef]
33. Mauro, R.; Branco, F. Comparative analysis of compact multilane roundabouts and turbo-roundabouts. *J. Transp. Eng.* **2010**, *136*, 316–322. [CrossRef]
34. Emami, A.; Sarvi, M.; Asadi Bagloee, S. A review of the critical elements and development of real-world connected vehicle testbeds around the world. *Transp. Lett.* **2020**, *14*, 49–74. [CrossRef]
35. PTV Planung Transport Verkehr AG. *PTV VISSIM User Manual*; PTV Planung Transport Verkehr AG: Karlsruhe, Germany, 2017.

smart cities

MDPI

Article

User Experience of Navigating Work Zones with Automated Vehicles: Insights from YouTube on Challenges and Strengths

Melika Ansarinejad [1,*], Kian Ansarinejad [2], Pan Lu [3] and Ying Huang [1,*]

1 Department of Civil, Construction and Environmental Engineering, North Dakota State University, Fargo, ND 58105, USA
2 Department of Computer Science, North Dakota State University, Fargo, ND 58105, USA; kian.ansarinejad@ndsu.edu
3 Department of Transportation, Logistics, and Finance, North Dakota State University, Fargo, ND 58105, USA; pan.lu@ndsu.edu
* Correspondence: melika.ansarinejad@ndsu.edu (M.A.); ying.huang@ndsu.edu (Y.H.)

Highlights

What are the main findings?

- Automated vehicles (AVs) show a near-equal mix of strengths and challenges in navigating real-world construction zones, with 117 strengths and 105 challenges documented from 38 user-recorded cases.
- Tesla's FSD system demonstrates adaptability and dynamic response in complex work zones, while Waymo shows better handling of human interactions and structured environments.

What is the implication of the main finding?

- Continued AV development requires improvements in perception of temporary infrastructure, interaction with human actors, and navigation in unmapped or irregular layouts.
- User-generated video content offers valuable insights into real-world AV behavior, informing infrastructure design, policy development, and public trust in AV deployment.

Abstract

Understanding automated vehicle (AV) behavior in complex road environments and user attitudes in such contexts is critical for their safe and effective integration into smart cities. Despite growing deployment, limited public data exist on AV performance in construction zones; highly dynamic settings marked by irregular lane markings, shifting detours, and unpredictable human presence. This study investigates AV behavior in these conditions through qualitative, video-based analysis of user-documented experiences on YouTube, focusing on Tesla's supervised Full Self-Driving (FSD) and Waymo systems. Spoken narration, captions, and subtitles were examined to evaluate AV perception, decision-making, control, and interaction with humans. Findings reveal that while AVs excel in structured tasks such as obstacle detection, lane tracking, and cautious speed control, they face challenges in interpreting temporary infrastructure, responding to unpredictable human actions, and navigating low-visibility environments. These limitations not only impact performance but also influence user trust and acceptance. The study underscores the need for continued technological refinement, improved infrastructure design, and user-informed deployment strategies. By addressing current shortcomings, this research offers critical insights into AV readiness for real-world conditions and contributes to safer, more adaptive urban mobility systems.

Keywords: work zone; automated vehicles; transportation infrastructure; lane closure; urban planning; object detection; construction zone

1. Introduction

Highway work zones in the United States have become increasingly hazardous for both workers and motorists. With numerous work zones established nationwide, transportation agencies are focusing on enhancing communication within these areas while also encouraging safer driving habits to reduce the risk of accidents [1]. Between 2013 and 2022, U.S. work zone fatalities rose by 50% [2]. In 2022 alone, crashes in these areas led to 891 deaths and 37,701 injuries. These incidents occurred within work zones or on roads leading to or from them due to traffic-related activities, behaviors, or controls. Of the 891 fatalities, 528 took place in construction zones, 305 were in work zones of unspecified type, 49 occurred in maintenance zones, and nine happened in utility zones [3]. The rapid advancement of automated vehicle (AV) technology holds great potential for transforming various aspects of urban and rural road transportation, including mobility, safety, and environmental impact under various road conditions [4,5]. According to the Society of Automotive Engineers (SAE), driving automation systems are classified into six levels, from Level 0 to Level 5, based on the degree of automation and the role of the human driver. Level 0 represents no automation, where the human driver performs all driving tasks. At Level 5, the system is fully automated, capable of handling driving tasks under all conditions without any human input. Levels 2 and 3 provide partial and conditional automation, respectively, where the vehicle can manage certain driving functions, but the human driver must remain attentive and ready to intervene. Levels 4 and 5 eliminate the need for driver intervention in specific (Level 4) or all (Level 5) driving environments. This taxonomy is essential for understanding the functional expectations and limitations of AV systems operating in complex scenarios such as work zones [6]. Work zones are expected to remain a challenging environment for automated vehicle (AV) systems due to their dynamic and unpredictable nature. These zones often involve temporary changes to lane configurations, vary in duration (ranging from short-term to long-term), may affect a single lane or multiple lanes, and can be present year-round. Additionally, irregular signage, pavement markings, fluctuating speed limits, and the presence of construction equipment and workers require AVs to exhibit heightened awareness and adaptability to navigate safely [1].

Numerous research studies have evaluated and are still assessing the performance, benefits, and challenges of connected and/or automated vehicles in urban and rural transportation settings using field tests, real-world data, simulations, surveys and interviews [7,8], or based on user experience on video sharing platforms [9]. Various datasets exist for AVs, including disengagement and incident data from the California Department of Motor Vehicles [10,11], NHTSA AV incident data [12], and Waymo's diverse automated driving datasets [13]. However, there is a significant lack of studies employing any systematic method aimed at assessing AV performance and rider experience, specifically in work zone environments; consequently, a shortage of publicly available datasets on this topic is observed. This is a crucial gap, as AVs are expected to account for a notable share of vehicles in U.S. cities by 2030 [14]. Understanding the challenges and potential of AVs in work zones, along with public perception, acceptance, and willingness to ride in AVs or along with AVs despite the possibility of encountering construction zones, is essential for guiding infrastructure owners and operators on necessary physical and digital modifications. This will inform manufacturers on required AV technology development and assist policymakers in establishing regulations that support the safe and efficient integration of AVs in future urban and rural environments, ensuring that individuals across different financial backgrounds, geographic locations, physical abilities, technological literacy levels, and health conditions not only have the economic means but also feel confident and willing to consider AVs as a viable transportation option.

This paper analyzes the performance of AVs navigating work zones, drawing on real-world case studies shared by users on YouTube between 2018 and 2024. The analysis focuses on strengths and challenges related to safety, mobility, and reliability to identify areas where AV technology and infrastructure can improve to better support future deployment. The study examines how two major AV systems, Tesla's Full Self-Driving (FSD) and Waymo, responded to diverse construction environments. It is worth noting that throughout this study, any reference to FSD refers to "supervised" Full Self-Driving systems. According to the SAE definitions of autonomy levels, there are currently no commercially available Level 5 vehicles; none that can operate under all conditions without human oversight or restriction.

Observations are organized into six key categories: Perception & Sensing, Decision Making & Planning, Control & Speed Management, Human Interaction & Override, Infrastructure Compliance & Navigation, and Environmental & Weather Conditions. In addition to technical performance, the study offers insights into user experience and perception. It explores how AV disengagements and errors are interpreted by different road users, including passengers, pedestrians, construction workers, and human drivers, and how these responses reflect moments of trust, frustration, and stress. The findings help evaluate whether AV systems like supervised FSD reduce the cognitive burden on drivers and riders in complex work zones or inadvertently intensify it. They also reveal whether users are more likely to trust the system or initiate disengagement, seek remote assistance, or even prefer human-driven vehicles in these situations. This study is guided by the following research questions:

(1) What types of challenges do AV users report encountering when navigating construction zones, and what strengths do they demonstrate in such complex environments?
(2) To what extent do current AV technologies meet user expectations regarding safety, reliability, and automated performance in work zones, and do the strengths outweigh the challenges?
(3) How do users respond to AV behavior in construction zones, and what do their reactions reveal about trust, acceptance, and willingness to adopt the technology in these contexts?
(4) What insights can be extracted from user-shared videos to guide future improvements in AV technology, transportation systems, urban infrastructure, and policy development related to AV deployment in work zones?

2. Materials and Methods

This study adopts a qualitative exploratory approach, incorporating elements of a scoping review to analyze user-shared YouTube videos and examine how AVs navigate construction zones, while also assessing user opinions about AV performance. The analysis focuses on videos recorded by individuals who were either AV users (riders/passengers) or external observers, such as drivers or passengers in nearby vehicles, or pedestrians who captured AV behavior in real-world settings. These publicly accessible videos provide valuable insights into AV performance, human-AV interaction, and user perception that are often absent from formal datasets or require the integration of multiple research methodologies to fully uncover.

The analysis process began by developing a series of keyword combinations aligned with the research objectives to find videos containing terms such as "Work Zone" or "Construction Zone" in the title, in conjunction with at least one keyword representing automated driving behavior, including "self-driving," "automated vehicle," "automated car," or "autopilot." Table 1 outlines the conceptual categories and associated synonyms and terms used in constructing the search queries. For selecting AV manufacturers, we focused on companies that are recognized as major players in the AV industry and have conducted frequent public road testing. Since the goal of this study was to compare AV performance and user experiences over time, we targeted companies for which user-documented experiences were available consistently across multiple years and covered various software versions, updates, and geographic locations across different U.S. states and, in some cases, other countries. Based on these criteria, Waymo and Tesla were selected as the primary focus in this study. Other AV manufacturers were considered during the video screening process. However, these companies did not have a sufficient volume of user-generated content that met our inclusion criteria.

Table 1. Key concepts and synonyms used for search query development.

Concepts	Synonyms/Related Terms
Automated vehicle	Self-driving car, automated vehicle, automated car, driverless vehicle
Work zone	Construction zone, roadwork, road construction
Navigation challenges	Lane shift, obstacle, detour, cones, temporary signage
Major AV companies with public road testing	Tesla, Waymo

In the initial phase of our methodology, video acquisition was executed through a purpose-written Python 3.11.9 script that interacts with the official YouTube Data API v3. For each search cycle, the script iterated through a predefined set of Boolean keyword strings (listed in Table 1) and submitted each query to the API with the "publishedAfter" and "publishedBefore" parameters set to 1 January 2018 and 31 December 2024, respectively. The upload date range was selected to capture both early and recent AV deployments, providing a longitudinal view of technological evolution and public response.

The API returned the title, publication date, and video identifier for each query. Two case-insensitive regular expressions were then applied: one confirming the presence of work-zone terminology (construction zone, work zone, roadwork), and the other ensuring reference to an AV concept (automated vehicle, self-driving, Tesla, Waymo, automated vehicle, or full self-driving). Items satisfying both patterns were retained, and a URL-based check was used to remove duplicates arising from overlapping queries. No restrictions on video duration were applied, allowing inclusion of all content types (including standard videos and shorts). As a result, the selected videos ranged from less than one minute to over 20 min in length. Next, all extracted videos were individually watched in full to ensure their relevance to the topic and alignment with the study's criteria and objectives. Some videos resulting from the search were manually excluded from the analysis, despite being titled as demonstrations of AV navigation in work zones and have relevant contents. The excluded videos met one or more of the following criteria: (1) they were released by AV companies, which were omitted to focus exclusively on user-generated content and ensure an unbiased representation of real-world experiences; (2) they lacked narration, captions, or subtitles, and were excluded to avoid potential misinterpretation by the authors. To facilitate the summarization and accurate interpretation of AV behavior and user experiences, the selected videos were then categorized into two groups for analysis: (1) those containing spoken narration (supported by captions, descriptions, or subtitles), and (2) those without spoken audio, relying on written captions or descriptions. For videos and shorts that included only on-screen text or descriptions without spoken audio, all content was read, analyzed, and summarized. For videos with spoken narration, transcripts were extracted using YouTube's built-in "Show transcript" feature to access the official, time-stamped dialog. Each line of speech was reviewed for emotional cues, such as confidence, hesitation, surprise or frustration, and descriptive comments regarding AV performance. User sentiment was inferred through a multimodal qualitative approach that combined three layers of analysis: verbal content, paralinguistic cues, and behavioral observations. Verbal content included emotionally charged phrases such as "I don't feel safe" or "that was smooth." Paralinguistic cues such as tone of voice, inflection, pacing, and audible signals like laughter or tension were used to interpret emotional states with greater accuracy. Behavioral observations, including hand movements, facial expressions, or manual disengagements, were also considered when available to support or clarify the emotional tone of the narration. A calm voice paired with a positive comment indicated confidence, while strained speech and physical signs of discomfort reflected anxiety or loss of trust. This approach ensured a comprehensive and reliable analysis across different content types.

In the next step, thematic analysis was conducted with the assistance of OpenAI's API to identify key themes related to the strengths and challenges presented in the collected references. This approach aligns with recent research that utilizes advanced large language models (LLMs) to analyze AV disengagement reports, demonstrating the models' effectiveness in identifying patterns, classifying causes, and informing infrastructure improvements based on complex, unstructured data [15]. Original transcripts and video summaries were programmatically analyzed to extract recurring patterns, with the API supporting pattern recognition and clustering, from which categories and subcategories of AV performance were developed.

To assess the accuracy of the model's thematic categorization of AV strengths and challenges, we conducted a manual evaluation using the full set of extracted events. From the dataset of 222 events (117 strengths and 105 challenges) identified across 38 user-shared videos, all events were manually reviewed and categorized based on their root causes through detailed analysis of video content, captions, and subtitles. We then compared the manually assigned root cause category to the one assigned by the model. A prediction was considered correct if the model's assigned category matched the manual classification. Out of the 117 strength events, the model's categorization matched the manual classification in 104 cases. For the 105 challenge events, the model matched the manual classification in 93 cases. This results in an overall accuracy of:

$$\text{Accuracy} = (\text{Number of Correct Predictions}/\text{Total Evaluated Events}) \times 100$$

$$\text{Accuracy} = ((104 + 93)/(117 + 105))) \times 100 = 88.67\%$$

This indicates that the model correctly categorized approximately 89% of the evaluated events. This process led to the establishment of six main thematic categories: Perception & Sensing, Decision Making & Planning, Control & Speed Management, Human Interaction & Override, Infrastructure Compliance & Navigation, and Environmental & Weather Conditions.

3. Results

This section synthesizes 38 user-documented experiences involving two AV companies, Waymo and Tesla, between 2018 and 2024. Figure 1 illustrates the distribution of YouTube video content related to AV navigation in work zones during this period. From 2018 to

2021, the percentage of available English-language video content remained relatively low and stable, with each year contributing approximately 5% of the overall distribution. This suggests that, in the early stages, AV navigation in work zones was either less frequently documented or had not yet become a major public focus. A notable rise occurred in 2020 with the share increasing to about 16%, likely reflecting expanded AV testing and growing efforts to demonstrate performance under complex conditions. Although a slight decline followed in 2021, content volume rebounded in 2022 to around 13%, coinciding with the resumption of broader testing activities after pandemic-related slowdowns. The most significant growth was observed in 2023, which accounted for nearly 29% of the total content, highlighting intensified AV activity in construction zones and reflecting technological advancements, heightened regulatory interest, and greater public scrutiny. In 2024, while the share slightly decreased to approximately 26%, it remained substantially higher than in earlier years, indicating sustained engagement and interest. The growth in video content underscores the increasing importance of evaluating AV performance in unpredictable work zone environments as the technology continues to mature.

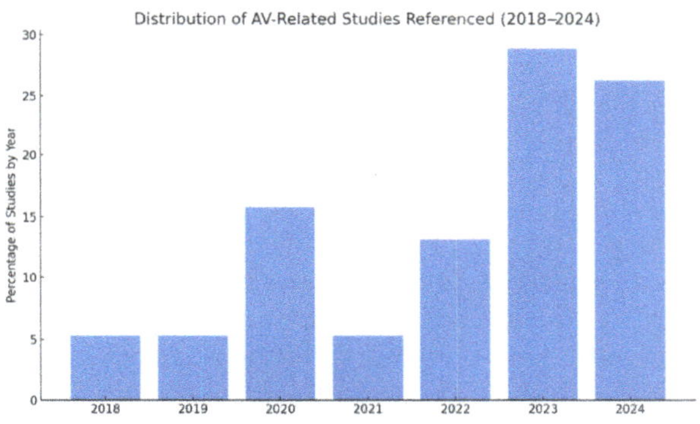

Figure 1. Trend of studies on AV deployment in construction zones (2018–2024).

The synthesis is organized into two subsections, beginning with Waymo and followed by Tesla. Within each subsection, user experiences are arranged chronologically based on the video upload dates, allowing for an examination of how user perceptions and AV system performance have evolved over time. Synthesizing user-generated content on YouTube played a critical role in revealing both the progress and persistent challenges of AV performance in complex and dynamic construction zone environments. These videos offer authentic glimpses into AV behavior under real-world conditions, capturing moments of user confidence, discomfort, hesitation, and fear that simulation studies, surveys, and interviews alone often fail to replicate. An important feature of this study is the broad appeal of its content. The insights captured in the results section are valuable to a wide audience, including transportation professionals and industry specialists seeking technical observations, as well as general users who are interested in gaining a better understanding of AV performance in real-world settings.

3.1. Waymo's Navigation Experience

Waymo is an automated driving technology company and a subsidiary of Alphabet Inc., Google's parent company. Established in 2009 as the Google Self-Driving Car Project, it was rebranded as Waymo in 2016. The company's mission is to make it safe and easy for people and goods to move around. Waymo develops the "Waymo Driver," the world's first autonomous ride-hailing service that combines hardware and software, including LiDAR, cameras, radar, and a powerful artificial intelligence (AI) computer platform, to provide a 360-degree view of the driving environment. This technology aims to enable Waymo's vehicles to navigate complex urban environments without human intervention. Waymo continues to expand its services and test its technology in various environments to improve road safety and redefine mobility [16]. Waymo's performance in construction zones, as observed through user-recorded videos, reflects a blend of technological advancements and persistent challenges.

In 2021, a Waymo vehicle in Chandler, Arizona, encountered difficulty navigating around construction cones placed at an intersection. The vehicle came to a stop and requested remote assistance. However, similar difficulties occurred later, indicating limited capability in adapting to temporary construction barriers [17]. In contrast, Waymo demonstrated notable improvement in 2022. During a test in downtown Phoenix, Arizona, the AV successfully maneuvered through road barriers, navigated around jaywalkers, and operated effectively in wet conditions. The trip, which was arranged via the Waymo app, was completed without major incidents, showcasing the system's growing ability to handle complex urban conditions [18]. However, the year 2023 introduced new concerns. In Tempe, Arizona, one Waymo vehicle accelerated from a stop to 50 mph within a construction zone where the posted speed limit was 35 mph. It failed to stop at a red light and exited the work zone traveling at 45 mph, which raised questions about the system's compliance with temporary speed and signal regulations [19]. That same year in Los Angeles, California another Waymo vehicle entered two separate work zones. Although it used turn signals correctly, it became confused by abrupt lane changes and required remote assistance to navigate safely. Additionally, dense fog during the drive complicated the vehicle's ability to interpret its surroundings, though it ultimately avoided any unsafe behavior. Despite these challenges, some 2023 tests highlighted the system's strengths [20]. In one instance, a Waymo vehicle navigated around active construction equipment in the main road without human intervention, earning praise from passengers [21]. In another real-world encounter, Waymo vehicle was observed driving into an active construction site without any human occupants. As it approached the area, the vehicle seemed confused by the irregular layout, including an open trench and scattered construction cones. Its perception system failed to identify a clear path, and the vehicle came to a stop near the edge of the trench, unable to determine how to proceed. The situation drew the attention of bystanders, who noted the vehicle's hesitation and lack of response. To redirect it, a construction worker placed cones in front of the vehicle, which eventually prompted it to reverse and steer away from the hazard [22]. During a set of tests in San Francisco that same year, the Waymo vehicle encountered multiple construction and obstruction-related events in an urban setting. It began its route by preparing to make a right turn onto a street obstructed by cones and a stationary motorcycle. The cones and signage were clearly visible on the passenger display. The vehicle maintained steady behavior as traffic progressed slowly due to a temporary stop sign held by a construction crew. Once the sign changed to "slow," the vehicle responded appropriately and continued forward. Further along the route, the AV approached another zone impacted by utility or road maintenance work. Instead of continuing straight as initially planned, the system automatically rerouted with a smooth right turn, seamlessly re-entering the flow of traffic. Later, it encountered a double-parked vehicle with an open door and cones surrounding it. Despite the limited space, the vehicle identified the hazard and executed a precise maneuver around the obstruction. This moment even drew a smile from a nearby pedestrian who witnessed the AV's careful behavior. Toward the end of the drive, the vehicle adjusted its route again in response to ongoing tree maintenance. Arborists had placed cones and equipment along the lane, prompting the system to detect the obstruction and alter its path accordingly. Throughout this journey, the AV's actions remained steady, predictable, and minimally disruptive, contributing to the user's positive impression of its ability to manage real-world, cone-dense environments with composure [23]. Nonetheless, in 2024, Waymo experienced another setback. In Chandler, California, a Waymo vehicle became immobilized within a work zone. Nearby construction workers attempted to assist by waving and moving cones to create a clearer path. Despite these efforts, the vehicle remained unresponsive and unable to proceed, highlighting an ongoing weakness in dealing with unexpected layout changes and a lack of clear external guidance [24]. Table 2 presents a summary of Waymo performance and user perceptions in various construction zone conditions between 2021 and 2024, highlighting instances of major disengagements or assistance and the corresponding overall user sentiment.

Table 2. Waymo performance and user perception in construction zones.

Year	Location	Work Zone Condition	Major Disengagement/ Assistance	Waymo Overall User Perception
2021 [17]	Chandler, Arizona	Construction cones at an intersection.	Yes	Negative
2022 [18]	Phoenix, Arizona	Barriers, jaywalkers, wet conditions, red light.	No	Positive
2023 [19]	Tempe, Arizona	Posted speed limit of 35 mph.	No	Negative
2023 [20]	Los Angeles, California	Required abrupt lane changes, light rain and dense fog.	Yes	Mixed
2023 [21]	N/A	Active construction equipment.	No	Positive
2023 [22]	N/A	Active construction site with irregular layout, open trench, scattered cones, presence of construction workers.	Yes	Negative
2023 [23]	San Francisco, California	Multiple work zones, street cones, stationary motorcycle, temporary stop sign held by a construction crew, signage change from "Stop" to "Slow," double-parked vehicle with an open door, presence of pedestrian, ongoing tree maintenance site, construction equipment.	No	Positive
2024 [24]	Chandler, California	Unexpected layout changes, cones, presence of construction workers.	Yes	Negative

Figure 2a,b present word clouds visualizing user-reported challenges and strengths of Waymo in work zones, respectively, highlighting key themes in negative and positive feedback.

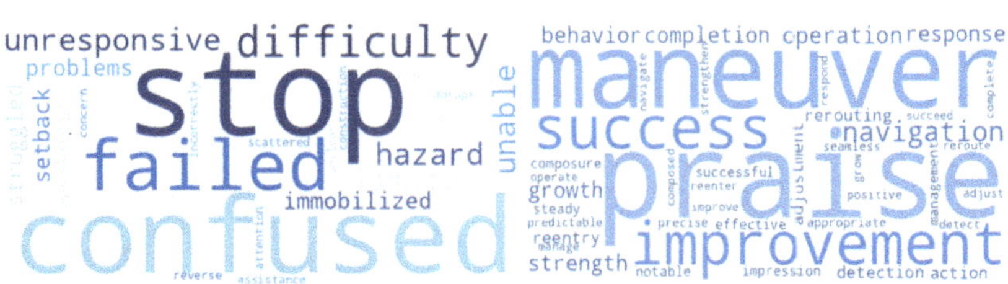

(**a**) Challenge-related Keywords (**b**) Strengths-related Keywords

Figure 2. (**a**,**b**): Word Clouds of user-reported challenges and strengths of Waymo in work zones.

3.2. Tesla's Navigation Experience

Tesla, founded in 2003, is a global leader in electric vehicles and clean energy, aiming to accelerate the transition to sustainable energy [23]. In addition to producing EVs, Tesla develops Autopilot and supervised Full Self-Driving (FSD) technologies, which use a vision-based system powered by cameras, neural networks, and AI to perceive and interpret driving environments. Although these systems require active driver supervision, they are continuously refined through over-the-air updates and large-scale data collection [25]. Since 2018, users have tested Tesla's systems in real-world construction zones, highlighting both progress and challenges. Early tests showed that Autopilot could follow cones when lane markings were absent [26] and use aggressive settings like "Mad Max" to manage merges effectively [27]. In 2019, Autopilot version 2019.40.1.1 handled obstructed lanes by smoothly changing lanes without visualizing the barrels on-screen [28], and another test demonstrated its ability to maintain stable speed and direction through cone-defined paths [29]. In 2020, Autopilot handled rainy construction zones well at first but struggled when lane visibility declined, triggering emergency braking and requiring driver intervention [30]. Another Model 3 test showed strong lane-following in tight zones, aided by visualization updates, though merging still required driver readiness [31]. Autopilot 2020.16.2.1 showed improved rendering of signs and cones but failed at speed adjustment and made unsafe lane attempts in construction-diverted areas [32]. A comprehensive 2020.4.1 test highlighted both near-collisions with cones and later success in fully navigating an unmarked, cone-only lane; even using a lead vehicle as a dynamic guide or performing the route solo [33]. Autopilot version 2020.40.3 represented a leap in capability with improved object detection, night vision, and traffic light recognition in city work zones [34]. It also handled complex maneuvers across solid lines and cones with high precision [35]. By 2021, FSD Beta 10.5 could stop workers holding signs but misinterpreted movement cues, prompting multiple driver interventions [36]. In 2022, version 10.69.2.2 struggled in blocked-lane scenarios and required manual control [37], while FSD 0.2.2 failed consistently in a simple cone-lined path despite succeeding in highway conditions [38]. Other users noted jerky corrections, wide turns, and hesitation under debris-strewn conditions, though the system remained stable [39]. FSD Beta 10.11.2, tested in 2022, performed well through a complex construction route with cyclists, stop signs, and ambiguous intersections, requiring brief driver input [40]. In 2023, testing in Dallas showed FSD identifying a lane closure, slowing to 30 mph, and executing a clean lane change [41]. Other tests that year revealed better cooperation with human drivers [42], stable barrier navigation [43], and dynamic human awareness, such as safe clearance during tree trimming operations [44]. However, the system still exhibited uncertainty, hugging lane edges near potholes [45], and occasionally drove too slowly, requiring manual throttle input to keep pace with traffic [46]. In 2024, the release of FSD version 12.3.3 marked significant advancements. In complex work zones with shifting cones and signage, the system struggled initially but adjusted well by following a lead vehicle and mimicking human-like behavior with only minor intervention [47]. Another test showed confident adaptation to construction cones and lane shifts [48], and Beta 12 rerouted automatically at a fenced closure, succeeding where Beta 11 had failed [49]. Version 12.3.6 also handled uneven pavement, sticker-only lane markings, tight merges, and heavy cone density with composure, correcting its own mistaken lane entries without disengagement [50]. A separate test in mixed rain conditions showed the system reducing speed from 77 to 70 mph and managing misaligned signs, construction cones, and parking without intervention [51]. A Cybertruck test saw the system reroute around a road closure and self-park while recognizing pedestrians and adapting to residential speeds [52]. Tesla's Basic Autopilot also showed reliable stop-and-go control and predictive behavior in adaptive traffic near temporary lane closures [53]. Near Arthur Ashe Stadium, FSD 12.3.6 successfully scanned traffic gaps and executed a smooth lane change through a narrow work zone, demonstrating enhanced real-time decision-making [54]. Despite these gains, a real-world 2024 test revealed recurring limitations in urban construction zones: misinterpreting inactive signals, failing to identify closures, poor rerouting, hesitation at intersections, and visual inconsistencies in cone detection; all contributing to the need for frequent driver intervention [55]. Table 3 presents a comprehensive summary of user-documented evaluations of Tesla Autopilot and supervised Full Self-Driving (FSD) systems from 2018 to 2024, detailing vehicle models and software versions tested, construction zone conditions, observed outcomes, major disengagements events, and users' overall perceptions.

Table 3. Tesla performance and user perception in construction zone.

Year	Vehicle Model/ Software Version	Work Zone Condition	Outcome	Major Disengage-ment/Assistance	Tesla User Overall Perception
2018 [26]	Model 3	Construction zone without lane markings, used cones.	• Adapted to atypical conditions centered itself based on cones.	No	Positive
2018 [27]	Model X on Mad Max mode	Nighttime construction, Mad Max Lane changes, dynamic layout.	• Executed smooth lane shifts and merged into traffic. • Used turn signals automatically and enhanced driver awareness with UI cues. • Hesitated occasionally during lane changes and struggled in tight merges. • Prioritized caution over aggressiveness, even in assertive mode.	No	Positive
2019 [28]	Autopilot 2019.40.1.1 (HW2.5)	Construction barrels obstructing lanes, limited visibility.	• Performed smooth lane change to maintain a safe buffer. • Demonstrated cautious but effective behavior without object visualization. • Detected and reacted to barrels. • Transitioned back to the original lane after passing the obstruction.	No	Positive
2019 [29]	Model 3 (HW 3.0)	Construction zone with cones, no visible lane markings.	• Maintained consistent speed. • Followed temporary boundaries created by cones instead of painted pavement markings.	No	Positive
2020 [30]	Autopilot 2020.8.1	Rainy construction zone, low visibility, emergency braking triggered.	• Detected construction cones. • Adjusted speed appropriately. • Maintained lane position in light traffic. • Navigated the initial portion of the work zone smoothly. • Followed the flow of surrounding traffic. • Exhibited instability as the roadway narrowed. • Struggled with reduced visibility and poorly marked lanes. • Drifted toward the edge of the road near a reflective barrier. • Triggered emergency braking system to avoid impact.	Yes	Mixed
2020 [31]	Model 3	Unclear road markings, cones present.	• Demonstrated strong lane-following performance. • Maintained lane integrity using cameras and sensors. • Detected nearby obstacles. • Used enhanced visualization updates to detect cones and construction elements. • Struggled with lane changes in unpredictable sections. • Faced difficulty merging in complex work zone layouts. • Required ongoing driver readiness.	No	Positive

Table 3. *Cont.*

Year	Vehicle Model/ Software Version	Work Zone Condition	Outcome	Major Disengagement/Assistance	Tesla User Overall Perception
2020 [32]	Autopilot 2020.16.2.1	Motorway, roundabout in UK, construction zones with cones and temporary markings.	• Failed to change lanes automatedly on a dual carriageway. • Attempted to enter a closed lane diverted by cones. • Hesitated around large vehicles sometimes misjudging safe gaps for lane changes. • Executed abrupt lateral shifts at junctions. • Failed to adjust speed to temporary speed limits in construction zones. • Aborted a turn unexpectedly in a car park.	Yes	Mixed
2020 [33]	Autopilot 2020.4.1 (HW 3.0)	Complex construction zone, cones, newly paved unmarked lanes.	• Executed a lane shift. • Veered dangerously close to a reflector cone, nearly struck the vehicle's mirror. • Triggered a strong driver reaction and increased caution. • Displayed a 40-mph speed despite a 35-mph construction zone. • Adjusted speed later in response to conditions, slowed down noticeably in cone-dense areas, even if traffic moved faster. • Followed another car through the unmarked zone, as a dynamic guide. • Navigated the freshly paved lane using only cone placement. • Made the correct turn and maintained proper spacing. • Demonstrated capability for fully visual-based navigation, surprised and impressed the user.	No	Positive
2020 [34]	Autopilot 2020.40.3	Urban nighttime construction zones, narrow shifting lanes.	• Recognized cyclists. • Navigated through narrow and shifting construction lanes. • Used high-resolution cameras for object identification. • Executed careful and cautious lane changes in dynamic environment. • Detected signal changes from 150 to 200 m away. • Came to a complete stop when needed. • Required driver confirmation to proceed through intersections. • Integrated real-time map data. • Made accurate navigational decisions even when road cues were missing or unclear.	No	Positive
2020 [35]	Model 3 2020.40.3	A work zone, where cones directed the vehicle to cross a solid center line in construction.	• Narrowly avoided a cone by less than one foot. • Stayed within the temporary cone-defined path. • Smoothly transitioned back to the original lane after passing the obstruction. • Demonstrated high precision. • Showed strong spatial awareness and precision. • Received user praise for handling the challenge effectively.	No	Positive

Table 3. *Cont.*

Year	Vehicle Model/ Software Version	Work Zone Condition	Outcome	Major Disengage-ment/Assistance	Tesla User Overall Perception
2021 [36]	Model 3 FSD Beta 10.5	Construction-heavy area in Houston, Texas, worker with stop sign, potholes	• Recognized and responded to a worker holding a stop sign, came to a safe stop. • Misinterpreted the worker's slight forward movement as a signal to proceed. • Attempted to follow the stop sign holder, requiring driver intervention. • Upon re-engagement, tried to steer around the worker, prompting another manual takeover. • Unexpectedly followed the walking worker with the stop sign. • Disengaged when a pedestrian entered the intersection. • Struggled to handle a T-intersection without occasional disengagements. • Maintained cautious behavior near barricades and construction workers. • Had difficulty on pothole-ridden roads, requiring frequent driver takeover. • Failed to slow down at a railroad crossing; driver had to apply brakes.	Yes	Mixed
2022 [37]	Model Y FSD Beta 10.69.2.2	Nighttime, construction cones, blocked lane, confusion at intersections.	• Navigated basic turns smoothly, handled quiet streets without issue. • Became confused at an intersection, made a wrong turn. • Hesitated and failed to identify a navigable path, when encountered cones and a blocked lane in a construction zone, came to a full stop.	Yes	Negative
2022 [38]	FSD	Five consecutive trials through a temporary right-hand lane defined by construction cones, slight uphill grade with clear lane marking and consistent cone placement	• Hesitated at the approach to cones in each trial. • Backed up slightly at the fourth cone during every run. • Stalled and failed to proceed past the cones. • Misinterpreted a construction worker's "Slow" sign gesture as a cue to steer toward a cone. • Repeated failures occurred despite clear and consistent conditions. • Performed reliably in structured freeway environments (executed 85 mph lane changes, operated well even in heavy traffic and nighttime conditions).	Yes	Negative

Table 3. *Cont.*

Year	Vehicle Model/ Software Version	Work Zone Condition	Outcome	Major Disengagement/Assistance	Tesla User Overall Perception
2022 [39]	FSD	Gravel-strewn lane, uneven surfaces, cautious navigation.	• Briefly hesitated before navigating the lane. • Made a small, jerky steering correction. • Proceeded cautiously through the gravel-strewn path. • Made wide turns at intersections. • Occasionally drifted near curbs. • Responded slowly to changes in speed. • A stable and safe passage due to conservative driving behavior. • Successfully navigated uneven terrain and scattered debris. • Resumed normal operation outside the work zone without further incident.	No	Positive
2022 [40]	FSD Beta 10.11.2	Active construction zone with stop signs, cyclists, intersections, and traffic congestion, low visibility drone mounted on the vehicle captured the drive	• Navigated complex routes confidently. • Smoothly handled a notoriously confusing intersection by following another vehicle. • Demonstrated adaptive perception and interpreted an ambiguous trailer correctly. • Made well-timed directional decisions, even under pressure from tailgating vehicles. • Maintained control in low-visibility conditions with unclear lane markings.	No	Positive
2023 [41]	FSD	Construction zone in Dallas, Texas with lane closure and residential navigation	• Identified a lane closure. • Automatically reduced speed to 30 mph. • Executed a smooth lane change to avoid the obstruction. • Performed a slightly abrupt right turn into a residential neighborhood.	No	Positive
2023 [42]	FSD	Narrowed road section due to construction	• Approached a narrowed road section safely, yielded to oncoming traffic. • Exchanged a visual acknowledgment with a human driver (indicative of cooperative behavior). • Maintained a smooth trajectory throughout the interaction. • Sustained reliable lane positioning. • Ensured consistent obstacle detection under physical and social driving conditions.	No	Positive
2023 [43]	FSD Beta	Construction zone with barriers.	• Navigated around construction barriers with stable control. • Adjusted its path safely and smoothly. • Maintained appropriate following distances.	No	Positive

Table 3. *Cont.*

Year	Vehicle Model/ Software Version	Work Zone Condition	Outcome	Major Disengagement/Assistance	Tesla User Overall Perception
2023 [44]	Model 3 FSD Beta	Tight construction zone in Silicon Valley, California with human workers and construction equipment trimming trees	• Displayed strong situational awareness. • Dynamically adjusted to the presence of humans and roadside equipment. • Maintained safe clearance from the active work area.	No	Positive
2023 [45]	Model 3	Road narrowing with large pothole	• Exhibited uncertainty as the road narrowed. • Created its own lane by closely hugging the right edge. • Approached a large pothole unsafely.	Yes	Negative
2023 [46]	Model Y FSD Beta	Cone-dense construction zone, conservative speed (California).	• Recognized and reacted to each cone. • Maintained overly cautious speed. • Required light manual acceleration by the driver to maintain appropriate traffic flow.	No	Positive
2024 [47]	Model 3 FSD 12.3.3	Complex work zone with shifting lanes, scattered cones, and temporary signage.	• Struggled to interpret faded or missing lane markings. • Had difficulty with unexpected obstacles early in the test. • Identified and began reliably following a lead vehicle. • Quickly adjusted its trajectory to match the traffic flow. • Exhibited human-like decision-making (adjusted speed and path). • Required only minimal driver intervention (Inputs were limited to occasional steering corrections or system acknowledgments).	No	Positive
2024 [48]	Model 3 FSD 12.3.3	Construction zone with cones and lane shifts	• Demonstrated impressive performance in construction zones, addressed challenges that traditional autopilot systems struggled with. • Adapted to dynamic scenarios. • Executed smooth lane changes despite construction obstacles like cones and lane shifts.	No	Positive
2024 [49]	FSD Beta 12	Fenced-off construction area with road closure and adjacent equipment	• Immediately recognized the road closure and nearby equipment. • Steered into and followed the detour without hesitation. • Adjusted correctly to altered lane markings. • Maintained a steady and controlled pace through the work zone.	No	Positive

Table 3. *Cont.*

Year	Vehicle Model/ Software Version	Work Zone Condition	Outcome	Major Disengagement/Assistance	Tesla User Overall Perception
2024 [50]	FSD 12.3.6	Dynamic construction zone, uneven pavement, cone density.	• Detected need for a lane change due to road work well in advance, executed smooth lane change despite moderate surrounding traffic. • Crossed a double yellow line confidently. • Navigated uneven pavement. • Maintained proper spacing around cones and curbs. • Stayed within path marked by temporary stickers (not paint), with only minor wobble. • Responded properly at a red light and proceeded smoothly on green. • Adjusted trajectory effectively around new curbs and tight traffic sections. • Managed areas with no lane lines on the navigation map using real-time perception. • Adapted to lane shifts and merged seamlessly with traffic. • Maintained safe clearance in narrow, cone-dense zones with roadside activity. • Mistakenly entered a left-turn lane but corrected smoothly and re-entered the appropriate lane. • Successfully navigated fake turn lanes, obstructive equipment, and tight lane shifts. • Reacted appropriately to audible alerts from nearby objects. • Maintained composure and continued operation without hesitation.	No	Positive
2024 [51]	FSD 12.3.6	City streets, highways, and an active construction zone; Light, intermittent rain	• Operated with minor performance warnings despite visibility issues from rain. • Adjusted speed from preset 77 mph to 70 mph in response to weather conditions. • Encountered a new construction layout with cones and yellow lane markings, recognized ongoing drainage work. • Hesitated at a stop line with triangular markings, required a tap on the accelerator to proceed through heavy cross-traffic. • Managed through misaligned stop signs and construction cones. • Merged efficiently into a tight traffic gap, adjusted speed according to surrounding vehicle flow. • Paused at a red right-turn arrow until prompted by driver tap (due to vehicle pressure behind).	No	Positive

Table 3. *Cont.*

Year	Vehicle Model/ Software Version	Work Zone Condition	Outcome	Major Disengagement/Assistance	Tesla User Overall Perception
2024 [52]	Cybertruck FSD	Road closure due to construction on the main road	• Detected the road closure on the main road, rerouted automatically through local streets. • Encountered workers and traffic cones on the alternate route, slowed down appropriately near active work. • Recognized nearby pedestrians and vehicles. • Adapted well to residential neighborhood conditions by maintaining appropriate speeds, adjusted braking naturally.	No	Positive
2024 [53]	Model 3 Autopilot	Urban driving in Pittsburgh, Pennsylvania with stop-and-go traffic near temporary lane closures in active construction areas.	• Monitored behavior of nearby vehicles continuously, and adjusted its own speed based on surrounding traffic dynamics. • Maintained a safe following distance. • Reduced overall driver workload. • Demonstrated predictive modeling for enhanced comfort and safety.	No	Positive
2024 [54]	FSD 12.3.6	Complex, constrained construction work zone in Queens, New York City.	• Scanned for an available gap in traffic or lane, executed a smooth lane change to the left within a construction zone. • Successfully navigated through challenging and constrained road conditions.	No	Positive
2024 [55]	FSD	Urban construction zone with complex (inactive signals, road closures, construction barriers)	• Misinterpreted inactive traffic signals as green lights. • Drove 8–10 mph below posted speed limits. • Failed to recognize road closures and construction barriers. • Struggled with real-time reroute, caused confusion and unsafe maneuvers. • Occasionally entered repeated loops in dead ends. • Hesitated at stop signs and blinking red lights. • Creeped forward without confidently determining right-of-way. • Misidentified objects inconsistently displayed cones and construction elements. • Handled basic lane keeping well in clearly marked areas. • Performance was unreliable in construction-heavy scenarios.	Yes	Negative

Figure 3a,b present word clouds visualizing user-reported challenges and strengths of Tesla in work zones, respectively, highlighting key themes in negative and positive feedback.

(a) Challenge-related Keywords (b) Strength-related Keywords

Figure 3. (**a**,**b**): Word Clouds of user-reported challenge and strengths of Tesla in work zones.

4. Discussion

4.1. AV Performance Analysis

Analyzing the synthesis of 38 video contents, 222 events were identified, comprising 105 challenges and 117 strengths. This near-even split, with strengths representing 52.7% and challenges 47.3% of the total, reflects a transitional phase in AV development, where systems are becoming increasingly capable but still require further refinements to achieve consistently reliable performance when navigating dynamic and unpredictable work zone environments. Thematic analysis of user-documented encounters with AVs in construction environments led to a deeper understanding of distinct challenge and strength categories along with their distribution, as shown in Figure 4.

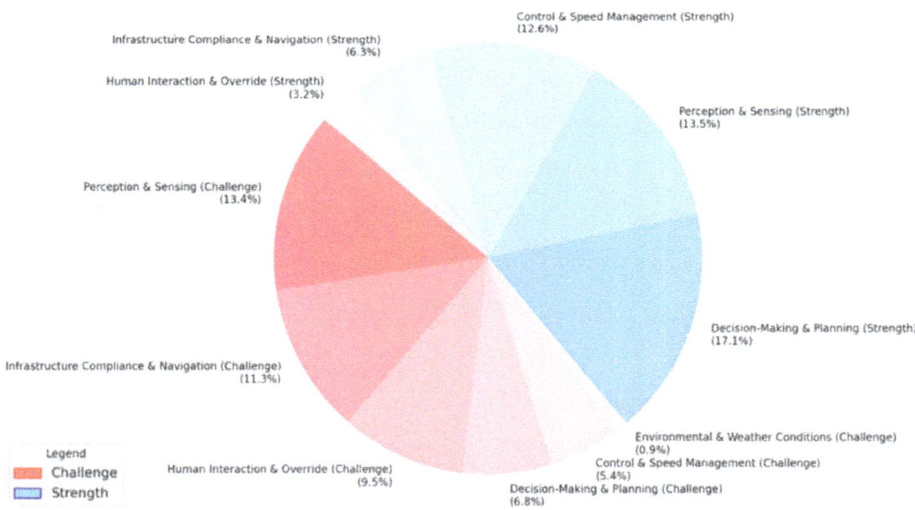

Figure 4. Distribution of strength and challenges in AV work zone navigation.

4.1.1. Challenges

In Table 4, the challenge categories and subcategories are ranked by frequency, with Perception & Sensing emerging as the most reported issue, followed by Infrastructure Compliance & Navigation, and Human Interaction & Override. These results indicate that perception-related challenges remain a primary concern as AVs often struggle to detect, interpret, and respond to temporary, unexpected, or uncommon road features such as construction signs, cones, and irregular lane markings. Infrastructure compliance also presents a significant challenge, with vehicles failing to recognize traffic signals, respond to missing or unclear lane guidance, or properly navigate detours. Furthermore, difficulties in human interaction, such as responding appropriately to pedestrians, construction workers, and other unpredictable human behaviors, contribute to disengagement events. Although fewer challenges were reported in Decision-Making & Planning, Control & Speed Management, and Environmental & Weather Conditions, occasional failures in these areas suggest that further advancements are needed to ensure reliable AV operation in dynamic and complex construction environments. In particular, the low frequency of events related to environmental and weather conditions is likely attributed to the limited availability of real-world testing conducted under adverse weather scenarios.

Table 4. Ranking and total number of reported AV work zone navigation challenges.

Rank	Challenge Category	Challenge Subcategories	Tesla	Waymo	Total Frequency
1	Perception & Sensing	• Non-standard Markings; • Temporary Signs; • Occluded Visibility; • Unexpected Obstacles.	28	2	30
2	Infrastructure Compliance & Navigation	• Traffic Signals; • Construction Zone Entry and Exit; • Unclear/Missing Lane Markings.	19	6	25
3	Human Interaction & Override	• Pedestrian/Worker Interaction; • System Disengagements/Human Intervention Required.	17	4	21
4	Decision-Making & Planning	• Complex Intersection/Turning movement; • Lane Prediction/Planning; • Road Closures or Detours.	13	2	15
5	Control & Speed Management	• Speed Modulation.	11	1	12
6	Adverse Weather Conditions	• Presence of Precipitation/Fog.	1	1	2
All	All	All	89	16	105

A comparative breakdown of challenge categories for 38 user-recorded videos involving Tesla and Waymo (Figure 5) reveals key differences in the operational performance of the two companies.

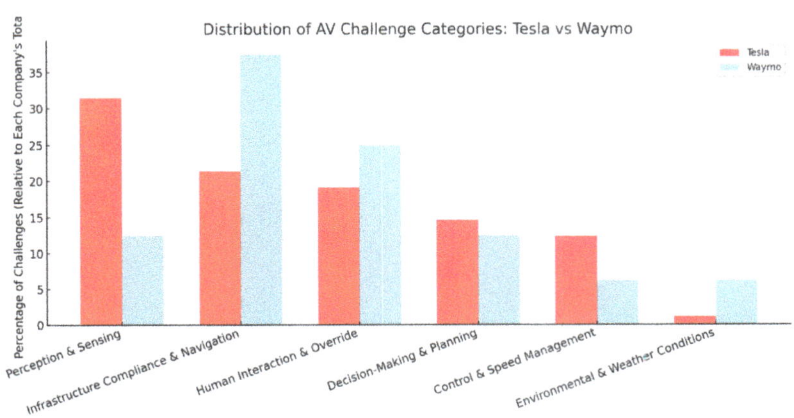

Figure 5. Comparison of AV challenge types encountered by Tesla and Waymo.

Tesla showed a higher rate of perception and sensing challenges (31.46%) compared to Waymo (12.50%), likely due to Tesla's camera-only system versus Waymo's multi-sensor fusion using LiDAR, radar, and cameras. Waymo's richer sensor suite appears more robust in detecting and interpreting the driving environment. Conversely, Waymo faced more issues with infrastructure compliance and navigation (37.50%) than Tesla (21.35%), especially in poorly marked roads. Tesla's neural-network-based system seems better at generalizing to dynamic or unmapped environments, while Waymo's map-dependent planning may struggle in these contexts. Human override needs were present in both systems (Waymo 25%, Tesla 19.10%). Decision-making and planning issues were slightly more common in Tesla (14.61%) than in Waymo (12.50%). Similarly, speed or control challenges occurred more frequently in Tesla (12.36%) than in Waymo (6.25%), suggesting that Tesla exhibited less smooth control under certain conditions. Weather-related events were rare overall, with only one reported incident for each company.

4.1.2. Strengths

In Table 5, the strength categories highlight key areas where AVs demonstrated notable performance in construction zones. Decision-Making & Planning emerged as the most notable strength, driven by the AVs' ability to follow lead vehicles near construction zones, reroute around obstacles, and adapt to dynamic environments. Following lead vehicles near construction zones particularly reflected human-like decision-making, allowing AVs to navigate through complex and changing conditions. Perception & Sensing ranked next, with AVs demonstrating improved capabilities in detecting obstacles and interpreting road signage, although occasional inconsistencies still highlight areas for further enhancement. Control & Speed Management also showed strong performance as AVs were able to execute smooth lane changes in detours and adjust speed cautiously near cones and other temporary elements. Lower rankings were observed in Infrastructure Compliance & Navigation, where strengths included lane re-centering and estimation, and in Human Interaction & Override, where AVs demonstrated some ability to handle pedestrian interactions. The relatively lower frequencies in these categories suggest that, while progress has been made, further development is necessary to achieve consistent and reliable AV behavior in construction environments.

Table 5. Ranking and total number of AV work zone navigation strengths.

Rank	Strength Category	Strength Subcategories	Tesla	Waymo	Total Frequency
1	Decision-Making & Planning	• Following Lead Vehicle; • Rerouting and Path Planning; • Adaptation to Dynamic Environment.	32	6	38
2	Perception & Sensing	• Road Signs; • Obstacles.	27	3	30
3	Control & Speed Management	• Smooth Lane Change in Detours; • Slowdown/Cautious Speed near Cones.	25	3	28
4	Infrastructure Compliance & Navigation	• Lane Re-centering and Estimation.	13	1	14
5	Human Interaction & Override	• Handling of Pedestrian Interactions; • No Major Human Intervention.	5	2	7
Total	All	All	102	15	117

The comparative strength distribution of Tesla and Waymo across five categories (Figure 6) reveals important distinctions when interpreted alongside their corresponding challenge frequencies. These distributions highlight not only where each system excels, but also how those strengths relate to observed limitations in construction zone navigation.

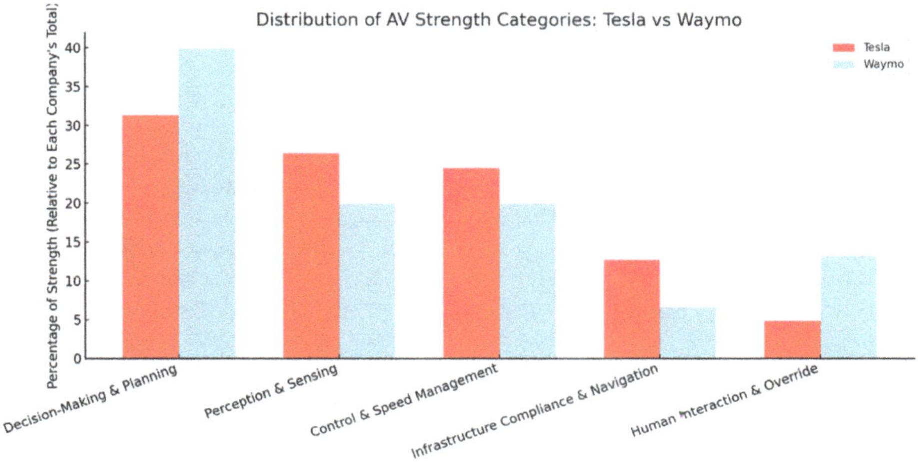

Figure 6. Distribution of strength categories for Tesla and Waymo in work zones.

Tesla demonstrates a broader distribution of strengths, with notable performance in Perception & Sensing (26.47%) and Control & Speed Management (24.51%). These align with Tesla's camera-based system, which frequently succeeded in identifying cones, signs, and other road elements while maintaining cautious speed adjustments. Interestingly, these areas were also among the most challenging categories for Tesla; Perception & Sensing (31.46%) and Control & Speed Management (12.36%); suggesting that while Tesla encounters issues in these domains, it has also shown repeated improvements and system adaptability in response. This dual presence of strengths and challenges reflects Tesla's iterative learning model, where real-world data informs software updates that gradually improve performance. Tesla's strength in Infrastructure Compliance & Navigation (12.75%) also aligns with a relatively moderate challenge rate (21.35%), indicating that the system, despite some navigation difficulties in construction layouts, often successfully handles temporary infrastructure changes using cone guidance and visual cues. Meanwhile, Decision-Making & Planning (31.37%) is another strong area for Tesla, supported by real-world examples of smooth detours and dynamic responses, although its challenge rate in this category (14.61%) suggests occasional hesitation or misjudgment in high-stakes environments. However, Human Interaction & Override represents Tesla's weakest area of strength (4.90%) and corresponds with a notably high challenge frequency (19.10%). This can be primarily attributed to Tesla's supervised Full Self-Driving (FSD) system that assumes that a human will intervene when necessary, unlike Waymo, which is built to operate without driver input. Waymo's strengths are more concentrated, with the majority falling under Decision-Making & Planning (40%) and Human Interaction & Override (13.33%). These strengths correspond well to Waymo's low challenge percentages in Decision-Making (12.5%) and Control (6.25%), showing that its rule-based, map-supported system excels in structured planning and cautious interaction with pedestrians and construction workers. However, Waymo's greatest challenge area, Infrastructure Compliance & Navigation (37.5%), is also where it shows the least strength (6.67%), pointing to a significant gap in adapting to temporary or unmapped construction layouts. Similarly, its relatively low strength in Perception & Sensing (20%) reflects modest ability to respond fluidly to visual changes, which matches a moderate challenge rate (12.5%) in that category.

4.1.3. Comparison Between Strengths and Challenges

A comparative analysis of overall strength and challenge categories also reveals notable trends in the evolving performance of AV systems within construction zone environments (Figure 7). Decision-making & Planning demonstrated the greatest improvement, with 38 strength events recorded compared with 15 challenge events. While AVs occasionally struggled with complex intersections as well as lane and detour prediction, the substantially higher number of strengths suggests that capabilities in rerouting, path planning, and adaptation to dynamic environments have matured considerably. In contrast, Perception & Sensing presented a more balanced outcome, with both strengths and challenges recorded in 30 events each. Although AVs have demonstrated improved abilities in obstacle detection and road sign recognition within work zones, persistent difficulties in interpreting temporary signs, uncommon markings, occluded visibility, and unexpected obstacles highlight ongoing inconsistencies in perception. This may indicate two key insights: perception and sensing remain a central focus of AV performance, representing the most frequently observed category with a total of 60 events; and user feedback on AVs' perception and sensing capabilities in work zones is evenly divided. Control & Speed Management exhibited a favorable trend, with 28 strength events reported versus 12 challenge events, indicating that AVs are increasingly capable of managing speed modulation and executing smooth lane changes under variable conditions. However, Infrastructure Compliance & Navigation continues to present notable challenges, with 25 challenge events recorded compared with 14 strengths related to lane re-centering and estimation. Human Interaction & Override also remain critical areas for improvement, with 21 challenge events and only seven strength events reported. This highlights the complexity of responding effectively to pedestrian interactions and unexpected human behavior and emphasizes that human intervention, either by the driver or through remote assistance, is still necessary in certain situations. Environmental & Weather conditions were cited infrequently, with only two challenge events and no corresponding strengths, suggesting limited exposure to adverse conditions in the summarized YouTube content.

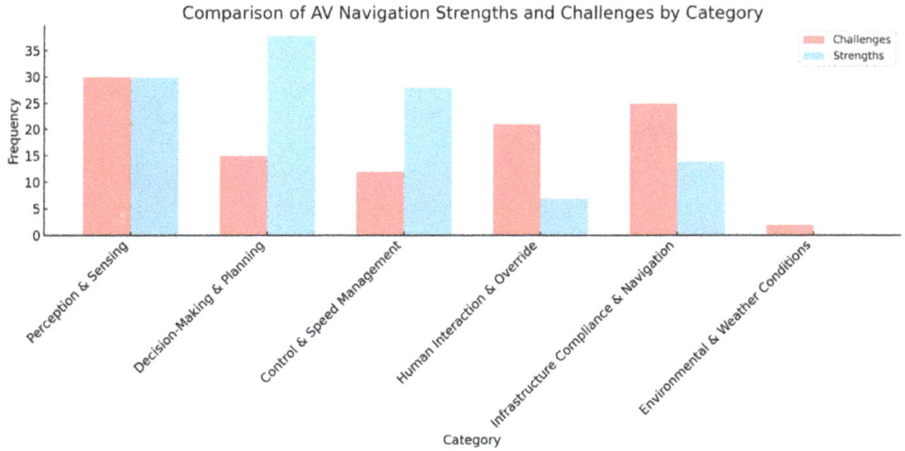

Figure 7. Comparison of challenges and strengths of AV navigation in work zones.

4.2. User Perception Analysis

The analysis of user-documented experiences reveals an overall positive perception of AVs navigating work zones, although notable concerns persist. Among the 38 experiences analyzed between 2018 and 2024, approximately 74% of user perceptions were positive, 11% were mixed, and 15% were negative (Figure 8).

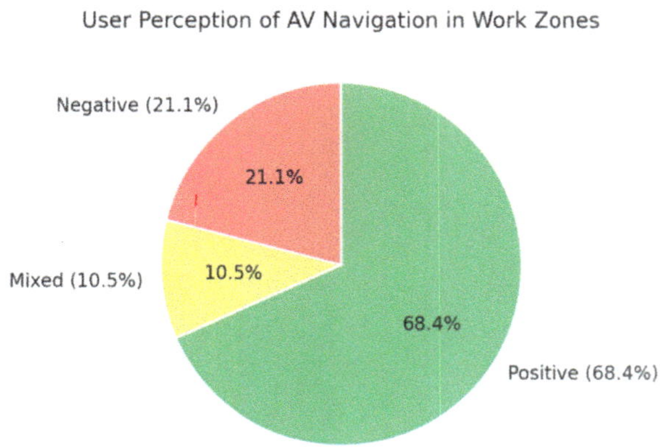

Figure 8. Overall user perception of AV navigation in work zones.

Users frequently praised AVs' strengths in structured tasks such as smooth lane changes, cautious speed modulation, and obstacle avoidance in dynamic construction environments. However, as Tables 2 and 3 showed, negative and mixed perceptions were associated with cases where major disengagements or external interventions were necessary. Specifically, negative or mixed feedback was often reported when AVs encountered issues such as confusion with temporary infrastructure layouts (e.g., open trenches and scattered cones), failure to correctly interpret construction signs or lane closures, hesitation or incorrect responses to pedestrians and construction workers, difficulties navigating poorly marked lanes under low-visibility conditions, and abrupt or unsafe maneuvers requiring manual takeover. These events highlight that while AV systems like Tesla's supervised FSD and Waymo's driverless technology have made progress in controlled scenarios, challenges in interpreting complex and rapidly changing construction environments continue to affect user trust and acceptance.

An analysis of the user perception chart comparing Tesla and Waymo (Figure 9) further illustrates these dynamics. Among Tesla's recorded experiences (30 events), approximately 77% were positive, 13% negative, and 10% mixed, indicating a predominantly favorable response from users. In contrast, Waymo's user feedback (8 events) was more divided, with 38% positive, 50% negative, and 13% mixed perceptions. Waymo's higher proportion of negative responses, despite the smaller sample size, may reflect either more challenging test conditions or more frequent system limitations during complex navigation tasks. This disparity may also be partly attributed to operational differences: Waymo's system functions in a fully automated mode without a human driver present at the wheel, whereas Tesla's supervised FSD trials involve a safety driver capable of intervening. The absence of a human driver in Waymo vehicles may heighten user sensitivity to system behavior, especially in uncertain or high-risk environments like work zones.

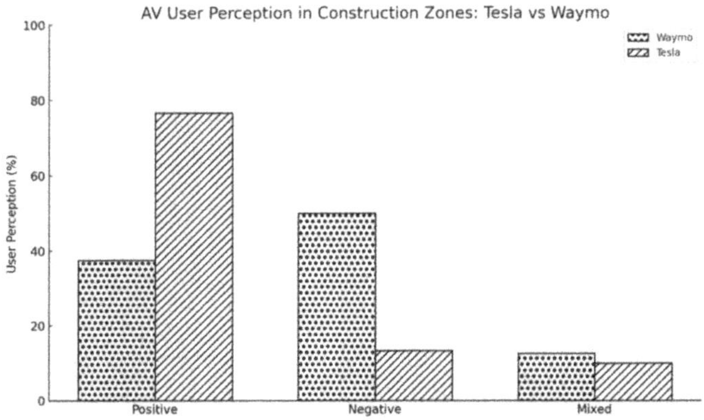

Figure 9. Comparison of user perceptions of Tesla and Waymo navigation in work zones.

4.3. Safety Implications and Regulatory Considerations of AV Failures in Work Zones

The safety implications of observed AV failures, such as misinterpretation of stop signs, failure to detect construction cones or workers, and difficulty navigating detours; warrant explicit discussion due to their potential impact on public safety and regulatory oversight. AV misbehavior in construction zones can introduce serious real-world risks. For instance, incidents where AVs failed to stop at red lights or exceeded reduced speed limits in work zones [19] highlight the danger posed to both road workers and other road users. Similarly, Tesla FSD's confusion when interpreting a worker's stop sign, or hesitation near pedestrian activity [36], could result in unintended vehicle movement that jeopardizes human safety. These failures are not just technical issues; they are potential regulatory triggers. In many jurisdictions, AVs are expected to meet stringent functional safety standards and comply with local traffic laws. Disengagements requiring human override suggest that current systems fall short of SAE Level 4 autonomy expectations and may influence how policymakers classify and certify AV readiness for public roads. Moreover, regulatory agencies such as NHTSA may view persistent errors in work zones as indicative of insufficient real-world training data or poor generalization in AV algorithms. These shortcomings could necessitate new guidelines around AV testing and require manufacturers to demonstrate system robustness in temporary and irregular environments. The implications also extend to liability; if an AV fails to detect a worker or misreads a stop sign leading to a collision, questions of fault, whether attributable to the manufacturer, operator, or infrastructure, become legally and ethically complex. Therefore, addressing these challenges through improved AV perception models, clearer infrastructure standards, and cross-sector collaboration is vital.

5. Research Findings

This study set out to examine the performance of AVs in navigating work zones by analyzing real-world user-documented experiences shared on YouTube. Between 2018 and 2024, AV navigation in work zones evolved from a relatively limited area of focus into a prominent subject of documentation and analysis. Through a comprehensive assessment of user narratives, on-screen captions, subtitles, and transcribed observations, the research highlighted the progress and limitations of AV systems developed by Tesla and Waymo when confronted with the construction environments. A distinctive value of this study lies in its use of user-generated content, which not only documented AV performance but also captured the emotional and cognitive responses of AV users. These insights offered a more complete evaluation of AV deployment by revealing the "human factor" behind technological performance. These first-hand narratives offered insights into moments of confidence, such as when an FSD system navigated a complex cone layout with limited human intervention, and moments of anxiety or loss of trust, such as when the system followed a construction worker too closely or came to an unexplained stop in a confusing detour, which required human assistance either remotely or on-site. The observations revealed how users perceive, interpret, and respond to AV behavior in real-world contexts. Real-world demonstration tests are often accompanied by post-experiment surveys, while traffic simulation studies are frequently paired with driving simulator experiences to assess the human comfort level and perception on automated systems. This combined approach enables researchers to capture insights across three critical dimensions: system performance, infrastructure interaction, and rider comfort. Similarly, analyzing user-generated YouTube content in our study served a comparable purpose, which offered a window into how AVs perform in real-world settings, how they interact with road infrastructure, and how comfortable or confident passengers and bystanders feel during these encounters.

To support reproducibility and enable other researchers to conduct similar studies, we propose a standardized framework for annotating AV behavior using public video content. Our approach involved collecting user-generated YouTube videos through a custom Python script that interacted with the YouTube Data API, applying Boolean keyword combinations such as "construction zone" and "self-driving" across a defined date range, as detailed in Table 1. Videos were retained if they were created by users; and included narration, captions, or subtitles, ensuring sufficient contextual information for analysis. AV behavior within these videos, was annotated as either a "Strength" or "Challenge was classified into six thematic categories: Perception & Sensing, Decision-Making & Planning, Control & Speed Management, Human Interaction & Override, Infrastructure Compliance & Navigation, and Environmental & Weather Conditions. These categories were established using the OpenAI API to assist with thematic clustering and pattern recognition, followed by manual validation to ensure accuracy and consistency. The comparison between automated and manual classifications yielded an accuracy of 88.67% across 222 annotated events, demonstrating strong alignment between model-assisted and human-coded results.

From this analysis, a total of 117 strengths and 105 challenges were identified across 38 user-documented AV operation cases in work zones spanning different locations. These findings indicate that AVs demonstrate a balanced but still-developing performance in complex and irregular road environments. While these strengths highlight meaningful progress toward greater autonomy, they are not sufficient on their own to achieve SAE Level 4 or 5 autonomy. Tesla's Full Self-Driving system offers a range of advanced driver assistance features, while still requiring continuous driver supervision and responsibility in the work zone. Waymo's autonomous driving system is designed to operate vehicles independently within specific environments. While it does not rely on a safety driver during deployment, real-world observations indicate that it occasionally required remote or on-site assistance when navigating complex or irregular road conditions. While the naming of such systems may reflect intended future capabilities, their actual functionality corresponds to differing levels of automation when viewed through the lens of the SAE automation framework, and neither system meets the criteria for full autonomy as defined at Level 5, which entails operation under all roadway and environmental conditions without human oversight.

To further contextualize these findings, user perception of AV behavior in work zones was analyzed. Approximately 68% of documented reactions were positive, 21% were negative, and 11% were mixed. The analysis of specific AV capabilities showed that AV perception and sensing were the most frequently referenced aspects, with 30 strengths and 30 challenges. This balance reflects ongoing inconsistencies in how AVs interpret cones, lane markings, signage, and nearby actors. In contrast, decision-making and speed modulation functions showed greater reliability, with more strengths than challenges. Notably, human interaction and override functions emerged as key weaknesses, having the lowest strength counts and the highest challenge frequency, underscoring the complexity of engaging effectively with human agents in dynamic work zone environments.

Expanding this comparison, insights from users also shed light on the differences between Tesla and Waymo systems. Tesla's supervised FSD system exhibited strengths in real-time learning, maneuver execution, and adaptability to irregular geometry, though it was more prone to perception-related challenges and required more frequent driver intervention, likely due to its reliance on a camera-only vision system. Waymo, which employs LiDAR, radar, and cameras, demonstrated fewer perception errors and improved handling of human actors but was occasionally less flexible in responding to unmapped detours or temporary infrastructure changes. User perception was more positive for Tesla (77% of reactions) than Waymo (38%), possibly influenced by Tesla's semi-supervised mode allowing safety drivers to intervene, while Waymo's fully driverless model exposed the system to more critical observation during unexpected events. These system-specific insights were reinforced by a number of observed technical challenges. Common limitations included misinterpretation of temporary or inconsistently placed cones and barriers, difficulty detecting or reacting to faded or occluded lane markings and signage, and failure to interpret pedestrian gestures or construction worker hand signals in certain instances. Furthermore, some systems exhibited hesitation or inability to reroute around temporary detours or physical work zone obstructions. In parallel, infrastructure-related challenges emerged as a significant contributor to AV disengagements. These issues were frequently associated with zones featuring nonstandard geometry, inadequate signage, or ambiguous lane boundaries. AV confusion was prevalent in environments lacking high-contrast or reflective materials, or where cones were irregularly placed, especially under low-light or visually obstructed conditions. These findings emphasize the interdependence between AV design and infrastructure readiness.

To contextualize these findings, the 2018 NHTSA report [56] on Automated Driving Systems (ADS) presented performance metrics from public road testing that closely align with the challenges identified in this study. The report documented safety-critical events, including crashes and disengagements, most commonly occurring in complex environments such as intersections and non-standard road configurations. These conditions are consistent with the work zones analyzed in this study, where AVs exhibited uncertain or inconsistent behavior. Furthermore, the NHTSA's emphasis on the need for a "fallback-ready user" reflects this study's observations of frequent human intervention and user discomfort when AV systems failed to confidently navigate work zones. Furthermore, the findings of this study are well-aligned with the priorities and concerns highlighted in the 2021 final report of Autonomous/Connected Vehicles Assessment study in Work Zones [57]. First, the frequent user-reported confusion and system disengagements in work zones reflect the report's identification of work zones as high-risk, complex environments requiring targeted AV adaptation and infrastructure support. Second, users' calls for clearer communication, signage, and system transparency reinforce the report's emphasis on improving human–machine interface design and driver awareness. Third, the paper's insight into mixed public perception echoes the report's recognition of the importance of public education, outreach, and trust-building. Lastly, both our study and the report [57] stress the need for continuous learning from real-world deployments, stakeholder feedback, and collaborative efforts to advance AV safety and integration. Additionally, the 2024 report by the U.S. Department of Transportation [58] reinforces our findings by identifying construction zones as a key non-ADS factor contributing to disengagements, specifically citing scenarios such as blocked lanes. This consistency highlights a shared understanding that current AV systems face significant challenges operating reliably in construction zones, underscoring the need for enhanced system adaptability, improved infrastructure support, and targeted safety measures in these complex environments.

6. Future Directions

In response to the combined technical and infrastructure challenges, several targeted directions for improvement are proposed. Infrastructure enhancements should prioritize high-contrast, retroreflective, and machine-readable elements that function under varying weather and lighting conditions. Integrating Bluetooth beacons and dynamic signage could also help guide AVs during temporary layout changes. On the technology front, efforts should focus on strengthening perception models to recognize temporary and occluded signage, improving the recognition of human gestures, and enhancing system adaptability in diverse environmental conditions, including inclement weather. Complementing these technical and infrastructural strategies, policy and user-centered considerations must also be addressed. Regulators should develop and enforce AV-compatible work zone design standards and require manufacturers to report detailed system performance data under diverse scenarios. In terms of rider experience, incorporating real-time, context-aware in-vehicle messages, such as "Yielding to road worker" or "Construction detected; adjusting path"; can help reassure passengers and build trust. Structured educational tools, including mobile apps or onboarding sessions, can further support user understanding of AV behavior, particularly in work zone scenarios. To support safe and consistent AV operation in dynamic environments, improvements in human interaction protocols are also essential. Training for construction personnel on standardized AV interaction cues and emergency disengagement procedures could foster safer and more predictable engagements between AVs and human workers. Furthermore, enhancing data diversity is crucial. Future research should include a broader range of AV manufacturers, geographic regions, and digital platforms beyond YouTube to improve the generalizability of findings. Building on the insights identified in this study, future work will also focus on grounding observed AV decision-making in construction zones within formal algorithmic frameworks. Specifically, AV behaviors will be analyzed through the lens of established planning and optimization strategies, such as A*, D*, RRT, and constrained decision-making models. This approach will enable stronger connections between real-world AV performance and theoretical models of decision-making under uncertainty.

7. Limitations

While the findings from this study offer valuable insights, several limitations must be acknowledged. The reliance on user-generated YouTube videos, while offering real-world depth, may introduce selection bias or subjective interpretations. The dataset predominantly featured videos recorded in clear daylight conditions, with limited representation of challenging conditions such as fog, snow, rain, or high wind, as well as nighttime driving. The absence of internal AV data, such as sensor inputs, decision logs, and software versions, also limited the ability to determine the precise causes of system behavior.

Additionally, user demographic and psychological variables (e.g., age, gender, driving experience, stress level, familiarity with AVs) were not captured, though they likely influence the way AV behavior is perceived and reported. In addition, the sample of YouTube content creators and viewers may not be representative of the broader public. Individuals who post or engage with AV-related content on YouTube may differ in terms of technological interest, risk tolerance, communication style, or geographic and socio-economic background. These factors could influence which AV experiences are shared and how they are interpreted, potentially introducing bias in the observed patterns.

Another limitation lies in the geographic concentration of the dataset, which is predominantly composed of videos from the USA. Road infrastructure, signage conventions, construction practices, and driver behavior can vary across countries; all of which influence how AVs perceive and respond to work zone environments and shape how users interpret and evaluate AV performance in such contexts. For instance, a maneuver perceived as cautious or appropriate in one country might be viewed as confusing or unsafe in another due to differing norms and expectations. Expanding future datasets to include more diverse international content would enhance both the global relevance of AV behavior analysis and our understanding of cross-cultural differences in user perception.

In addition to expanding data diversity, future research could benefit from a hybrid approach that combines annotated AV-manufacturer footage with user-generated videos. In this study, we intentionally focused on user-generated videos to prioritize authentic, real-world experiences that reflect user interactions and spontaneous AV behavior. While manufacturer-produced content often offers structured insights into system behavior during edge cases, such as rare failure modes or complex construction scenarios; user-generated footage provides authentic context around real-world usage and human responses. Integrating both sources would allow for a more comprehensive analysis that leverages the technical precision of controlled tests and the experiential depth of spontaneous, real-world encounters. This blended methodology contributes to a more holistic evaluation of system robustness, safety, and user trust.

Given these limitations, future research should pursue several important directions. These include comparing the infrastructure and technological adaptations needed for safe AV navigation in rural versus urban areas, as well as exploring work zone personnel's perceptions of AV operations. Further investigation is also needed to assess the effectiveness of external AV indicators, such as lights or signage; in improving worker trust and safety. Additionally, studying AV performance under adverse weather conditions in conjunction with complex work zone layouts will be critical. Finally, expanding the dataset to include insights from AV users in cold-climate regions may inform strategies for safe navigation during road closures and hazardous weather events.

Author Contributions: Conceptualization, M.A., K.A., P.L. and Y.H.; methodology, M.A. and K.A.; software, M.A. and K.A.; validation, M.A., K.A., P.L. and Y.H.; formal analysis, M.A. and K.A.; investigation, M.A.; resources, M.A., K.A., P.L. and Y.H.; data curation, M.A. and K.A.; writing—original draft preparation, M.A. and K.A.; writing—review and editing, M.A., K.A., P.L. and Y.H.; visualization, M.A. and K.A.; supervision, P.L. and Y.H.; project administration, M.A., P.L. and Y.H.; funding acquisition, P.L. and Y.H. All authors have read and agreed to the published version of the manuscript.

Funding: This research was funded by the Upper Great Plains Transportation Institute at North Dakota State University and the Center for Multimodal Mobility in Urban, Rural, and Tribal Areas (CMMM), a Tier 1 University Transportation Center funded by the U.S. Department of Transportation. The authors are responsible for the content and accuracy of the information presented.

Data Availability Statement: Dataset available on request from the authors.

Acknowledgments: The authors would like to thank the Upper Great Plains Transportation Institute at North Dakota State University and the Tier 1 University Transportation Center for Multi-Modal Mobility in Urban, Rural, and Tribal Areas (CMMM) for their support.

Conflicts of Interest: The authors declare no conflicts of interest.

References

1. National Highway Traffic Safety Administration. Work/Construction Zones. Available online: https://www.nhtsa.gov/sites/nhtsa.gov/files/workzones.pdf (accessed on 5 March 2025).
2. Work Zone Barriers. Work Zone Fatalities, Injuries, & Crashes. Available online: https://www.workzonebarriers.com/work-zone-crash-facts.html (accessed on 22 April 2025).
3. National Safety Council. Motor Vehicle Safety Issues, Work Zone. Available online: https://injuryfacts.nsc.org/motor-vehicle/motor-vehicle-safety-issues/work-zones/ (accessed on 22 April 2025).
4. Ansarinejad, M.; Huang, Y.; Qiu, A. Impact of Fog on Vehicular Emissions and Fuel Consumption in a Mixed Traffic Flow with Autonomous Vehicles (AVs) and Human-Driven Vehicles Using VISSIM Microsimulation Model. In Proceedings of the International Conference on Transportation and Development 2023: Transportation Safety and Emerging Technologies, Austin, TX, USA, 14–17 June 2023. [CrossRef]
5. Maddineni, V.K.; Ansarinejad, M.; Ahmed, M.M. Safety and operational impacts of different Autonomous Vehicle operations on freeway work zones. *Adv. Trans. Stud.* **2024**, *4*, 117–126.
6. SAE International. SAE Levels of Driving Automation. Available online: https://www.sae.org/blog/sae-j3016-update (accessed on 28 September 2024).
7. Ansarinejad, M.; Ansarinejad, K.; Lu, P.; Huang, Y.; Tolliver, D. Autonomous Vehicles in Rural Areas: A Review of Challenges, Opportunities, and Solutions. *Appl. Sci.* **2025**, *15*, 4195. [CrossRef]
8. Makahleh, H.Y.; Ferranti, E.J.S.; Dissanayake, D. Assessing the Role of Autonomous Vehicles in Urban Areas: A Systematic Review of Literature. *Future Trans.* **2024**, *4*, 321–348. [CrossRef]
9. Das, S.; Dutta, A.; Lindheimer, T.; Jalayer, M.; Elgart, Z. YouTube as a Source of Information in Understanding Autonomous Vehicle Consumers: Natural Language Processing Study. *J. Trans. Res. Board* **2019**, *2673*, 242–253. [CrossRef]
10. State of California Department of Motor Vehicles. Disengagement Reports. Available online: https://www.dmv.ca.gov/portal/vehicle-industry-services/autonomous-vehicles/disengagement-reports/ (accessed on 22 April 2025).
11. State of California Department of Motor Vehicles. Autonomous Vehicle Collision Reports. Available online: https://www.dmv.ca.gov/portal/vehicle-industry-services/autonomous-vehicles/autonomous-vehicle-collision-reports/ (accessed on 22 April 2025).
12. National Highway Traffic Safety Administration(NHTSA). Incident Report Data. Available online: https://www.nhtsa.gov/data (accessed on 22 April 2025).
13. Waymo. Waymo Open Dataset. Available online: https://waymo.com/open/ (accessed on 22 April 2025).
14. Zhang, W.; Wang, K. Parking futures: Shared automated vehicles and parking demand reduction trajectories in Atlanta. *Land Use Policy* **2020**, *91*, 103963. [CrossRef]
15. Ansarinejad, M.; Ahmed, M.; Gaweesh, S. Assessing the Efficacy of Pre-Trained LLMs in Analyzing Autonomous Vehicles Field Test Disengagements. In *Assessing the Efficacy of Pre-Trained LLMs in Analyzing Autonomous Vehicles Field Test Disengagements*; Transportation Research Board (TRB): Washington, DC, USA, 2025; Available online: https://annualmeeting.mytrb.org/OnlineProgramArchive/Details/23204 (accessed on 11 May 2025).
16. Waymo. Waymo Driver. Available online: https://waymo.com/waymo-driver/ (accessed on 22 April 2025).
17. 12 News. Waymo Driverless Car Unable to Navigate Traffic Cones in Chandler. Available online: https://www.youtube.com/watch?v=B4O9QfUE5uI (accessed on 25 April 2025).
18. Chen, K. Waymo Robotaxi vs. Downtown Construction | Waymo Ride Along #1. Available online: https://www.youtube.com/watch?v=Dk4qc_4IgZQ (accessed on 22 April 2025).
19. GNAV TV. WAYMO Autonomous Driving—Driverless Car Speeding Through a Construction Zone—Tempe Arizona. Available online: https://www.youtube.com/watch?v=njfSKMcM5-w (accessed on 22 April 2025).
20. Chen, K. Driverless Car Confused by Construction Zone | Waymo Ride Along #14. Available online: https://www.youtube.com/watch?v=wWZGZWuUx-Y (accessed on 22 April 2025).
21. Baruchbrett. Waymo vs. Construction Equipment. Available online: https://www.youtube.com/watch?v=vcw48xJXZKg (accessed on 22 April 2025).
22. meme me. Fully Autonomous Car Glitches in a Construction Zone and Gets Stuck. Available online: https://www.youtube.com/watch?v=i96nws0MtDg (accessed on 22 April 2025).
23. Maya. 15 Clips of Waymo Vehicles Navigating Roadwork/Construction/Cones. Available online: https://www.youtube.com/watch?v=K08nfCxNKUM (accessed on 22 April 2025).
24. Marcel Irnie (Irnieracing). Google's Self Driving Electric Car Waymo Taxi Got Lost in California Construction Zone. Available online: https://www.youtube.com/watch?v=0mFZEuwIceQ (accessed on 22 April 2025).
25. TESLA. Impact Report 2024. Available online: https://www.tesla.com/impact (accessed on 22 April 2025).
26. Land, M. Tesla Model 3 Autopilot Perfection Through a Construction Zone. Available online: https://www.youtube.com/watch?v=clR994Jg4O0 (accessed on 22 April 2025).

27. Teslavangelist. First Try of Tesla Drive on Navigation: Construction Zone, Solid White Lines, etc. Available online: https //www.youtube.com/watch?v=bqbOEKZa2vg (accessed on 22 April 2025).
28. Tesla, D. Tesla Autopilot Avoids Construction Barrel in My Lane | Navigate on Autopilot | Full Self Driving |. Available online https://www.youtube.com/watch?v=SS10B51qAz0 (accessed on 22 April 2025).
29. CfTesla. Tesla Autopilot Creates It's Own Path Through Construction Zone! AMAZING!!! | Tesla Model 3. Available online https://www.youtube.com/watch?v=3wzkKM6RmVc (accessed on 22 April 2025).
30. RodrickEV. Can Autopilot Navigate A Construction Zone In The Rain? (HW3 2020.8.1). Available online: https://www.youtube com/watch?v=w2VWXwMyLIw (accessed on 22 April 2025).
31. JermEV. Tesla Model 3 Autopilot Through Construction Zone Jan 2020! Available online: https://www.youtube.com/watch?v= 4Ifa7-Caat0 (accessed on 22 April 2025).
32. Tesla Driver. Lorry SWERVES into My Lane SELF DRIVING Through Construction Zone!—Tesla Autopilot 2020.16.2.1. Available online: https://www.youtube.com/watch?v=WLI_ydsrrZM (accessed on 22 April 2025).
33. RodrickEV. Tesla Autopilot Can Drive in Construction Zone Lanes? (HW3 2020.4.1). Available online: https://www.youtube com/watch?v=5NdpgXef4sw (accessed on 22 April 2025).
34. Tesla Life Europe. Impressive Tesla Autopilot Performance at Night & in Construction Zone with EU Update 2020.40.3. Available online: https://www.youtube.com/watch?v=bVkpbNV4uEA (accessed on 22 April 2025).
35. Teslajake. Tesla Model 3 Autopilot Construction Zone. Available online: https://www.youtube.com/watch?v=mGV6_o1gHsY (accessed on 22 April 2025).
36. Tesla Shree. Tesla FSD Beta Version 10.5 | During Road Work | Moving Stop Sign! Available online: https://www.youtube.com/ watch?v=2UAGNrmlLGE (accessed on 22 April 2025).
37. TeslaNerd 360. Tesla FSD Beta 10.69.2.2 Construction Zone Confusion—Full Self Driving on Model Y. Available online: https: //www.youtube.com/watch?v=kXh-7gnq_OY (accessed on 22 April 2025).
38. The Megawatts. Tesla Model Y Full Self Driving FSD Almost Perfect! FSD Construction Zone Not So Much! MUST WATCH! Available online: https://www.youtube.com/watch?v=JEPzLnxdQvw (accessed on 22 April 2025).
39. Huey L. Tesla Full Self Driving Construction Zone (1). Available online: https://www.youtube.com/watch?v=NNZpVJEJjwY (accessed on 22 April 2025).
40. Minimal Duck. Tesla FSD—Construction Zone!! (Full Self Driving Beta 10.11.2). Available online: https://www.youtube.com/ watch?v=1QLEY4pT5HA (accessed on 22 April 2025).
41. ByTesla237. Tesla Full Self Driving Handles a Construction Zone Like a CHAMP!! Available online: https://www.youtube.com/ shorts/vGA21V30v_Y (accessed on 22 April 2025).
42. Spicy Tech. Tesla FSD Beta: Real-World Traffic Test—Discovering Opportunities for Improvement During Rush Hour. Available online: https://www.youtube.com/watch?v=49Wd0beBxUg (accessed on 22 April 2025).
43. Tesla Owners Silicon Valley. FSD Beta Handles Construction Zone. Available online: https://www.youtube.com/shorts/ oWbkN33POEs (accessed on 22 April 2025).
44. Torque News. Tesla FSD Beta Demonstrates Incredible Progress in Tight Construction Zone. Available online: https://www. youtube.com/watch?v=-Lcu4BEFQfA (accessed on 22 April 2025).
45. Winston's Garage. Tesla Autopilot VS Construction Zone! Available online: https://www.youtube.com/shorts/-mfgMf5PyjE (accessed on 22 April 2025).
46. It's Binh (Been) Reviewed & More. Tesla Model Y FSD Full Self Driving Beta Navigating Construction Zone Lots of Cones. Available online: https://www.youtube.com/watch?v=DWviwLfvVWI (accessed on 22 April 2025).
47. FSD604. Learning From the Construction Zone: A Tesla Model 3 Experience with FSD 12.3.3. Available online: https://www. youtube.com/watch?v=Qpe7LA4K5ho (accessed on 22 April 2025).
48. FSD604. FSD 12.3.3 in Construction Zone Navigating the Construction Zone: A Tesla Model 3 with FSD #TeslaFSD. Available online: https://www.youtube.com/watch?v=rxSNDqGJm2E (accessed on 26 April 2025).
49. Tesla FSD (Full Self Driving). Tesla FSD Beta 12: The ONE with the Construction Zone. Available online: https://www.youtube. com/watch?v=MNi-vsAIqD8 (accessed on 24 April 2025).
50. Iowa Tesla Guy. Tesla Full Self Driving (FSD) vs Heavy Construction Zone. Available online: https://www.youtube.com/watch? v=RR5jM3L-ad8 (accessed on 22 April 2025).
51. Jakub Kudlacz. Tesla FSD in the Rain Doing Its Thing! (Construction Zone, City Streets, Highway, Parking Lot). Available online: https://www.youtube.com/watch?v=fbnhWYgJgLY (accessed on 26 April 2025).
52. my tesla cybertruck. Tesla fsd. It's Smarter than Me. It's Taking Shortcuts to Avoid Road Construction, and Parks Itself. Available online: https://www.youtube.com/watch?v=e57jTMEI0WE (accessed on 22 April 2025).
53. Frunk to Trunk. Navigate Construction Zones with Ease! Tesla's Magic! Available online: https://www.youtube.com/shorts/ gLKc51bg4Cs (accessed on 22 April 2025).

54. FSD Observer. FSD 12.3.6 Lane Change to Passing Lane in Work Zone. Arthur Ashe Stadium Test #fsd #fsdbeta. Available online: https://www.youtube.com/shorts/_bwYe_8xnYM (accessed on 22 April 2025).
55. FSDDriver. Tesla FSD Nightmare: Construction Zone Chaos & No Fixes in Sight! Available online: https://www.youtube.com/watch?v=I9fPBwhTf9A (accessed on 22 April 2025).
56. National Highway Traffic Safety Administration; U.S. Department of Transportation. *A Framework for Automated Driving System Testable Cases and Scenarios*; National Highway Traffic Safety Administration: Washington, DC, USA, 2018. Available online: www.ntis.gov (accessed on 22 April 2025).
57. Sun, C.; Edara, P.; Adu-Gyamfi, Y.; Reneker, J.; Zhang, S. Investigation of Autonomous/Connected Vehicles in Work Zones. December 2021. Available online: https://swzdi.intrans.iastate.edu/ (accessed on 22 April 2025).
58. Khan, M.; Lemaster, D.; Najm, W.G. Understanding Safety Challenges of Vehicles Equipped with Automated Driving Systems (ADS)-Analysis of ADS Disengagements. 2024. Available online: https://www.transportation.gov/sites/dot gov/files/2024-08/HASS_COE_Understanding_Safety_Challenges_of_Vehicles_Equipped_with_ADS_Aug2024.pdf (accessed on 22 April 2025).

smart cities

MDPI

Article

A Driving Simulator-Based Assessment of Traffic Calming Measures at High-to-Low Speed Transition Zones

Ali Pirdavani [1,2,*], Mahdi Sadeqi Bajestani [2], Maarten Mantels [1] and Thibaut Spooren [1]

[1] UHasselt, Faculty of Engineering Technology, Agoralaan, 3590 Diepenbeek, Belgium; maarten.mantels@hotmail.com (M.M.); thibaut.spooren@gmail.com (T.S.)

[2] UHasselt, The Transportation Research Institute (IMOB), Martelarenlaan 42, 3500 Hasselt, Belgium; mahdi.sadeqibajestani@uhasselt.be

* Correspondence: ali.pirdavani@uhasselt.be

Highlights

What are the main findings?

- Physical traffic calming measures, such as chicane and road narrowing, reduced speeds and altered the deceleration behaviors at high-to-low speed transition zones.
- Psychological traffic calming measures such as avenue planting and transverse markings have minimal standalone effects on speed and deceleration but can support smoother transitions when combined with physical interventions.

What is the implication of the main finding?

- Integrating gateway designs into urban entry zones can enhance smart mobility strategies by reducing reliance on enforcement and promoting self-explanatory roads.
- The study offers actionable insights for integrating simulator-based testing in traffic engineering, allowing early-stage evaluation of road design interventions.

Abstract

Effective speed management at urban entry points is essential for ensuring traffic safety and supporting sustainable mobility in smart cities. This study contributes to urban mobility planning by using a high-fidelity driving simulation to evaluate gateway designs that enhance safety and behavioral compliance at built-up entry zones. Seven gateway configurations, comprising physical (i.e., chicanes, road narrowing) and psychological (i.e., transverse markings, avenue planting) speed calming measures, were evaluated against a reference scenario. A total of 54 participants completed a 14 km simulated route under standardized conditions, with vehicle speed, acceleration/deceleration, and lateral position continuously recorded. The strongest effects were observed in designs featuring chicanes, which achieved the largest speed reductions but also induced abrupt deceleration. In contrast, the combination of road narrowing and transverse markings resulted in a smoother and more gradual deceleration, minimizing driver discomfort and lateral instability. Psychological measures alone, such as avenue planting, had a limited impact on speed behavior. These findings highlight the importance of combining physical and psychological traffic calming measures to create effective, perceptually engaging transitions that promote safer and more consistent driver responses.

Keywords: speed transition zones; gateway design; traffic calming measures; driving simulator; urban mobility

1. Introduction

Managing vehicle speed during transitions from high-speed to low-speed zones is a critical challenge in the design of urban mobility systems. These transition zones, commonly situated at the entry to urban areas, often experience elevated crash risks due to immediate changes in driver expectations and insufficient adaptation to posted speed limits [1,2]. Traditional reliance on traffic signage alone has proven inadequate, as studies show that many drivers fail to perceive or comply with speed limit changes when entering built-up areas [3,4]. Addressing this challenge is essential for smart cities that prioritize traffic safety and sustainable urban mobility.

To enhance safety and compliance, policymakers have increasingly adopted gateway treatments, with physical and perceptual interventions designed to prompt earlier deceleration and improve driver awareness at the urban threshold [5]. These include physical (e.g., chicane, road narrowing) as well as psychological (e.g., transverse marking and avenue planting) traffic calming measures (TCMs) [6]. Such treatments aim to shape driver behavior through spatial constraints and the manipulation of visual and cognitive cues, aligning with the principles of behavioral design in urban systems [7,8].

Smart cities provide an opportunity to rethink the planning and evaluation of such interventions through data-driven approaches and simulation-based tools [9,10]. Driving simulators, in particular, enable the controlled testing of infrastructure layouts under repeatable conditions, offering detailed insight into driver behavior without real-world risks [11]. They are increasingly integrated into smart urban planning workflows, including digital twin platforms and scenario-based decision support systems.

This study contributes to this emerging direction by using a high-fidelity driving simulator to evaluate seven gateway designs, comprising both physical and psychological elements, for their effectiveness in moderating driver speed, acceleration/deceleration, and lateral positioning during the transition from rural to urban zones. A reference scenario without any TCMs serves as a baseline. In addition to objective behavioral metrics, subjective driver evaluations were collected to assess comfort and perceived effectiveness.

The aim is to identify which configurations most effectively promote safe and consistent deceleration before entering lower-speed zones, without inducing abrupt or uncomfortable maneuvers. By embedding these findings within the context of smart city planning and simulation-driven infrastructure design, this research offers practical insights for policymakers, engineers, and mobility planners seeking to enhance safety at critical urban thresholds.

2. Background and Related Works

Urban mobility highlights the importance of ensuring traffic safety on different road segments and transportation networks, specifically focusing on challenging conditions [12]. The transition from high-speed to low-speed road environments presents a critical challenge for traffic engineers seeking to enhance road safety and ensure drivers' compliance with posted speed limits. Various speed calming strategies have been developed and evaluated to address this, ranging from physical interventions to psychological treatments [13]. This section discusses three principal types of TCMs: vertical, lateral, and psychological. Emphasis is placed on how these measures influence driver behavior at transition zones and their comparative effectiveness in reducing speed and improving lane discipline under simulated conditions.

2.1. Vertical Traffic Calming Measures

Vertical TCMs play a central role in managing speed reduction at transitions from high-speed (e.g., 70 km/h) to low-speed (e.g., 50 km/h) road zones [14]. These interventions rely on physical elevation changes in the roadway to prompt deceleration. The most commonly applied vertical measures include speed humps, speed tables, and raised pedestrian crossings or intersections. Speed humps and bumps are short, abrupt elevations that force drivers to reduce speed to avoid discomfort or vehicle damage [15]. In contrast, speed tables offer a broader, flat-topped design that moderates speed while preserving ride comfort [16]. Raised crosswalks serve the dual function of enhancing pedestrian safety and reducing vehicle speed [17]. In practice, these devices are often deployed in combination with horizontal elements or visual markings to amplify their impact.

Driving simulator studies have extensively assessed the effectiveness of these vertical measures in speed transition zones. The results consistently demonstrate significant reductions in mean vehicle speed and notable changes in driver behavior at the location of vertical interventions. However, a common observation across studies is that drivers frequently accelerate after passing the calming device, diminishing the long-term effectiveness unless additional measures are implemented downstream [18]. To address this, simulations have tested vertical elements in series, spaced strategically to maintain reduced speeds over longer road segments. For example, placing speed humps or raised platforms every 75 to 200 m can prevent drivers from regaining high speeds too quickly [19].

The effectiveness of vertical measures is further enhanced when combined with other types of TCMs, such as road narrowing, horizontal deflections, or changes in pavement texture. Driving simulator studies have shown that such combinations lead to more sustained reductions in speed, as they increase cognitive load and force drivers to remain attentive throughout the transition zone. Even low-cost or temporary vertical treatments, such as portable crash barriers or marked elevations, have demonstrated meaningful impacts in simulator-based experiments, particularly at intersections or gateways into urban areas [20].

Design considerations derived from simulator-based research are crucial for optimizing the deployment of vertical calming devices [21]. Appropriate spacing between successive devices is essential to sustain the desired speed profile, with a recommended maximum distance of 200 m for maintaining speeds below 50 km/h, and closer spacing (around 75 m) for zones targeting 40 km/h or lower speeds. Additionally, differences in vehicle type significantly influence the perceived impact of vertical measures, emphasizing the need for tailored designs based on the expected traffic mix [14]. Overall, driving simulator studies affirm the utility of vertical speed calming devices in transition zones and offer empirical guidance for their strategic implementation to ensure both safety and effectiveness.

2.2. *Lateral Traffic Calming Measures*

Lateral TCMs aim to reduce vehicle speeds by modifying the horizontal alignment of the roadway or creating visual cues that alter a driver's perception of the driving environment. These interventions compel drivers to adjust their lateral path, thereby promoting speed reduction through steering adjustments and increased cognitive demand. Common lateral measures include chicanes, gate constructions, horizontal curves, and visual narrowing through pavement markings [22]. Driving simulator studies offer valuable insights into how these designs influence driver behavior during speed transitions from high-speed to low-speed zones.

Among the most prominent lateral interventions, chicanes have demonstrated consistent effectiveness in both simulator and field studies [23]. By introducing alternating shifts in road alignment, chicanes force drivers to reduce speed and maneuver carefully through the deflections [24]. Their efficacy is highly dependent on geometric design factors such as curvature, spacing, and deflection angle. Horizontal curves similarly sustain speed reductions over longer distances, as they increase driver workload and attentiveness. Simulator results suggest that although these features may slightly increase lateral position variability, they contribute meaningfully to lower speeds across the transition zone [22,25].

Gateway constructions, physical or visual structures placed at the boundary of speed zones, are another proven lateral measure. These gateways serve as strong psychological signals that cue drivers to decelerate upon entering a lower-speed area. Driving simulator studies indicate that gate constructions lead to localized speed reductions and modest increases in lateral movement, but these do not pose significant safety risks [26]. Similarly, peripheral hatched markings, which visually narrow the perceived width of the road, encourage drivers to maintain a central lane position while decreasing travel speed [27]. These visual treatments have been shown to reduce lateral wandering and enhance lane discipline, especially in transition zones where abrupt speed changes are expected.

In sum, lateral speed calming measures, particularly when designed with attention to human perception and roadway context, offer a powerful set of tools to manage driver behavior in speed transition zones. Empirical findings from simulator studies should guide the selection and placement of such measures to ensure optimal effectiveness and safety.

2.3. Psychological Traffic Calming Measures

Psychological TCMs are non-physical interventions that rely on visual stimuli and perceptual manipulation to influence driver behavior. Unlike vertical or lateral speed calming devices, these measures alter how drivers perceive the road environment rather than the road itself. They are instrumental in high-speed to low-speed transition areas where subtle yet effective cues can encourage earlier and more consistent deceleration. Driving simulator studies provide a controlled environment to isolate and assess the impact of such psychological interventions on speed regulation and lane discipline.

One of the most studied psychological interventions involves modified road signage. These include larger, more vividly colored, or uniquely shaped signs that draw the driver's attention to upcoming speed reductions [28]. Simulator experiments have shown that such signs can be as effective as physical road markings in reducing vehicle speeds at the entry to lower-speed zones [29]. Enhanced signage functions by capturing the driver's cognitive focus earlier, enabling smoother deceleration and greater compliance with posted limits. The early engagement it prompts is particularly important in transitions where conventional signage might otherwise be overlooked or filtered out cognitively. In addition, strategically placed vegetation along the roadway creates an optical narrowing effect [30], contributing to a reduction in driving speed. The amount of vegetation and its distance from the road are important factors [31].

Another set of psychological interventions includes road surface markings, such as transverse lines, colored pavement sections, and peripheral hatched markings. These visual cues create the illusion of speed or narrowing, which in turn prompts drivers to reduce their speed. Though technically tactile, transverse rumble strips contribute visually by emphasizing deceleration zones [32]. Driving simulator research has shown that these markings not only reduce average speed but also improve lateral control, particularly when peripheral markings are used. Gateways, whether painted or physically framed, also serve as strong psychological indicators of a changing traffic environment, leading to immediate deceleration at the transition threshold. Additionally, optical pavement treatments, such as converging lines, gradient markings, or brightness-modulated textures, have proven effective in manipulating speed perception [33]. These techniques rely on visual illusions that suggest acceleration or narrowing, prompting drivers to slow down reflexively. Simulator studies confirm that these markings successfully reduce speeds without compromising lateral control [34].

While psychological measures consistently produce significant reductions in speed at the point of transition, their effects tend to diminish once drivers pass beyond the treated zone. Simulator studies indicate that although drivers slow down in response to visual cues, many begin accelerating again shortly afterward. This rebound effect highlights the importance of continuity or repetition in applying psychological calming strategies. These measures are best used in conjunction with physical devices or as part of a sequenced intervention plan that maintains driver attention and reinforces speed regulation across the transition area to achieve more sustained impact.

Despite extensive research on TCMs, notable gaps persist in the literature regarding the combined effects of physical and psychological interventions in high-to-low speed transition zones. Previous studies have predominantly examined isolated TCMs without evaluating their synergistic potential when implemented in coordinated gateway designs. Moreover, there is limited empirical understanding of how such combinations simultaneously influence longitudinal (speed, acceleration) and lateral (deviation) driving behavior. This study addresses these gaps by systematically testing seven distinct gateway configurations that integrate both physical and psychological elements using a high-fidelity driving simulator. This research offers insights into the effectiveness of combined TCM strategies by capturing a wide range of behavioral indicators, including speed, acceleration/deceleration, and lateral positioning. The findings directly inform the design of smarter, more self-explanatory urban entry points, aligning with the broader objectives of sustainable and data-informed urban mobility.

3. Materials and Methods

The primary aim of this study was to assess the effectiveness of seven distinct gateway designs using a driving simulator. Each gateway design comprises a combination of physical and psychological TCMs. These designs were evaluated by analyzing participant driving behavior in terms of speed, acceleration/deceleration, and lateral position. This allowed for a comprehensive assessment of how each design influenced driver responses and maneuver execution. To determine the impact of each gateway design, comparisons were made against a reference scenario without any intervention. The systematic collection and analysis of relevant data enabled a direct comparison of each design's performance and behavioral effects.

3.1. Overview of Gateway Designs

Figure 1 provides a general overview of all gateway designs implemented in this research. It shows the distances within each design and the position of the 'built-up area' sign, which signals the drivers that they are entering the low-speed zone.

The transition zone is designed with a fixed length of 100 m, allowing for a smooth and comfortable deceleration from 70 km/h to 50 km/h [35,36]. Based on existing literature, this distance effectively facilitated the intended speed reduction [37]. In four of the tested configurations, transverse road marking and avenue planting were implemented before the transition zone. In all seven designs, physical TCMs (i.e., chicane, road narrowing, or a combination of both) were employed in the transition zone. Additionally, a reference design without any intervention served as the baseline condition.

Each design scenario comprised a total roadway length of 650 m and simulated driving in the direction from a rural to an urban environment. Both the 70 km/h and 50 km/h segments were configured as two-way roads with one lane per direction and equipped with a centerline and edge markings. The minimum required pavement widths for these zones were 6.40 m and 6.10 m, respectively, with corresponding lane widths of 3.00 m and 2.85 m.

For simulation purposes, a uniform pavement width of 7.00 m was applied to the 70 km/h zone, and 6.50 m to the 50 km/h zone. The edge markings were 0.15 m wide, and the centerline consisted of broken segments measuring 2.5 m in length and 0.15 m in width, spaced at 10 m intervals. A continuous curb, measuring 10 cm in height and 50 cm in width, was added along the transition zone, as prior studies indicated its potential to double the speed-reducing effect compared to zones without curbs [26]. The built-up area sign was positioned precisely at the end of the 100 m transition zone. All driving scenarios were performed under daytime conditions with good visibility to ensure consistency across all experimental designs.

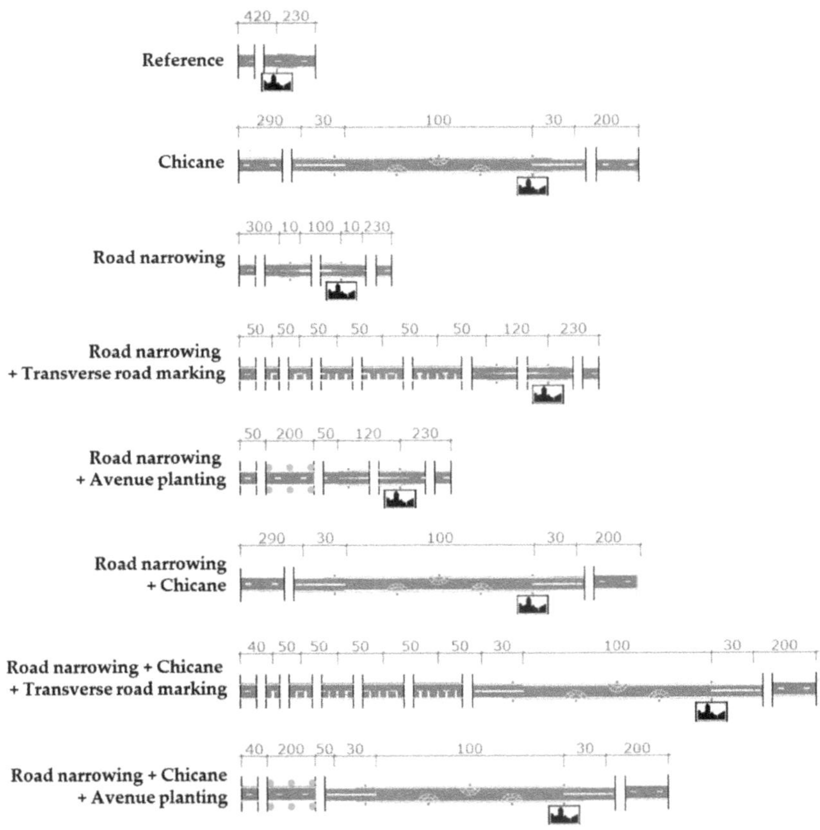

Figure 1. General overview of all gateway designs.

In the road narrowing design, the roadway width was visually and physically reduced. The total pavement width was narrowed to 6.70 m, with each lane measuring 2.85 m, corresponding to the minimum allowable lane width for a speed zone of 50 km/h [36]. This layout preserved sufficient paved space to accommodate wider vehicles while maintaining the narrowing effect. The construction was reinforced through road markings, enhancing its visual impact. Appropriate signage was implemented to inform drivers: a warning sign for the upcoming narrowing was placed 200 m in advance, followed by a "no overtaking" sign at 100 m before the narrowing. At 100 m beyond the end of the narrowed road section, a final sign indicated that overtaking was once again permitted.

3.2. Overview of Interventions

In this study, the term 'chicane' refers to a staggered arrangement of three half-circle speed cushions that primarily function as a lateral deflection measure. Figure 2 provides further clarification and shows the dimensions within the transition zone. The cross-section AA′ shown in this figure is further elaborated in Figure 3.

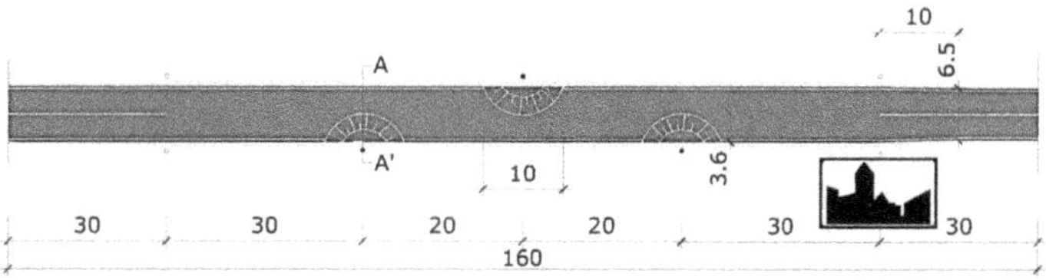

Figure 2. Dimensions of the chicane in meters with indication of cross-section AA′.

Each speed cushion features a linear incline, sloping upward from the center toward the edge of the roadway, reaching a maximum height of 14 cm at the outer edge. This configuration slightly deviates from the conventional standard, typically consisting of a uniform elevation of 12 cm across the full roadway width [38]. Due to technical constraints within the simulation software, it was not feasible to implement a consistent 12 cm height across the entire surface. The cushions were constructed in each single passing lane with a 4% slope along the cross-section to accommodate this. To enhance visibility and alert drivers, delineator posts were placed 75 cm from the edge of the paved surface at the location of each raised element. A cross-section of the roadway at the location of a cushion is shown in Figure 3.

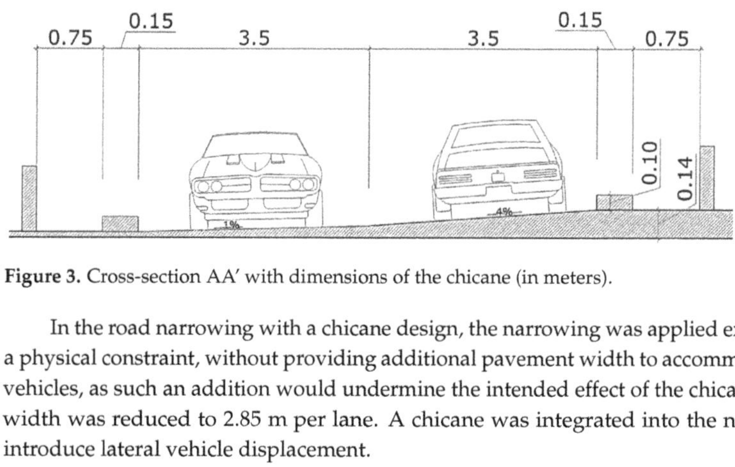

Figure 3. Cross-section AA′ with dimensions of the chicane (in meters).

In the road narrowing with a chicane design, the narrowing was applied exclusively as a physical constraint, without providing additional pavement width to accommodate wider vehicles, as such an addition would undermine the intended effect of the chicane. The lane width was reduced to 2.85 m per lane. A chicane was integrated into the narrowing to introduce lateral vehicle displacement.

The designs incorporating transverse road markings were applied over a 200 m stretch, with a new group of markings every 50 m. Each marking measured 0.5 m in width, with a minimum spacing of 4 m between successive lines [39]. The number of markings per group increased progressively, starting with one and culminating in five transverse lines in the final group.

Finally, visual narrowing was introduced in designs that included avenue planting by placing tall trees with high trunks, approximately 13 m in height. These were spaced at 10 m intervals along a 200 m stretch and planted at a distance of 2 m from the edge of the roadway. This configuration was intended to enhance the psychological narrowing effect and reinforce the perception of entering a lower-speed environment.

3.3. Simulator Setup

Simulations were conducted using an in-house fixed-base driving simulator (STISIM Drive® 3; Systems Technology Inc., Hawthorne, CA, USA). The simulator consisted of a Ford Mondeo chassis, integrated with a 180-degree projection system (utilizing three projectors) and Fanatec hardware components, which provided a wide field of view and high visual immersion. The simulator is shown in Figure 4.

Figure 4. Driving simulator with 180° projection system.

3.4. Participants and Procedure

A total of 54 participants (13 women and 41 men) took part in the experiment. Before participation, all individuals signed an informed consent form and completed a pre-survey. Before starting the experimental trials, each participant performed a brief familiarization drive to become accustomed to the simulator environment. Eligibility criteria included possessing at least a provisional category B driving license and a minimum of 20 h of driving experience. Participants were excluded in cases of pregnancy, illness, or substance impairment (e.g., alcohol or drug use). Participation was entirely voluntary and could be withdrawn at any time. The experimental drives comprised sequential gateway designs distributed across a 14 km driving route. Each scenario featured a combination of TCMs interspersed with neutral road segments, referred to as filler pieces, to mitigate carryover effects. The scenarios specifically targeted the transition from rural to urban environments.

In total, four unique driving scenarios were developed, each presenting the gateway designs in a different order. This variation was intended to minimize the learning effect—the potential change in driving behavior due to increasing familiarity with the simulation environment. By alternating both the structure and sequencing of scenarios across participants, predictability was reduced, and habituation was effectively minimized. Although these steps reduced the likelihood of behavioral learning unrelated to the interventions, it cannot be ruled out entirely that repeated exposure influenced some responses.

The present study did not perform subgroup analyses by age, gender, or driving experience, as the main objective was to isolate the effects of gateway designs rather than demographic variability. While informative, such analyses would have extended the scope considerably beyond the intended focus of this research. We aim to identify relative differences between gateway designs under controlled conditions, not to estimate population-level effects for all driver groups. It should also be noted that with 54 participants, our sample size is relatively large for driving simulator experiments, which typically involve fewer participants due to the resource-intensive nature of such studies.

3.5. Data Collection and Processing

Data collection was conducted for driving behavior metrics. The collected data were analyzed using IBM SPSS Statistics version 29.0.1.1. Objective variables, including speed, acceleration/deceleration, and lateral position, were recorded at a sampling frequency of 0.014 s across the whole 650 m test trajectory and predefined measurement points. Each gateway design was treated as a separate data block. A comprehensive set of measurements was gathered during the simulation runs and subsequently subjected to parameter-specific analysis to minimize data loss.

The processing of participant data began with identifying and excluding statistical outliers. If a participant was flagged as an outlier for any of the main parameters (i.e., speed, lateral deviation, or acceleration/deceleration), they were excluded from the dataset entirely. Outliers were determined based on the interquartile range (IQR) method, with values exceeding 1.5 times the IQR above the third quartile or below the first quartile classified as extreme [40,41]. These thresholds were consistently applied across all parameters. Participants exhibiting outlier values in more than 15% of the measurement points were excluded from the analysis, which aligned with common practices reported in the literature [42]. As a result, no outliers were detected in the collected dataset, indicating a consistent and reliable data distribution.

4. Results

This section presents the findings of the driving simulator study, structured around three key behavioral indicators: speed, acceleration/deceleration, and lateral deviation. Each parameter was analyzed across the eight gateway designs to assess the effectiveness of individual and combined TCMs. The results are derived from data captured at predefined measurement points throughout the gateway. Measurement points were carefully selected for each parameter to ensure relevance and accuracy so that the influence of each design, as well as each TCM, can be interpreted.

Given the unbalanced design of our driving simulator experiments, we employed a Linear Mixed-effect Model (LMM) approach, which is well-suited for analyzing repeated measures and data structures with unequal group sizes or missing observations. Unlike traditional ANOVA, which assumes a balanced design and homogeneity of variance, LMMs can account for both fixed and random effects, offering greater flexibility and robustness in real-world experimental setups [43,44]. Type III tests of fixed effects were used within the LMM framework to identify statistically significant differences in driving behavior across the tested designs.

The following subsections discuss the results in detail, highlighting where and how the TCMs influenced driver performance. First, speed and acceleration/deceleration data are presented from measuring points across all designs. Then, the effectiveness of each TCM, if implemented as a standalone design, is also statistically studied. Finally, the drivers' behavior when encountering chicanes, as the most influential TCMs, is investigated.

4.1. Speed

Speed data were collected at six fixed measurement points to assess how each gateway design influenced speed. These points include: the beginning of the design segment, the end of any pre-transition measure (e.g., markings or avenue planting), three locations evenly spaced within the 100 m transition zone (i.e., start, middle, and end), and a final point 100 m downstream of the built-up entry sign. Figure 5 shows the locations of the measuring points for speed recording.

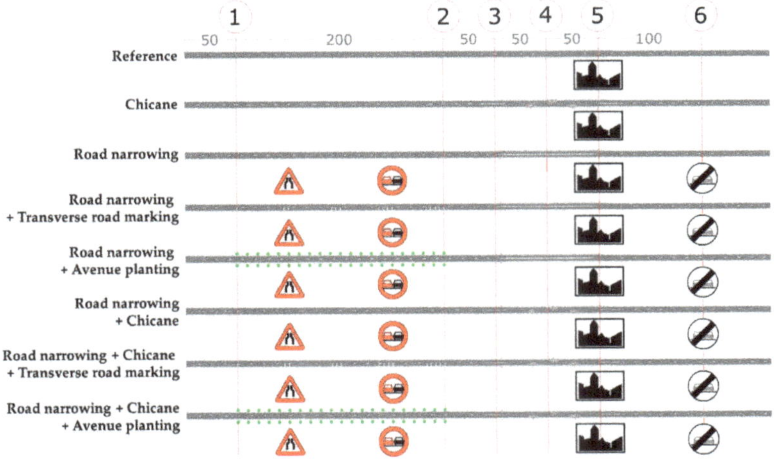

Figure 5. Measurement points for speed (distances are in meters).

Table 1 presents the results of the Type III tests of fixed effects with speed as the dependent variable. The analysis reveals that both the gateway's design and the measurement point's location had a statistically significant impact on vehicle speed. Specifically, the design yielded an *F*-value of 82.015 ($p < 0.001$), indicating that the different gateway designs resulted in substantial variations in average driving speed. Similarly, the effect of measurement point was significant ($F = 433.841$, $p < 0.001$), reflecting the expected variation in speed across the transition zone and into the urban segment. The significant intercept ($F = 4777.922$, $p < 0.001$) confirms that the estimated baseline level of the dependent variable—representing the expected value when all predictor variables are set to zero—is significantly different from zero, confirming that the model captures a meaningful starting point for the outcome measure.

Table 1. Type III tests of fixed effects with speed as the dependent variable.

Source	Numerator Df	Denominator Df	*F*	*p*
Intercept	1	52	4777.922	<0.001
Design	7	2409	82.015	<0.001
Measurement Point	5	2409	433.841	<0.001

In the LMM with speed as the dependent variable, the effect of the gateway design on average speed was systematically analyzed. The results are shown in Table 2. The model's intercept was estimated at 53.378 km/h, representing the expected mean speed in the reference design at measurement point 6, which is located within the built-up area where the legal speed limit is 50 km/h. All design-related coefficients were found to be negative and statistically significant ($p < 0.001$), indicating that the implemented measures resulted in significantly lower speeds compared to the reference scenario.

The largest reductions were observed in designs combining chicane and road narrowing. This gateway design led to an average speed decrease of more than 9 km/h relative to the reference. Designs featuring only a chicane produced reductions of over 7 km/h, while a standalone road narrowing yielded a moderate decrease of just over 3 km/h. The positive estimated coefficients for most measurement points, except for point 5 (i.e., the actual start point of the urban zone indicated by the built-up entry sign), reflect higher speeds at those locations compared to the reference point, and the least amount of speed at point 5. Collectively, the values reveal a gradual speed-reduction pattern across successive measurement locations. However, within individual designs, no statistically significant differences in average speed were found between measurement points 4, 5, and 6, suggesting a consistent speed level in that zone.

Table 2. Estimates of fixed effects with speed as the dependent variable.

	Parameter	Estimate	Std Error	Df	*t*	*p*
	Intercept	53.378	0.937	110.558	56.965	<0.001
Designs	Reference *	0	0	–	–	–
	Chicane	−7.315	0.608	2409	−12.032	<0.001
	Road narrowing	−3.032	0.608	2409	−4.987	<0.001
	Road narrowing + Transverse road marking	−3.925	0.608	2409	−6.455	<0.001
	Road narrowing + Avenue planting	−3.275	0.608	2409	−5.386	<0.001
	Road narrowing + Chicane	−10.020	0.608	2409	−16.481	<0.001
	Road narrowing + Chicane + Transverse road marking	−10.485	0.608	2409	−17.245	<0.001
	Road narrowing + Chicane + Avenue planting	−9.405	0.608	2409	−15.470	<0.001
Points	1	18.417	0.527	2409	34.977	<0.001
	2	12.246	0.527	2409	23.258	<0.001
	3	6.035	0.527	2409	11.461	<0.001
	4	0.636	0.527	2409	1.209	0.227
	5	−0.545	0.527	2409	−1.035	0.301
	6 *	0	0	–	–	–

* Baseline categories.

Figure 6 presents the speed profiles across the six measurement points for each gateway design. The graph demonstrates that the presence and type of TCMs significantly influence how drivers adapt their speed while approaching and traversing the transition zone. Designs incorporating a chicane have formed a cluster, associated with the most pronounced speed reductions, with average speeds dropping below 45 km/h before the built-up entry sign. In contrast, configurations lacking a chicane, such as the reference design or those featuring only visual cues or road narrowing, exhibit a more gradual decrease pattern, with speeds typically remaining above the 50 km/h threshold until after the transition is completed.

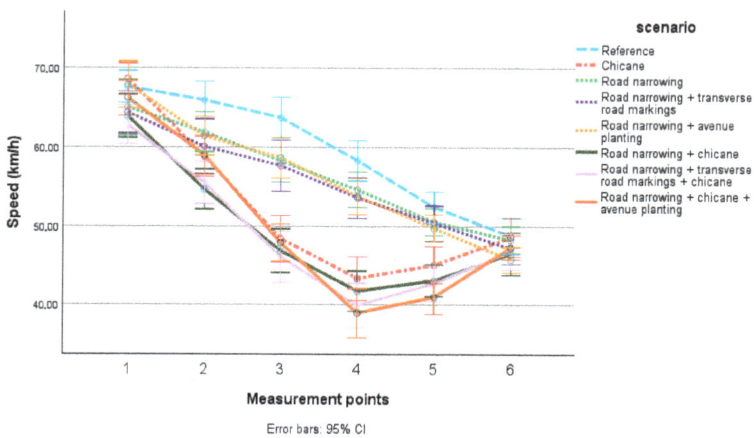

Figure 6. Speed at measurement points in different designs.

4.2. Acceleration/Deceleration

This section analyzes the effects of the different gateway designs on drivers' longitudinal acceleration and deceleration behavior. Acceleration/deceleration was measured at five specific points in each gateway design, capturing the dynamics before, during, and after the TCM. The spatial distribution of these measurement points is shown in Figure 7.

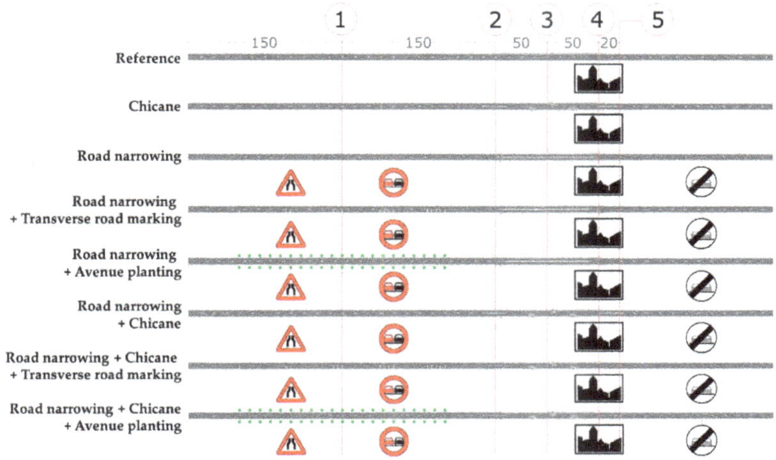

Figure 7. Measurement points for acceleration/deceleration (distances are in meters).

Table 3 presents the outcomes of the Type III fixed effects tests with acceleration/deceleration as the dependent variable. The results indicate that all three examined factors, intercept, design, and measurement point, have statistically significant effects on acceleration behavior. The design factor yielded an *F*-value of 6.355 with a *p*-value below 0.001, confirming that the specific gateway configuration has a meaningful influence on how drivers accelerate or decelerate. An even more substantial effect was observed for the measurement point, with an *F*-value of 75.909, reflecting a substantial variation in acceleration as drivers progress through different segments of the test environment. Additionally, the intercept was highly significant, confirming that the baseline acceleration/deceleration level, corresponding to the condition when all predictors are zero, is statistically different from zero.

Table 3. Type III tests of fixed effects with acceleration/deceleration as the dependent variable.

Source	Numerator Df	Denominator Df	*F*	*p*
Intercept	1	50	156.646	<0.001
Design	7	1978	6.355	<0.001
Measurement Point	4	1978	75.909	<0.001

Table 4 presents the fixed effects estimates for acceleration and deceleration behavior across various gateway design configurations and measurement points. The reference design showed a slight deceleration (-0.107 m/s^2), serving as the baseline. Designs featuring a chicane, either alone or in combination with road narrowing and psychological measures, demonstrated significantly higher positive estimates, indicating more pronounced acceleration events. The strongest effects were observed in configurations combining chicanes with transverse markings or avenue planting, each producing acceleration estimates around 0.23 m/s^2 ($p < 0.001$). This pattern reflects a rebound effect where the acceleration following the gateway can outweigh the earlier deceleration, resulting in net positive values.

To avoid this potential bias, the analysis in Section 4.3 focuses only on zones where the gateway design is located, thereby excluding the post-gateway segment from consideration. This allows the impact of each individual TCM to be assessed more directly. Moreover, road narrowing alone (0.086 m/s^2) was not statistically significant (p = 0.095), and the combination with avenue planting (0.04 m/s^2) showed no meaningful effect (p = 0.442).

Measurement point analysis revealed that points 1, 2, and 3 show significant negative estimates, with point 2 exhibiting the most pronounced deceleration (estimate = −0.594, $p < 0.001$), while points 4 and 5 (i.e., the reference point) do not significantly differ.

Table 4. Estimates of fixed effects with acceleration/deceleration as the dependent variable.

	Parameter	Estimate	Std Error	Df	t	p
	Intercept	−0.107	0.045	1699.260	−2.390	0.017
Designs	Reference *	0	0	–	–	–
	Chicane	0.161	0.051	1978	3.124	0.002
	Road narrowing	0.086	0.051	1978	1.673	0.095
	Road narrowing + Transverse road marking	0.130	0.051	1978	2.525	0.012
	Road narrowing + Avenue planting	0.040	0.051	1978	0.768	0.442
	Road narrowing + Chicane	0.229	0.051	1978	4.441	<0.001
	Road narrowing + Chicane + Transverse road marking	0.230	0.051	1978	4.468	<0.001
	Road narrowing + Chicane + Avenue planting	0.239	0.051	1978	4.649	<0.001
Points	1	−0.122	0.041	1978	−2.994	0.003
	2	−0.594	0.041	1978	−14.6	<0.001
	3	−0.300	0.041	1978	−7.366	<0.001
	4	0.000	0.041	1978	0.005	0.996
	5 *	0	0	—	—	—

* Baseline categories.

Figure 8 illustrates acceleration/deceleration patterns across the measurement points for gateway designs. A distinct variation in how drivers respond can be observed depending on the type of TCM encountered. Designs incorporating a chicane exhibit sharp deceleration just before the chicane, followed by a noticeable acceleration immediately afterward, highlighting a rebound effect. This rapid deceleration–acceleration sequence reflects a more abrupt and reactive driving style. In contrast, designs with road narrowing show a more gradual and linear deceleration behavior with a modest and delayed recovery in acceleration, suggesting a smoother and more predictable driver response. The reference design and those with only visual cues, such as transverse road markings or avenue planting, demonstrate relatively steady acceleration/deceleration profiles, indicating minimal behavioral adaptation.

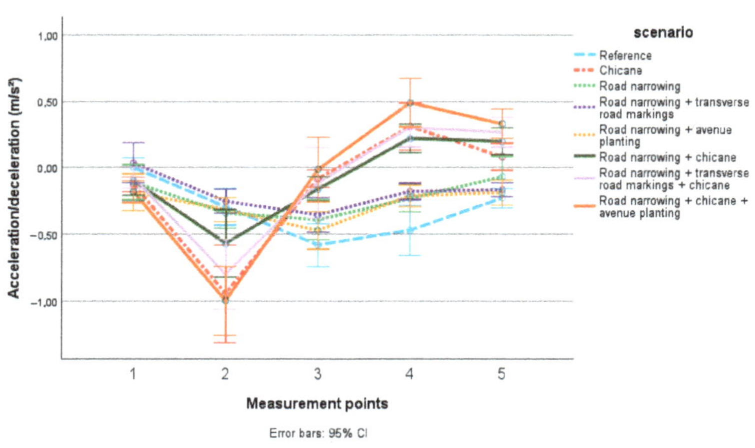

Figure 8. Acceleration/deceleration at measurement points in different designs.

4.3. Impact Assessment of Individual Traffic Calming Measures

We extended our analysis by performing additional LMM to gain a deeper understanding of how each individual TCM affects driving behavior. While the initial LMMs reported in previous sections focused on combined TCMs, these subsequent models isolate the independent impact of each treatment. The analysis was restricted to the segment where the gateway design was located to ensure that the observed effects reflect the direct influence of the TCMs rather than downstream driving responses. By excluding post-gateway segments where behavioral rebound effects may occur, this focused approach improves the precision of our estimates and allows for a more precise evaluation of each TCM's local effect. This enables us to identify which individual TCMs exert the most influence when controlling for other factors.

Table 5 presents the results of the Type III tests of fixed effects for speed, focusing on the independent contribution of each TCM. The chicane intervention had a statistically significant impact on driving speed ($F = 246.286$, $p < 0.001$), confirming its strong influence in reducing speed. The impact of road narrowing is also statistically significant ($F = 22.745$, $p < 0.001$), suggesting a less pronounced but still meaningful impact. In contrast, avenue planting ($F = 1.364$, $p = 0.243$) and transverse road markings ($F = 1.354$, $p = 0.245$) did not show statistically significant effects.

Table 5. Type III tests of fixed effects of individual TCMs with speed as the dependent variable.

Source	Numerator Df	Denominator Df	F	p
Intercept	1	84.976	3618.853	<0.001
Chicane	1	1585	246.289	<0.001
Road narrowing	1	1585	22.745	<0.001
Avenue planting	1	1585	1.364	0.243
Transverse marking	1	1585	1.354	0.245

Table 6 provides parameter estimates for the fixed effects with speed as the dependent variable. For each TCM, the reference category is where the intervention was present, while the effect estimate represents the mean difference in speed when the TCM is absent. The presence of a chicane corresponds to a significant reduction in speed, with an estimated increase of 8.122 km/h when the chicane is removed ($p < 0.001$), underscoring its strong traffic calming effect. Similarly, removing road narrowing results in a 3.491 km/h speed increase ($p < 0.001$), confirming its moderate but significant role in reducing speed. In contrast, the absence of avenue planting yields a negligible and non-significant effect (estimate = −0.855 km/h, $p = 0.243$), as does the removal of transverse markings (estimate = 0.852 km/h, $p = 0.245$).

Table 6. Estimates of fixed effects with speed as the dependent variable for each TCM.

Parameter	Estimate	Std Error	Df	t	p
Intercept	51.847	1.233	224.459	42.037	<0.001
Chicane = 0	8.122	0.518	1585	15.694	<0.001
Chicane = 1 *	0	0	—	—	—
Road narrowing = 0	3.491	0.732	1585	4.769	<0.001
Road narrowing = 1 *	0	0	—	—	—
Avenue planting = 0	−0.855	0.732	1585	−1.168	0.243
Avenue planting = 1 *	0	0	—	—	—
Transverse marking = 0	0.852	0.732	1585	1.163	0.245
Transverse marking = 1 *	0	0	—	—	—

* Baseline categories.

Table 7 presents the Type III fixed effects analysis outcomes using acceleration/deceleration as the dependent variable. The chicane exhibits the most substantial effect ($F = 4.155$, $p < 0.042$), indicating its influence on driving behavior by inducing deceleration. In contrast, all other TCMs, including road narrowing ($F = 1.197$, $p = 0.274$), avenue planting ($F = 2.344$, $p = 0.126$), and transverse road markings ($F = 0.134$, $p = 0.715$), did not significantly impact acceleration.

Table 7. Type III tests of fixed effects of individual TCMs with acceleration/deceleration as the dependent variable.

Source	Numerator Df	Denominator Df	*F*	*p*
Intercept	1	2.028	0.751	0.476
Chicane	1	1177	4.155	0.042
Road narrowing	1	1177	1.197	0.274
Avenue Planting	1	1177	2.344	0.126
Transverse marking	1	1177	0.134	0.715

Table 8 offers parameter estimates to further interpret the effect of each TCM on acceleration/deceleration. A negative estimate indicates stronger deceleration compared to the reference condition (i.e., presence of TCM). The presence of a chicane results in significantly higher deceleration, with a mean increase of 0.077 m/s^2 when the chicane is removed ($p = 0.042$). Road narrowing (estimate = -0.058 m/s^2, $p = 0.274$), avenue planting (estimate = 0.081 m/s^2, $p = 0.126$), and transverse markings (estimate = -0.019 m/s^2, $p = 0.715$) did not significantly differ from their reference conditions.

Table 8. Estimates of fixed effects with acceleration/deceleration as the dependent variable for each TCM.

Parameter	Estimate	Std Error	Df	*t*	*p*
Intercept	−0.377	0.392	2.115	−0.960	0.434
Chicane = 0	0.077	0.038	1177	2.038	0.042
Chicane = 1 *	0	0	—	—	—
Road narrowing = 0	−0.058	0.053	1177	−1.094	0.274
Road narrowing = 1 *	0	0	—	—	—
Avenue planting = 0	0.081	0.053	1177	1.531	0.126
Avenue planting = 1 *	0	0	—	—	—
Transverse marking = 0	−0.019	0.053	1177	0.366	0.715
Transverse marking = 1 *	0	0	—	—	—

* Baseline categories.

4.4. Lateral Deviation Behavior

Since scenarios without chicanes showed negligible differences in lateral deviation among themselves, including the reference condition, the analysis of lateral behavior was focused only on gateway designs including chicanes. While the chicane intervention was found to reduce vehicle speeds significantly, it also led to a notable increase in deceleration rates. Although speed reduction is generally associated with improved road safety, the accompanying sharp deceleration may reflect abrupt driver responses or discomfort, potentially introducing new safety risks. This seeming contradiction highlights the need for a more detailed examination of how drivers physically navigate the chicane. In particular, understanding lateral avoidance behavior is essential for interpreting whether drivers are swerving or adjusting their path abruptly in response to the physical constraints imposed by the chicane.

To further analyze the drivers' behavior when encountering chicanes, the lateral avoidance strategies of drivers at the location of the chicane were studied based on measurement points indicated in Figure 9. Each measurement point was taken at the center of the cushion, spaced 20 m apart.

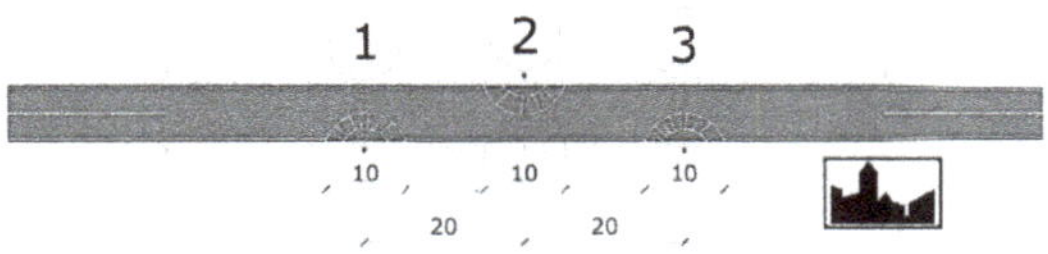

Figure 9. Measurement points for lateral deviation (distances are in meters).

Figure 10 visually illustrates the percentage of participants who used each driving strategy across the three chicane designs. Across all configurations, driving straight over the cushions remained the most prevalent behavior, indicating a general tendency among drivers to prioritize path continuity over evasive maneuvering. In a chicane design without road narrowing or oncoming traffic, almost 70% of participants drove straight over both cushions, making it the most dominant strategy, and only around 20% chose to perform a chicane maneuver. When the chicane was combined with road narrowing, the chicane maneuver strategy became more frequent, with 35%, and fewer drivers passed straight over both cushions. This indicates that the presence of a narrowed roadway heightens spatial awareness and prompts more lateral avoidance. On the other hand, when an oncoming vehicle is added to the design, the majority of participants (over 60%) again choose to drive over the cushions. This may reflect a reluctance to perform lateral maneuvers when facing dynamic conflict, possibly due to uncertainty or perceived risk.

Overall, these findings reveal that while physical or visual constraints can partially influence driver behavior, the tendency to drive over the cushions persists, even under more complex traffic conditions, raising questions about the chicane's actual effectiveness in enforcing safe lateral maneuvers. The combination of road narrowing and oncoming traffic, rather than encouraging evasive behavior, results in a reduction in maneuvering, with more participants choosing either to go straight over both cushions or only partially avoid them. However, although the chicanes did not consistently induce the expected lateral deviation, they still led to a noticeable reduction in speed, often accompanied by abrupt deceleration. This suggests that chicanes may function more effectively as speed-reducing elements than as tools for guiding smooth lateral repositioning, potentially affecting driver comfort and reaction patterns.

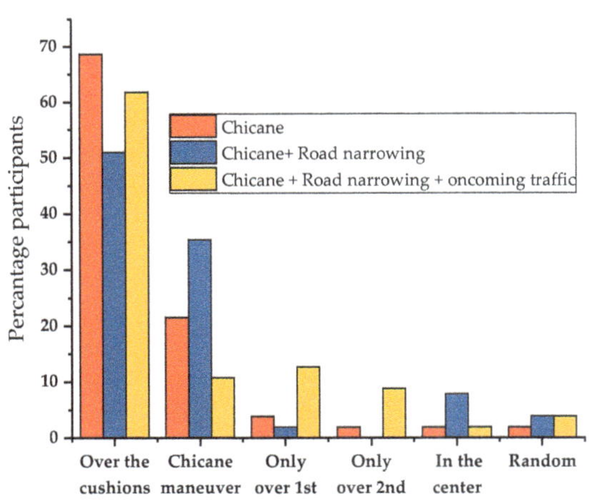

Figure 10. Overview of driving strategies at the chicane across different designs.

5. Discussion

The simulator results revealed a clear pattern: physical TCMs were substantially more effective at reducing vehicle speed during high-to-low speed transitions than psychological measures. Chicanes, both alone and in combination with road narrowing, consistently produced the most substantial speed reductions. These findings corroborate those of Sołowczuk et al. [22], who emphasized the efficacy of lateral deflections like chicanes in maintaining lower travel speeds in transition zones. Similarly, Lantieri et al. [26], demonstrated that curbs and lateral constraints significantly increase driver awareness and reduce approach speeds.

On the other hand, avenue planting and transverse markings failed to produce significant speed reductions when applied independently. This aligns with findings from Hussain et al. [34], who reported that while perceptual markings can influence speed momentarily, their effects dissipate quickly unless combined with physical features. However, Pazzini et al. [8], demonstrated that psychological TCMs, such as smart lighting at pedestrian crossings, can significantly reduce vehicle speeds and enhance yielding behavior. It should be noted that although Section 2 reviewed vertical traffic calming devices such as speed humps, tables, and raised crossings, these were not included in the tested designs. The raised cushions used in the chicane were primarily configured to enforce lateral deflection. As such, the findings of this study should be interpreted as focusing on lateral and psychological interventions rather than vertical ones.

Moreover, the speed data suggested that drivers only respond meaningfully to cues that demand cognitive and physical engagement. Designs that paired physical and perceptual elements, such as road narrowing combined with transverse markings, achieved smoother deceleration profiles without abrupt braking. This finding echoes Doomah and Paupoo [32], who noted that multimodal gateway treatments balance effectiveness with comfort and user acceptance.

Acceleration/deceleration profiles further reinforced the primacy of physical interventions. Chicanes led to abrupt deceleration followed by an immediate acceleration, creating a rebound effect. While this confirms their effectiveness in enforcing deceleration, it also raises concerns about driving comfort and consistency. Pérez-Acebo et al. [14] warned that such reactive maneuvers could lead to unpredictable driver behavior post-intervention, especially in mixed-traffic environments. In contrast, road narrowing designs prompted more gradual deceleration and less aggressive recovery in acceleration, indicating a smoother transition. These results are consistent with the findings of Melman et al. [45], who found that road narrowing, while not as forceful as chicanes, induces moderate but sustained deceleration through perceptual narrowing and steering tension.

The failure of avenue planting and transverse markings to produce significant changes in acceleration/deceleration confirmed earlier reports by Wu et al. [30], and Fitzpatrick et al. [31], who showed that vegetation and visual cues alone lack the tactile urgency to alter driver kinematics meaningfully. These measures might influence lane positioning or perceived safety, but are insufficient to alter longitudinal behavior without supporting physical constraints. The weak performance of psychological traffic calming measures in the present study may be attributed to several factors. First, the absence of tactile feedback (e.g., vibration from transverse rumble strips) likely reduced the salience of the markings. Second, the visual stimuli used in the simulator were uniform and predictable, potentially reducing their novelty effect. Third, the absence of other traffic participants or environmental variability limited the extent to which drivers needed to rely on perceptual cues. Such psychological measures may prove more effective in real-world settings, where drivers encounter multisensory feedback and more complex environmental contexts.

Therefore, simulator results should be interpreted cautiously when generalizing the impact of psychological interventions.

On the other hand, while chicanes maximized deceleration, their abruptness may compromise ride quality and safety, especially in areas where vulnerable road users might be present. Road narrowing, in contrast, offers a compromise, less dramatic in effect but more stable and predictable. It is important to note that the acceleration/deceleration estimates in the LMMs reflect an average over distinct driving segments; in some cases, the sharp rebound acceleration following a chicane, particularly after passing a gateway, outweighs the initial deceleration, resulting in a net positive estimate. Therefore, these values should be interpreted in conjunction with the full acceleration/deceleration profile, where these segmented patterns are clearly visible.

In addition, the analysis of lateral behavior at chicanes offers nuanced insight into how drivers physically negotiate TCMs. When an oncoming vehicle was added, most participants reverted to driving straight over the chicane cushions, prioritizing collision avoidance over ideal trajectory. This highlights a trade-off between safety and compliance, indicating that driver behavior becomes more risk-averse under perceived conflict conditions. It also aligns with Qin et al. [20], who showed that oncoming traffic can neutralize the behavioral effects of non-exclusive calming features. Therefore, lateral deviation behavior reveals the contextual sensitivity of calming interventions. Unless drivers are given compelling spatial or visual cues, and ideally, both, they tend to revert to default linear strategies. Effective gateway designs should thus limit ambiguity in maneuver paths while preserving options for larger vehicles and emergency access.

6. Conclusions

6.1. Summary

This study evaluated the effectiveness of seven gateway configurations, comprising both physical and psychological TCMs, at transition zones between 70 km/h and 50 km/h speed regimes. Key driving behavior parameters, including speed, acceleration/deceleration, and lateral deviation, were analyzed using a high-fidelity driving simulator and a within-subjects design. The results confirmed that physical interventions, particularly chicanes and their combinations with road narrowing, were the most effective in reducing vehicle speed and inducing significant deceleration. In contrast, when implemented independently, psychological measures such as avenue planting and transverse markings showed minimal behavioral impact. The study also demonstrated that driver response is context-dependent: spatial constraints and traffic presence significantly influence driving strategies and lane behavior at chicanes. Collectively, these findings reinforce the necessity of integrated gateway designs that blend physical constraints with perceptual cues to promote smoother, earlier, and more consistent speed reductions at urban entry points.

6.2. Limitations

While the simulator-based methodology offers controlled testing conditions and rich behavioral data, it also introduces limitations that must be acknowledged [46]. Firstly, the simulated environment may not fully replicate real-world conditions, particularly in terms of sensory input such as vibration, road texture, and dynamic interaction with other traffic participants [47]. This may influence the ecological validity of results, especially for psychological interventions that depend on visual or multisensory perception. Nonetheless, the relative validity of driving simulators for comparing driver behavior across design alternatives has been well established. Given the study's focus on identifying relative differences between gateway designs, the simulator provides a valid platform for this initial investigation. These results can guide future real-world studies to further assess the most promising interventions. Secondly, the participant pool included a diverse range of ages and driving experiences, providing valuable insights into relative effects across different driver profiles. While the sample size was modest, it still allowed us to capture meaningful trends. However, further studies could expand to include additional groups, such as older or professional drivers, to enhance generalizability. Nevertheless, the sample size of 54 participants is comparatively high for driving simulator research, providing a robust basis for detecting relative differences between designs. The findings should therefore be interpreted as relative effects across designs within a controlled environment.

Although the study accounted for learning effects by randomizing scenario sequences, repeated exposure to similar interventions in a controlled environment could still bias participant behavior. Another limitation is that the experiment captured only immediate driver responses during single-drive scenarios. However, examining these immediate effects is an essential first step in identifying the most promising gateway designs for follow-up research. It remains unclear whether the observed speed reductions and deceleration behavior changes would persist after repeated exposure or diminish as drivers become accustomed to the designs. This restricts the ability to generalize findings to long-term driver adaptation, highlighting the need for future longitudinal or field studies to confirm the durability of these behavioral effects.

Moreover, the experimental scenarios excluded other road users, such as pedestrians and cyclists, and higher traffic volumes. This simplification was intentional, as gateway locations are cognitively demanding points on the road network, and the primary focus was on measures to support drivers in adhering to speed changes. While other road users were included in the scenario design to enhance realism, they were not positioned to interact directly with drivers. This approach allowed for isolating the effects of the traffic calming measures. Finally, the simulator's fixed-base platform restricts physical movement, potentially dampening the driver's response to vertical interventions, such as speed cushions. Although no tactile feedback was provided, the raised cushions were visually rendered and perceived by drivers as elevated elements, ensuring that the key visual cue of the intervention was captured. This representation still elicited meaningful speed adjustments because visual information strongly guided driver behavior. While the absence of multisensory feedback may have reduced sensitivity to vertical features, previous research has consistently shown that simulators, even without full physical fidelity, yield reliable outcomes for relative comparisons of design alternatives. For instance, Hussain et al. [11] and Groeger et al. [48] demonstrated that simulators produce reliable behavioral outcomes consistent with real-world observations even when lacking full physical fidelity.

In addition, despite the limitations, the results remain valuable, particularly when analyzing relative differences across design conditions. The relative validity of the findings is strong, as the internal consistency and comparative trends offer meaningful insights into driver behavior in response to different interventions.

6.3. Potential Implications and Future Work

The results carry several inferences for policy and infrastructure design. For policymakers seeking to enhance safety at rural-urban transition points, this study provides clear evidence supporting the implementation of physically constraining measures like chicanes and road narrowing. In addition, the results have implications for smart city safety frameworks such as Vision [49], supporting the design of context-sensitive entry zones that proactively manage driver behavior. Gateway interventions evaluated through simulation can inform urban design guidelines and speed management policies at rural-urban boundaries, improving pedestrian and cyclist safety without relying solely on enforcement. As cities adopt digital twins for infrastructure planning [50,51], the tested gateway interventions can serve as validated modules in city-wide mobility simulations. Integrating behavioral data from simulators into such digital ecosystems supports smarter multi-criteria decision-making, enhances public engagement, and accelerates infrastructure optimization [52].

Future research should extend the findings through on-road experiments or augmented reality simulations that incorporate richer sensory feedback and dynamic traffic scenarios. Longitudinal studies are also needed to assess whether behavioral adaptations persist over time or diminish with driver familiarity. For instance, continuous analysis techniques on simulator time-series data can be employed. Such methods would allow the investigation of full speed and acceleration profiles, rather than discrete measurement points, thereby providing deeper insights into driver behavior dynamics across gateway designs. Moreover, integrating eye-tracking or biometric monitoring (e.g., heart rate variability) could yield more profound insights into the cognitive and emotional responses triggered by different gateway configurations. In addition, as smart city infrastructure evolves, exploring how digital interventions (e.g., in-vehicle alerts or adaptive signage) can complement physical gateways represents a promising direction for next-generation TCMs.

In addition, future studies should complement objective driving behavior data with detailed subjective evaluations, including perceived safety, comfort, and acceptance, to provide a more holistic assessment of gateway interventions. Future research with larger, more balanced samples should investigate whether demographic characteristics moderate design effects; targeted moderator analyses will be more informative once adequately powered. Vulnerable road users and higher traffic densities can also be incorporated into the experiments to better reflect real-world urban conditions and to evaluate the driver behavior in mixed-traffic environments. This was not performed in the current study to maintain focus on drivers' responses to speed changes at cognitively demanding gateway locations. Moreover, this study did not include a cost-effectiveness analysis, which lies beyond its scope, and future work can complement these findings with economic assessments to guide policymakers. Finally, this study can be integrated into a digital twin of the urban mobility infrastructure to design, assess, and iterate transportation policies in real-time.

Author Contributions: Conceptualization, A.P., M.M. and T.S.; methodology, A.P., M.S.B., M.M. and T.S.; software, A.P. and M.S.B.; validation, A.P. and M.S.B.; formal analysis, A.P., M.S.B., M.M. and T.S.; investigation, A.P., M.S.B., M.M. and T.S.; resources, A.P.; data curation, A.P.; writing—original draft preparation, A.P., M.S.B., M.M. and T.S.; writing—review and editing, A.P. and M.S.B.; visualization, M.S.B.; supervision, A.P.; project administration, A.P. All authors have read and agreed to the published version of the manuscript.

Funding: This study was partially supported by the European Union's Horizon Europe research and innovation programme through the CAMBER Project (Grant Agreement No. 101146800).

Institutional Review Board Statement: The study was conducted in accordance with the Declaration of Helsinki and approved by the Sociaal-Maatschappelijke Ethische Commissie (SMEC) of Universiteit Hasselt (confirmation ID REC/SMEC/2024-2025/52, approved on 28 April 2025).

Informed Consent Statement: Written informed consent has been obtained from the subjects to publish this paper.

Data Availability Statement: The data supporting the reported results are not publicly available due to GDPR considerations. However, anonymized data can be provided upon reasonable request to the corresponding author.

Acknowledgments: The authors would like to express their gratitude to Thomas Stieglitz from the Transportation Research Institute (IMOB) of UHasselt for his valuable assistance in setting up the driving simulator experiment.

Conflicts of Interest: The authors declare no conflicts of interest.

References

1. Karimpour, A.; Karimpour, A.; Kluger, R.; Liu, C.; Wu, Y.-J. Effects of speed feedback signs and law enforcement on driver speed *Transp. Res. Part F Traffic Psychol. Behav.* **2021**, *77*, 55–72. [CrossRef]
2. Mahmud, M.S.; Gates, T.J.; Johari, M.U.M.; Jashami, H.; Bamney, A.; Savolainen, P.T. Do Dynamic speed feedback signs impact drivers differently based on speeding tendencies? Insights from applications at select critical roadway contexts. *Transp. Res. Part F Traffic Psychol. Behav.* **2023**, *98*, 157–169. [CrossRef]
3. Costa, A.T.; Figueira, A.C.; Larocca, A.P.C. An eye-tracking study of the effects of dimensions of speed limit traffic signs on a mountain highway on drivers' perception. *Transp. Res. Part F Traffic Psychol. Behav.* **2022**, *87*, 42–53. [CrossRef]
4. Patel, D.; Alfaris, R.E.; Jalayer, M. Assessing the effectiveness of autism spectrum disorder roadway warning signs: A case study in New Jersey. *Transp. Res. Part F Traffic Psychol. Behav.* **2024**, *100*, 57–68. [CrossRef]
5. Lopoo, L.M.; Cardon, E.; Souders, S.; Dale, M.K.; Ngo, U. An evaluation of a Vision Zero traffic-calming intervention, an urban transportation safety policy. *J. Urban Aff.* **2024**, *47*, 3048–3069. [CrossRef]
6. Tumminello, M.L.; Macioszek, E.; Granà, A.; Giuffrè, T. Evaluating Traffic-Calming-Based Urban Road Design Solutions Featuring Cooperative Driving Technologies in Energy Efficiency Transition for Smart Cities. *Energies* **2023**, *16*, 7325. [CrossRef]
7. Bei, R.; Du, Z.; Huang, T.; Mei, J.; He, S.; Zhang, X. Analysis and regulation of driving behavior in the entrance zone of freeway tunnels: Implementation of visual guidance systems in China. *Accid. Anal. Prev.* **2024**, *202*, 107600. [CrossRef]
8. Pazzini, M.; Cameli, L.; Vignali, V.; Simone, A.; Lantieri, C. Video-Based Analysis of a Smart Lighting Warning System for Pedestrian Safety at Crosswalks. *Smart Cities* **2024**, *7*, 2925–2939. [CrossRef]
9. Dani, A.A.H.; Supangkat, S.H.; Lubis, F.F.; Nugraha, I.G.B.B.; Kinanda, R.; Rizkia, I. Development of a Smart City Platform Based on Digital Twin Technology for Monitoring and Supporting Decision-Making. *Sustainability* **2023**, *15*, 14002. [CrossRef]
10. Karmaker, A.K.; Islam, S.M.R.; Kamruzzaman, M.; Rashid, M.M.U.; Faruque, M.O.; Hossain, M.A. Smart City Transformation: An Analysis of Dhaka and Its Challenges and Opportunities. *Smart Cities* **2023**, *6*, 1087–1108. [CrossRef]
11. Hussain, Q.; Alhajyaseen, W.K.; Pirdavani, A.; Reinolsmann, N.; Brijs, K.; Brijs, T. Speed perception and actual speed in a driving simulator and real-world: A validation study. *Transp. Res. Part F Traffic Psychol. Behav.* **2019**, *62*, 637–650. [CrossRef]
12. Zare, N.; Macioszek, E.; Granà, A.; Giuffrè, T. Blending Efficiency and Resilience in the Performance Assessment of Urban Intersections: A Novel Heuristic Informed by Literature Review. *Sustainability* **2024**, *16*, 2450. [CrossRef]
13. Ambros, J.; Tomešová, L.; Jurewicz, C.; Valentová, V. A review of the best practice in traffic calming evaluation. *Accid. Anal. Prev.* **2023**, *189*, 107073. [CrossRef]
14. Pérez-Acebo, H.; Ziółkowski, R.; Linares-Unamunzaga, A.; Gonzalo-Orden, H. A Series of Vertical Deflections, a Promising Traffic Calming Measure: Analysis and Recommendations for Spacing. *Appl. Sci.* **2020**, *10*, 3368. [CrossRef]
15. Majer, S.; Sołowczuk, A.; Kurnatowski, M. Design and Construction Aspects of Concrete Block Paved Vertical Traffic-Calming Devices Located in Home Zone Areas. *Sustainability* **2024**, *16*, 2982. [CrossRef]
16. Solowczuk, A. Effect of traffic calming measures implemented on the approach to the Tempo–30 zone on the Degree of speed reduction. In *IOP Conference Series: Materials Science and Engineering*; IOP Publishing: Bristol, UK, 2019.
17. Sołowczuk, Comparative Analysis of the Effectiveness of Various Traffic Calming Measures Implemented on the Approach to the Tempo–30 Zone. *New Approaches Eng. Res.* **2021**, *4*, 35–45. [CrossRef]
18. Akbari, A.; Haghighi, F. Traffic calming measures: An evaluation of four low-cost TCMs' effect on driving speed and lateral distance. *IATSS Res.* **2020**, *44*, 67–74. [CrossRef]
19. Yeo, J.; Lee, J.; Cho, J.; Kim, D.-K.; Jang, K. Effects of speed humps on vehicle speed and pedestrian crashes in South Korea. *J. Saf. Res.* **2020**, *75*, 78–86. [CrossRef]
20. Qin, Y.; Wu, Y.; Guo, M. A Simulator Study on the Driving Impacts of Four Speed-Calming Measures at Unsignalized Intersections. *Appl. Sci.* **2024**, *14*, 3542. [CrossRef]
21. D'apuzzo, M.; Evangelisti, A.; Santilli, D.; Nardoianni, S.; Cappelli, G.; Nicolosi, V. Towards a New Design Methodology for Vertical Traffic Calming Devices. *Sustainability* **2023**, *15*, 13381. [CrossRef]
22. Sołowczuk, A.B.; Kacprzak, D. Identification of the Determinants of the Effectiveness of On-Road Chicanes in the Village Transition Zones Subject to a 50 km/h Speed Limit. *Energies* **2021**, *14*, 4002. [CrossRef]
23. Galante, F.; Mauriello, F.; Montella, A.; Pernetti, M.; Aria, M.; D'aMbrosio, A. Traffic calming along rural highways crossing small urban communities: Driving simulator experiment. *Accid. Anal. Prev.* **2010**, *42*, 1585–1594. [CrossRef]

24. Sołowczuk, A.; Kacprzak, D. Synergy Effect of Factors Characterising Village Transition Zones on Speed Reduction. *Energies* **2021**, *14*, 8474. [CrossRef]
25. Jassal, K.; Sharma, U. Replacing speed bumps with chicanes: Modernizing India's traffic calming strategy. *Proc. Inst. Civ. Eng. Munic. Eng.* **2024**, *177*, 197–207. [CrossRef]
26. Lantieri, C.; Lamperti, R.; Simone, A.; Costa, M.; Vignali, V.; Sangiorgi, C.; Dondi, G. Gateway design assessment in the transition from high to low speed areas. *Transp. Res. Part F Traffic Psychol. Behav.* **2015**, *34*, 41–53. [CrossRef]
27. Ariën, C.; Jongen, E.M.; Brijs, K.; Brijs, T.; Daniels, S.; Wets, G. A simulator study on the impact of traffic calming measures in urban areas on driving behavior and workload. *Accid. Anal. Prev.* **2013**, *61*, 43–53. [CrossRef]
28. Yang, Y.; Chen, Y.; Wu, C.; Easa, S.M.; Lin, W.; Zheng, X. Effect of highway directional signs on driver mental workload and behavior using eye movement and brain wave. *Accid. Anal. Prev.* **2020**, *146*, 105705. [CrossRef]
29. Calvo-Poyo, F.; de Oña, J.; Morcillo, L.G.; Navarro-Moreno, J. Influence of Wider Longitudinal Road Markings on Vehicle Speeds in Two-Lane Rural Highways. *Sustainability* **2020**, *12*, 8305. [CrossRef]
30. Wu, J.; Jiang, J.; Duffy, V.; Zhou, J.; Chen, Y.; Tian, R.; McCoy, D.; Ruble, T. Impacts of roadside vegetation and lane width on speed management in rural roads. *Proc. Hum. Factors Ergon. Soc. Annu. Meet.* **2023**, *67*, 2267–2273. [CrossRef]
31. Fitzpatrick, C.D.; Harrington, C.P.; Knodler, M.A., Jr.; Romoser, M.R.E. The influence of clear zone size and roadside vegetation on driver behavior. *J. Saf. Res.* **2014**, *49*, 97.e1–104. [CrossRef] [PubMed]
32. Doomah, M.Z.; Paupoo, D.P. Evaluation of the effectiveness of speed tables combined with other traffic calming measures and their community acceptance in Mauritius. *Case Stud. Transp. Policy* **2022**, *10*, 1550–1565. [CrossRef]
33. Awan, H.H.; Pirdavani, A.; Houben, A.; Westhof, S.; Adnan, M.; Brijs, T. Impact of perceptual countermeasures on driving behavior at curves using driving simulator. *Traffic Inj. Prev.* **2019**, *20*, 93–99. [CrossRef]
34. Hussain, Q.; Alhajyaseen, W.K.; Reinolsmann, N.; Brijs, K.; Pirdavani, A.; Wets, G.; Brijs, T. Optical pavement treatments and their impact on speed and lateral position at transition zones: A driving simulator study. *Accid. Anal. Prev.* **2021**, *150*, 105916. [CrossRef] [PubMed]
35. Hallmark, S.; Hawkins, N. *Evaluation of Gateway and Low-Cost Traffic-Calming Treatments for Major Routes in Small Rural Communities*; Institute for Transportation: Ames, IA, USA, 2007.
36. Agentschap Wegen en Verkeer. *Vademecum Vergevingsgezinde Wegen (VVW)—Deel Gemotoriseerd Verkeer*; Departement Mobiliteit en Openbare Werken: Graaf de Ferrarisgebouw, Brussel, 2020.
37. Mahmud, M.S.; Megat Johari, M.U.; Bamney, A.; Jashami, H.; Gates, T.J.; Savolainen, P.T. Driver response to a dynamic speed feedback sign at speed transition zones along high-speed rural highways. *Transp. Res. Rec.* **2023**, *2677*, 1341–1353.
38. Roelofs, T. *Evaluatie Overrijdbare Slingerremmers*; Megaborn, Rapport DWr1302; Waterschap Rivierenland: Tiel, The Netherlands, 2014.
39. Martindale, A.; Urlich, C. *Effectiveness of Transverse Road Markings on Reducing Vehicle Speeds*; NZ Transport Agency: Wellington, New Zealand, 2010.
40. Hussain, Q.; Alhajyaseen, W.K.; Pirdavani, A.; Brijs, K.; Shaaban, K.; Brijs, T. Do detection-based warning strategies improve vehicle yielding behavior at uncontrolled midblock crosswalks? *Accid. Anal. Prev.* **2021**, *157*, 106166. [CrossRef]
41. Dash, C.S.K.; Behera, A.K.; Dehuri, S.; Ghosh, A. An outliers detection and elimination framework in classification task of data mining. *Decis. Anal. J.* **2023**, *6*, 100164. [CrossRef]
42. Pirdavani, A.; Bajestani, M.S.; Bunjong, S.; Delbare, L. The Impact of Perceptual Road Markings on Driving Behavior in Horizontal Curves: A Driving Simulator Study. *Appl. Sci.* **2025**, *15*, 4584. [CrossRef]
43. West, B.T.; Welch, K.B.; Galecki, A.T. *Linear Mixed Models: A Practical Guide Using Statistical Software*; Chapman and Hall/CRC: Boca Raton, FL, USA, 2022.
44. Ali, Y.; Raadsen, M.P.; Bliemer, M.C. Effects of passing rates on driving behaviour in variable speed limit-controlled highways: Evidence of external pressure from a driving simulator study. *Transp. Res. Part F Traffic Psychol. Behav.* **2024**, *104*, 488–505. [CrossRef]
45. Melman, T.; Kolekar, S.; Hogerwerf, E.; Abbink, D. How road narrowing impacts the trade-off between two adaptation strategies: Reducing speed and increasing neuromuscular stiffness. In Proceedings of the 2020 IEEE International Conference on Systems, Man, and Cybernetics (SMC), Toronto, ON, Canada, 20 April 2020. [CrossRef]
46. Mullen, N.; Charlton, J.; Devlin, A.; Bedard, M. Simulator validity: Behaviours observed on the simulator and on the road. In *Handbook of Driving Simulation for Engineering, Medicine and Psychology*; CRC Press: Boca Raton, FL, USA, 2011; pp. 1–18.
47. Faschina, S.; Stieglitz, R.-D.; Muri, R.; Strohbeck-Kühner, P.; Graf, M.; Mager, R.; Pflueger, M.O. Driving errors, estimated performance and individual characteristics under simulated and real road traffic conditions—A validation study. *Transp. Res. Part F Traffic Psychol. Behav.* **2021**, *82*, 221–237. [CrossRef]
48. Groeger, J.A.; Murphy, G. Driver performance under simulated and actual driving conditions: Validity and orthogonality. *Accid. Anal. Prev.* **2020**, *143*, 105593. [CrossRef]
49. Mofolasayo, A. Towards 'Vision-Zero' in Road Traffic Fatalities: The Need for Reasonable Degrees of Automation to Complement Human Efforts in Driving Operation. *Systems* **2024**, *12*, 40. [CrossRef]

50. Wu, D.; Zheng, A.; Yu, W.; Cao, H.; Ling, Q.; Liu, J.; Zhou, D. Digital Twin Technology in Transportation Infrastructure: A Comprehensive Survey of Current Applications, Challenges, and Future Directions. *Appl. Sci.* **2025**, *15*, 1911. [CrossRef]
51. Fatorachian, H.; Kazemi, H.; Pawar, K. Enhancing Smart City Logistics Through IoT-Enabled Predictive Analytics: A Digital Twin and Cybernetic Feedback Approach. *Smart Cities* **2025**, *8*, 56. [CrossRef]
52. Fayyaz, M.; Fusco, G.; Colombaroni, C.; González-González, E.; Nogués, S. Optimizing Smart City Street Design with Interval-Fuzzy Multi-Criteria Decision Making and Game Theory for Autonomous Vehicles and Cyclists. *Smart Cities* **2024**, *7*, 3936–3961. [CrossRef]

smart cities

Article

Beyond Efficiency: Integrating Resilience into the Assessment of Road Intersection Performance

Nazanin Zare [1,*], Maria Luisa Tumminello [1], Elżbieta Macioszek [2,*] and Anna Granà [1]

[1] Department of Engineering, University of Palermo, Viale delle Scienze ed 8, 90128 Palermo, Italy; marialuisa.tumminello01@unipa.it (M.L.T.); anna.grana@unipa.it (A.G.)
[2] Department of Transport Systems, Traffic Engineering and Logistics, Faculty of Transport and Aviation Engineering, Silesian University of Technology, 40-019 Katowice, Poland
* Correspondence: nazanin.zare@unipa.it (N.Z.); elzbieta.macioszek@polsl.pl (E.M.)

Highlights

What are the main findings?

- Intersection resilience is as critical as efficiency, revealing that conventional performance metrics alone are insufficient under disruptive conditions.
- A combined heuristic–microsimulation framework effectively evaluates both operational performance and robustness, capturing vulnerabilities such as power outages.

What is the implication of the main finding?

- Urban planners and policymakers can prioritize intersection designs that preserve mobility during climate-related disruptions, strengthening urban resilience strategies.
- The integrated methodological approach provides a transferable tool for other cities to support evidence-based decisions on resilient intersection planning and investment.

Abstract

Extreme weather events, such as storms, pose significant challenges to the reliability and efficiency of urban road networks, making intersection design and management critical to maintaining mobility. This paper addresses the dual objectives of traffic efficiency and resilience by evaluating the performance of roundabouts, signalized, and two-way stop-controlled (TWSC) intersections under normal and storm-disrupted conditions. A mixed-method approach was adopted, combining a heuristic framework from the Highway Capacity Manual with microsimulations in AIMSUN Next. Three Polish case studies were examined; each was modeled under alternative control strategies. The findings demonstrate the superior robustness of roundabouts, which retain functionality during power outages, while signalized intersections reveal vulnerabilities when control systems fail, reverting to less efficient TWSC behavior. TWSC intersections consistently exhibited the weakest performance, particularly under high or uneven traffic demand. Despite methodological differences in delay estimation, the convergence of results through Level of Service categories strengthens the reliability of findings. Beyond technical evaluation, the study underscores the importance of resilient intersection design in climate-vulnerable regions and the value of integrating analytical and simulation-based methods. By situating intersection performance within urban resilience, this research provides actionable insights for policymakers, planners, and engineers seeking to balance efficiency with adaptability in infrastructure planning.

Keywords: road efficiency and resilience; urban mobility; level of service; heuristic approach; microsimulation

1. Introduction

The combination of hazards such as storms and floods, with vulnerabilities like rapid urbanization and inefficient road infrastructure, increases disaster risks in urban areas [1–3]. According to World Bank data [4], 56% of the global population (4.4 billion) currently lives in cities, a figure projected to rise to nearly 70% by 2050. At the same time, the intensity and frequency of disasters are increasing due to climate change and urban population growth [5–7]. Between 2000–2023, about 80% of major U.S. electricity outages (≥50,000 customers or ≥300 MW) were weather-related, with severe storms, winter weather, and tropical cyclones as the main causes, and their frequency nearly doubled in the last decade [8]. Similarly, long-term studies in Europe and Italy found that weather is the leading driver of large-scale power outages, particularly winter events affecting transmission lines [9]. These findings highlight the importance of adaptation and mitigation strategies for strengthening urban resilience, a priority emphasized by researchers, policymakers, planners, and stakeholders [10–12]. Road infrastructure, while vital for daily life and economic activities, remains highly vulnerable to disruptions [6]. Moreover, resilience is closely tied to the Sustainable Development Goals (SDGs), introduced in the 2030 Agenda [13], specifically SDGs 11 and 13, that underscore making cities resilient and sustainable, as well as taking actions to combat climate change and its effects [14].

In this context, resilience in urban systems refers to the capacity to maintain functionality under disruptions [6,15], while in transportation, it denotes the ability of road networks to restore service after disturbances [6,16]. Efficiency, by contrast, reflects the capacity of transport infrastructure to ensure reliable mobility of people and goods with minimal costs, time, energy use, and environmental impacts [6,17,18]. Road networks are thus central to urban resilience, with connectivity, geometry, and design strongly influencing efficiency and vulnerability [19,20]. Intersection performance, an indicator of efficiency, is typically evaluated using metrics such as the volume-to-capacity ratio and control delay time, which capture different aspects of service quality for road users [18]. The choice of specific metrics depends on the objectives of each study. Road network resilience is commonly assessed by comparing functionality before and after a disruptive event [20]. Traffic-related functionality metrics, including travel time and throughput, are typically examined at the road network level [21]; however, at the intersection level, control delay time can serve as a useful indicator of functionality.

This research proposes a heuristic approach supported by microsimulations for evaluating intersection efficiency with an emphasis on resilience. The method integrates both efficiency and resilience into a flexible framework, analyzing various intersection geometries and control modes under normal and disrupted conditions. The study aims to generate insights to support resilient and efficient intersection design and management, ultimately improving urban mobility and infrastructure stability. To define the scope, several assumptions are introduced. The heuristic draws on the analytical framework of the Highway Capacity Manual (HCM), 6th edition [18], which treats intersections as isolated systems, while microsimulations are conducted using AIMSUN Next [22]. Each case study begins with its specific geometry and layout, but is evaluated under three traffic control types (roundabout, signalized intersection, and two-way stop-controlled (TWSC) intersection) regardless of its current configuration. The primary metric for the heuristic approach is control delay time, while microsimulations consider delay time, travel time, speed, and average approach delay, all measured under pre- and post-storm conditions. Since analytical and simulation results are not directly comparable, the level of service (LOS) is included as a supplementary indicator for both methods. Together, these metrics provide a more comprehensive assessment of intersection performance and resilience. Storm duration was not considered a critical factor for this study, as it was not deemed necessary to evaluate its impact on intersection resilience. Instead, the consequence of a storm, namely, electricity outages, was considered a critical factor in this research. Therefore, any other aspects of storms, including intensity, frequency, and duration, are not important for the objectives of this study.

The primary objective of this research is to integrate the efficiency and resilience of intersections into a unified assessment framework, providing a novel approach to enhancing infrastructure adaptability and robustness against future disruptions. To evaluate intersection performance, traffic composed solely of Human-Driven Vehicles (HDVs) in various intersection layouts was analyzed to address the following key questions:

- What role does intersection design play in the efficiency and resilience of signalized intersections during traffic signal failures caused by extreme weather events?
- How do different traffic control types and geometries affect intersection performance and resilience under both normal and post-storm conditions?
- How can predictive resilience metrics be developed and applied to intersection design and evaluation to better anticipate and mitigate the impacts of storm-induced power outages?

Following this introduction, the paper is organized as follows Section 2 presents a brief literature review. Section 3 represents the study areas and methodology, followed by Section 4, which represents the findings of this study. Section 5 discusses the results of the heuristic approach and the microsimulations conducted using AIMSUN Next [22]. Section 6 provides the conclusions, highlights the contributions to the field, and outlines the study's limitations and potential future developments. The final section contains the references used in this research.

2. Literature Review

The literature review conducted in this study follows the step-based research method outlined in [23], using a hybrid approach that combined meta-synthesis and realist review. The review focused on the peer-reviewed papers published since 2013. Searches were conducted until 19 September 2025, in three phases.

In the first phase, the Scopus database was queried using the terms "resilient roads" and "urban mobility" in the Article title, Abstract, and Keywords fields, yielding 68 peer-reviewed articles. Expanding the query by adding "intersection efficiency" and "smart mobility" returned no additional results.

In the second phase, abstracts were screened for relevance. During this process, 52 papers were excluded for not meeting the selection criteria, leaving 16 relevant papers.

In the third phase, a hybrid search was conducted across relevant journal websites, using keyword-based and semantic search techniques to identify additional studies not captured in Scopus. This process yielded 24 further peer-reviewed articles.

In total, 40 studies were included. Although the literature cannot be divided into entirely independent categories, for the purposes of analysis and gap identification, the selected studies were grouped into three thematic areas: (i) resilient road and intersection design, (ii) smart and resilient urban mobility, and (iii) urban resilience planning and governance. These categories are outlined in Sections 2.1–2.3, with gaps discussed in Section 2.4.

2.1. Resilient Road and Intersection Design

Road networks are fundamental to urban design, shaping long-term growth and resilience through their structure and typology [19,20]. Given their long lifespans and interdependencies, resilience must be assessed at multiple scales, from intersections to corridors and entire networks, since vulnerabilities at one level can propagate across the system [24].

At the intersection level, Zhao et al. [25] developed a spatiotemporal framework to identify critical intersections with greater precision. Decision analysis and expert judgment have also been applied to prioritize regional infrastructure resilience [26]. Quantitative tools contribute further insights. Resilience indices highlight the redesign of vulnerable intersections, with roundabouts demonstrating superior robustness [27], while urban form design has been linked to improvements in both energy efficiency and resilience [28].

Studies have also examined intersection safety under rising traffic demands and technological transitions [29], and proposed flood-resilient infrastructure incorporating green design principles [30]. Research on pavement-vehicle interactions has identified implications for material degradation, operating costs, and environmental outcomes, particularly with the advent of electric vehicles [31]. In terms of operations, adaptive traffic signal systems have been shown to outperform fixed-timing controls, offering greater flexibility under dynamic traffic conditions [32].

Collectively, these contributions highlight that resilient road and intersection design requires an integrated approach that combines structural robustness, innovative design principles, and adaptive operational strategies.

2.2. Smart and Resilient Urban Mobility

The evolution of smart mobility emphasizes automation, data-driven methods, and adaptive technologies as key to urban resilience. Lv and Zheng [33] developed a framework addressing road operation, transport planning, and automated vehicles, while Ghaffarpasand et al. [34] demonstrated how telematics data can enhance safety, reduce emissions, and improve logistics and insurance policies. Studies in emerging economies highlight foresight and long-term planning as essential to resilient transport systems [35].

Forecasting and quantitative assessment are central to mobility resilience. Tang and Heinimann [36] proposed a resilience-based metric for recurrent congestion, providing a systemic way to quantify how traffic systems absorb disturbances and recover functionality. Recent approaches leverage traffic sensor networks and machine learning to improve speed prediction, system optimization, and traffic flow forecasting [37–39]. Hybrid deep learning models achieve high accuracy in spatiotemporal traffic forecasting [40], while extended long short-term memory (LSTM) models capture periodicity and variability in traffic profiles over long time horizons [41]. Together, these methods illustrate how predictive analytics and resilience metrics can support adaptive urban mobility.

Emerging technologies such as the Internet of Vehicles (IoV) and blockchain also contribute to mobility resilience. IoV enables layered communication architectures and supports autonomous systems, though it raises new challenges for security and resilience [42]. Blockchain-based systems facilitate transparency and decentralized coordination in mobility networks [43].

At the multimodal level, resilience benefits from real-time operational responses to cascading failures [44], while disaster-related studies underscore how mobility disruptions propagate across systems. Research highlights that traffic flows and spatial interactions, rather than hazard proximity alone, are key in shaping resilience [7]. Disruptions such as floods and hurricanes often create bottlenecks far beyond the directly affected zones [45], while pandemic-related research emphasizes dynamic recovery strategies, redeployment of public transport, and shifts toward walking and cycling as essential for maintaining urban mobility under stress [46–48].

Taken together, these studies demonstrate that smart and resilient urban mobility depends on the integration of predictive analytics, adaptive technologies, and multimodal strategies to maintain functionality under both routine and disruptive conditions.

2.3. Urban Resilience Planning and Governance

Resilience is not only a technical property but also a function of urban planning and governance. Socio-spatial perspectives illustrate how resilience evolves over time [49], while operational frameworks link design with adaptable urban forms, emphasizing both contextual sensitivity and the preservation of historic areas under development pressure [50].

System dynamics modeling across cities in the UK, Brazil, and India revealed that ambitious modal shifts toward active travel can significantly reduce mortality risks from flooding and infectious disease outbreaks, underscoring the critical role of forward-thinking transport design in building resilience [51]. In Bangkok, integrated flood–mobility modeling showed that inundation drives a shift from cars to rail, highlighting the importance of resilient transit systems and supportive policies such as fare incentives [52]. Fan et al. [53], applied AI models to guide post-disaster recovery, offering decision-support for emergency planning.

Beyond modeling, broader planning challenges remain. Urban resilience planning has gained urgency as cities face complex risks linked to globalization and climate change. Resilience is often seen as an inherent feature of cities [54], but emerging vulnerabilities require innovative solutions, including biophilic and smart design approaches [55]. Research from Latin America and Africa highlights the challenges of uncontrolled growth and limited infrastructure maintenance, which reduce resilience and exacerbate mobility issues [56,57]. Equitable access to essential services is also critical, as disaster scenarios in Latin America and Asia demonstrate how disruptions can restrict access to healthcare and public facilities [58,59].

Governance further shapes resilience outcomes. Mediterranean island case studies reveal that fragmented governance and limited coordination hinder holistic smart city development, underscoring the importance of aligning strategic aspirations with public discourse to ensure adoption of resilient innovations [60].

Overall, these findings show that urban resilience planning and governance require cross-sectoral integration across transportation, land use, energy, and telecommunications, along with targeted strategies to ensure equitable access for vulnerable populations.

2.4. Gaps in the Literature

This literature review focuses on studies assessing resilience procedures in urban road infrastructures, primarily at national, regional, and city levels [7,26,48,49,51–53]. The resilience of urban systems largely depends on road networks, whose connectivity, geometry, and design exert a strong influence on network efficiency and vulnerability [19,20]. Existing studies have explored resilience at multiple scales, from intersections to entire networks [24,25], but most research considers intersections as part of broader networks or focuses narrowly on physical characteristics and environmental conditions [20,29]. Advances in smart mobility, telematics, and machine learning have improved traffic management and forecasting [34,37], yet these approaches typically address day-to-day operations rather than extreme events. Although evidence suggests that disruptions can alter traffic patterns far beyond the affected zone [7,45], little attention has been given to how intersections—the most critical nodes of road systems—function and recover under such conditions. Furthermore, despite system-level modeling and case studies highlighting bottlenecks and recovery trajectories after shocks [51–53], empirical and simulation-based evidence on how intersection design and control influence these dynamics remains limited.

Overall, the literature indicates the need to move beyond resistance-based designs toward resilience-oriented approaches that integrate adaptability, dynamic recovery, and predictive metrics. Significant gaps remain in understanding how different intersection geometries and management strategies affect resilience and efficiency, particularly during climate-related hazards such as storm-induced power outages. To address these gaps, this research proposes a heuristic approach supported by microsimulations to evaluate intersection efficiency with an emphasis on resilience. The method integrates both efficiency and resilience into a flexible framework, analyzing various intersection geometries and control modes under normal and disrupted conditions. From a societal point of view, this study also aims to generate insights to support resilient and efficient intersection design and management, ultimately improving urban mobility and infrastructure stability.

3. Study Areas and Methodology

In this section, the studied intersections are first introduced, followed by a description of the methodologies applied in this research. Figure 1 presents a step-based chart illustrating the proposed framework.

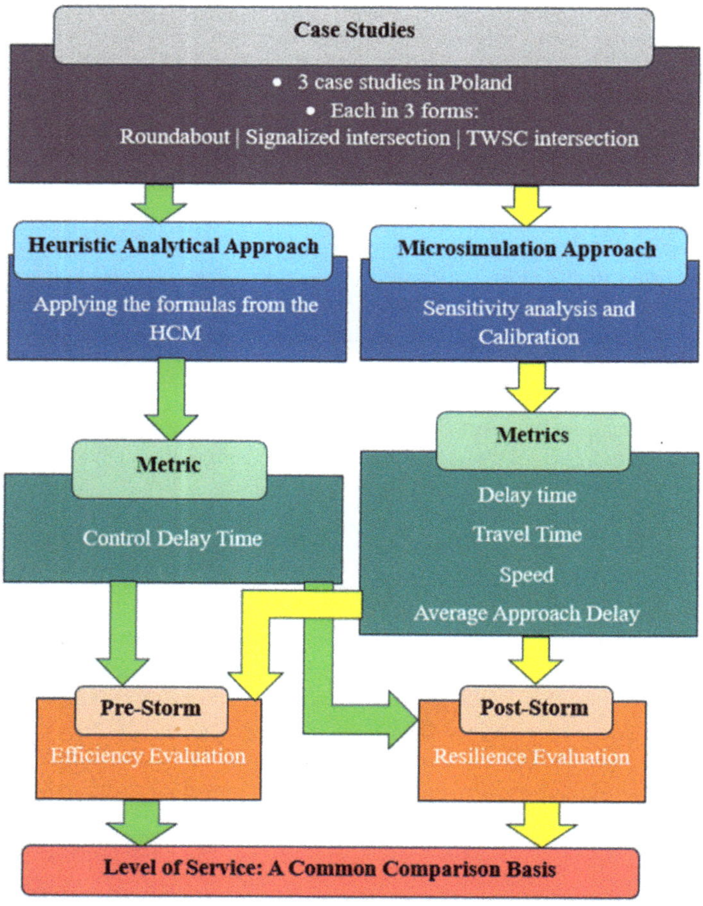

Figure 1. The methodological approach of the study.

3.1. Overview of Case Studies

This section is dedicated to introducing the case studies in their current situation, as well as further necessary information on the designed traffic control types for each of them. The views and geometric sketches of all examined intersections are illustrated in Figure 2.

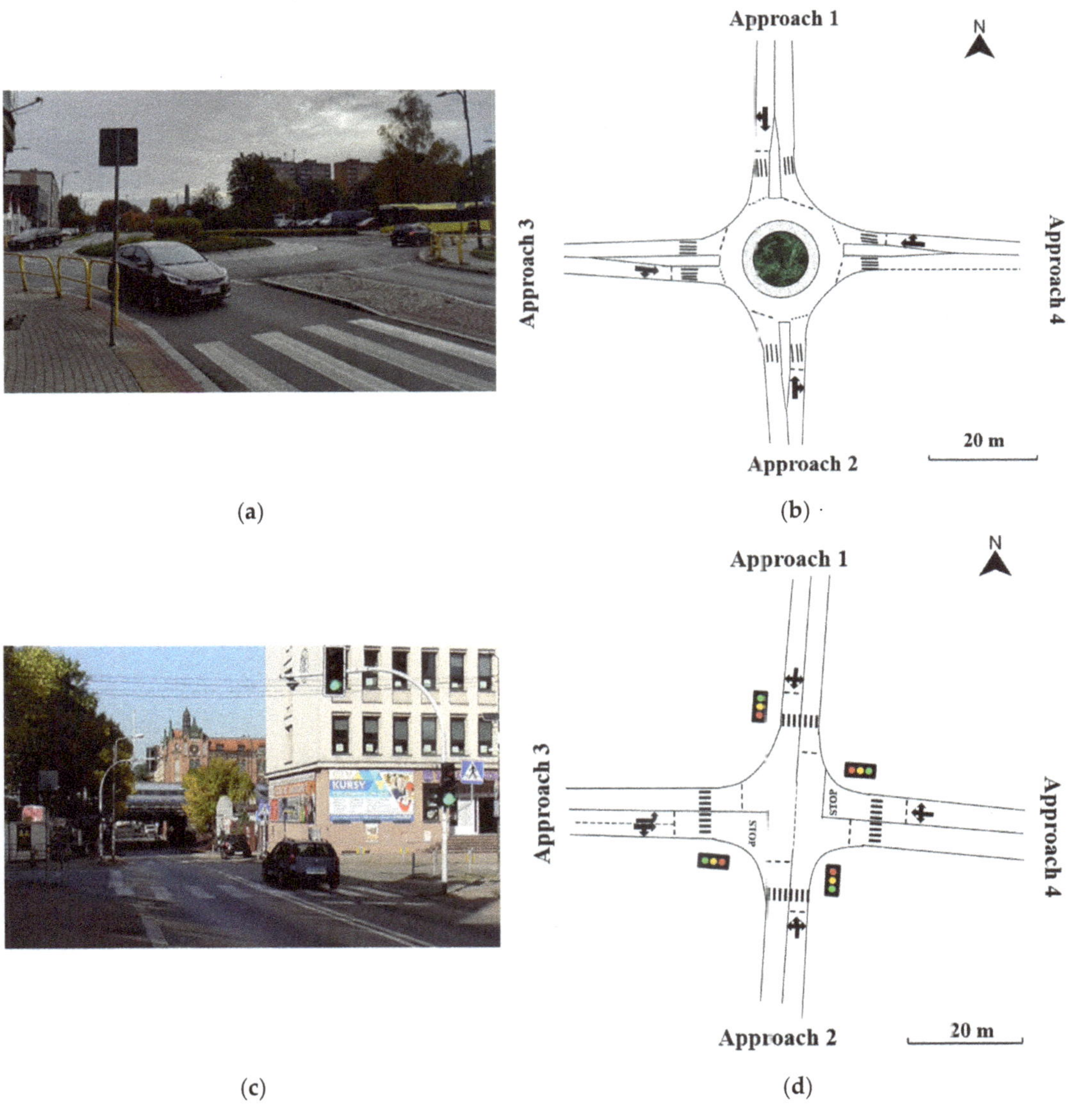

Figure 2. *Cont.*

(e)

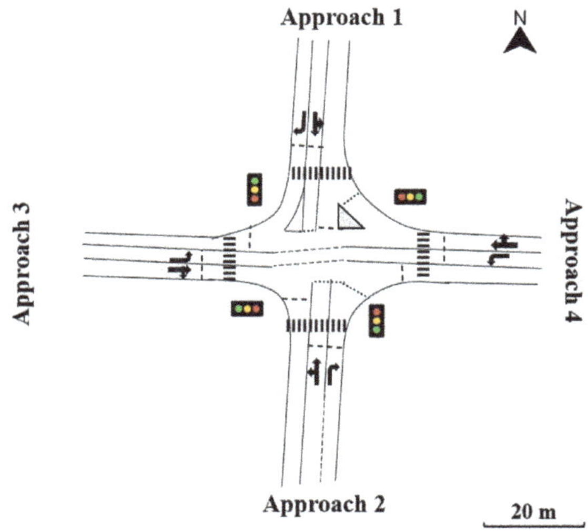

(f)

Figure 2. Examined case studies: (**a**) View from the west entry at Case Study 1 (latitude: 50.319716, longitude: 19.003237); (**b**) Geometric sketch of Case Study 1; (**c**) View from the south entry at Case Study 2 (latitude: 50.255549, longitude: 19.030781); (**d**) Geometric sketch of Case Study 2; (**e**) View from the south entry at Case Study 3 (latitude: 50.251557, longitude: 19.029674); (**f**) Geometric sketch of Case Study 3.

The selected intersections are typical urban road intersections; however, their geometries and layouts make them suitable for design as roundabouts, signalized intersections, or TWSC intersections, aligning with the objectives of this study. They were chosen based on their locations, available facilities, and compatibility with the study's aims. The intersections are situated near major trip generators, including a university, a cemetery, and several residential and administrative buildings. Field surveys confirmed that these sites experience traffic congestion and performance issues. Moreover, as described in the following paragraphs, the selected case studies differ in geometry, layout, number of lanes per approach, entry and exit traffic flows, and control type, enabling the analysis of a key research question.

The first case study is a single-lane four-leg roundabout [61], located in a suburban area of Siemianowice Śląskie. Each approach has one entry and one exit lane with a width of 3.50 m (see Figure 2b). It connects Kościelna and Oświęcimska, the major streets, with Maciejkowicka and Plac 11 Listopada, the minor streets. The roundabout has an outer diameter of 30.00 m, and the circulatory roadway is one lane with a width of 5.00 m. The second case study examines a four-leg signalized intersection situated in the urban area of Katowice. Each approach has one entry lane, except for approach 3, which has two entry lanes and one exit lane. All lanes are 3.50 m wide (see Figure 2d). This intersection is at the convergence of Konstantego Damrota, the major street, and Wojewódzka and Zygmunta Krasińskiego, the minor streets. All approaches have one shared lane for left-turn, right-turn, and through movements except approach 3, which has one shared lane for through and right-turn movements, and one exclusive left-turn lane. The third case study is a four-leg signalized intersection located in the urban area of Katowice. Each approach has two entry lanes and one exit lane, with each lane being 3.50 m wide (see Figure 2f). The intersection is at the junction of Konstantego Damrota, the major street, and Powstańców, the minor street. The major street entries consist of one shared lane for left-turn and through movements, and one exclusive right-turn lane. The minor street entries consist of one shared lane for right-turn and through movements, and one exclusive left-turn lane. All the intersections comply with Polish regulations [61,62]. The speed limit is 40 (km/h) in Case Study 1, and 30 (km/h) in Case Studies 2 and 3.

In line with the research objectives, each case study was conceptualized with three different configurations: its existing layout and two alternative designs. For example, Case Study 1, currently a roundabout, is also examined as a signalized intersection and a two-way stop-controlled intersection. Accordingly, the geometric characteristics of all configurations are described for each case study, and the current and designed layouts are presented in Table 1.

Table 1. The geometries of the current and designed configurations of case studies.

Case Study	No. of Entry (Exit) Lanes	Road Type [1]	Entry (Exit) Lane Width (m)	Width (No.) of Roundabout Circulatory Roadway (m)	Outer Diameter (m)
1	1 (1) 1 (1)	Major street Minor street	3.50 (3.50) 3.50 (3.50)	5.00 (1) [2]	30
2	1 (1) 1 (1) [3]	Major street Minor street	3.50 (3.50) 3.50 (3.50) [4]	3.50 (1)	20
3	2 (1) 2 (1)	Major street Minor street	3.50 (3.50) 3.50 (3.50)	3.50 (2)	25

[1] Approaches 1 and 2 are major streets and approaches 3 and 4 are minor streets in each case study (see Figure 2b,d,f) and are differentiated based on the number of entry lanes, the amount of entry (exit) traffic, or control mode. [2] The central island of the roundabout has a traversable apron. [3] one-lane entries and exits except for the entry from the west direction street, which has two entry lanes. [4] 3.50 m width for entries and exits, except for the entry from the west direction, which has 2.50 m width in each lane.

Traffic flow graphs for the case studies are illustrated in Figure 3. The total traffic flows for intersections 1, 2, and 3 are 1311, 1099, and 1452 vehicles per hour, respectively, with heavy vehicles accounting for 12%, 2%, and 3% of these flows. For the analysis, mean traffic volumes during peak hours were applied. Data were collected using two tripod-mounted digital cameras, primarily positioned on sidewalks to capture all vehicle movements, supplemented by on-site observations. Peak hours were identified in consultation with experts and locals and further validated through frequent field surveys conducted during the study period. These peak hours were 7:00 a.m. to 8:30 a.m., 12:00 p.m. to 1:00 p.m., and 3:00 p.m. to 5:00 p.m., with morning and afternoon showing heavier demand, and noon exhibiting a lighter peak. Data collection was performed on weekdays between September and November 2023, spanning at least five full days, from morning to evening at each site. Although the urban context may be regarded as generally conducive to pedestrian and cyclist activity, field observations revealed their presence to be minimal; therefore, their influence on traffic performance was considered negligible and excluded from the analysis.

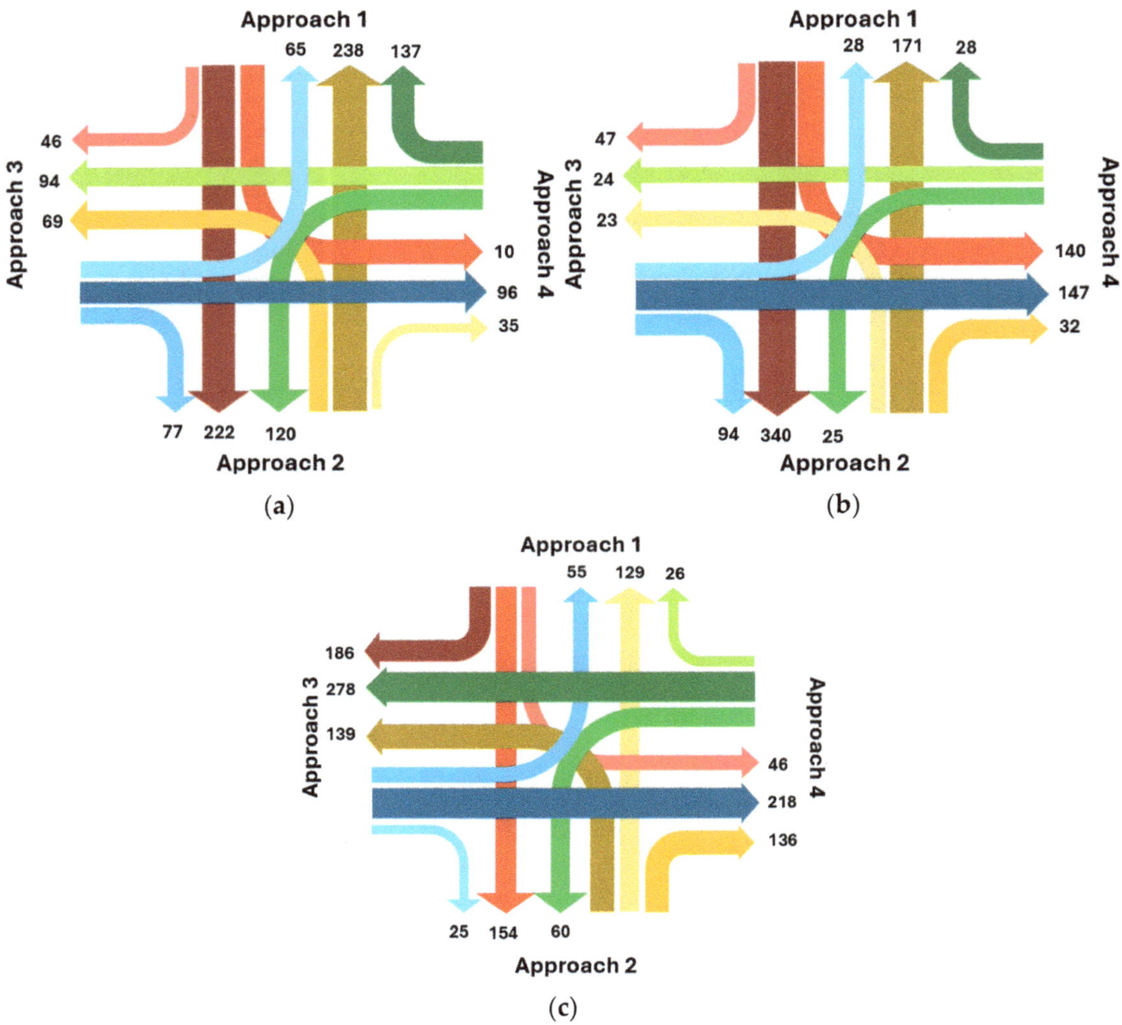

Figure 3. Traffic flow graphs for: (**a**) Case Study 1; (**b**) Case Study 2; (**c**) Case Study 3. Note: 1—All values are expressed in vehicles per hour. 2—In all traffic flow graphs, the red gradient represents flow from approach 1, the orange gradient from approach 2, the blue gradient from approach 3, and the green gradient from approach 4. 3—Approaches 1 to 4, for each case study, are the same as the approaches illustrated in Figure 2b,d,f.

For the signalized configurations, Case Study 1 was modeled with a cycle length of 90 s under pretimed, fixed timing control, following the design procedure in Chapter 31 of the HCM [18], to equalize the volume-to-capacity ratios for critical lane groups. The existing signalized intersections in Case Studies 2 and 3 operate with pretimed, fixed timings traffic signal control and have cycle lengths of 80 s. In Case Studies 1 and 2, the cycle length is divided into two phases, whereas in Case Study 3, it is divided into four phases. Phase details are presented in Figure 4.

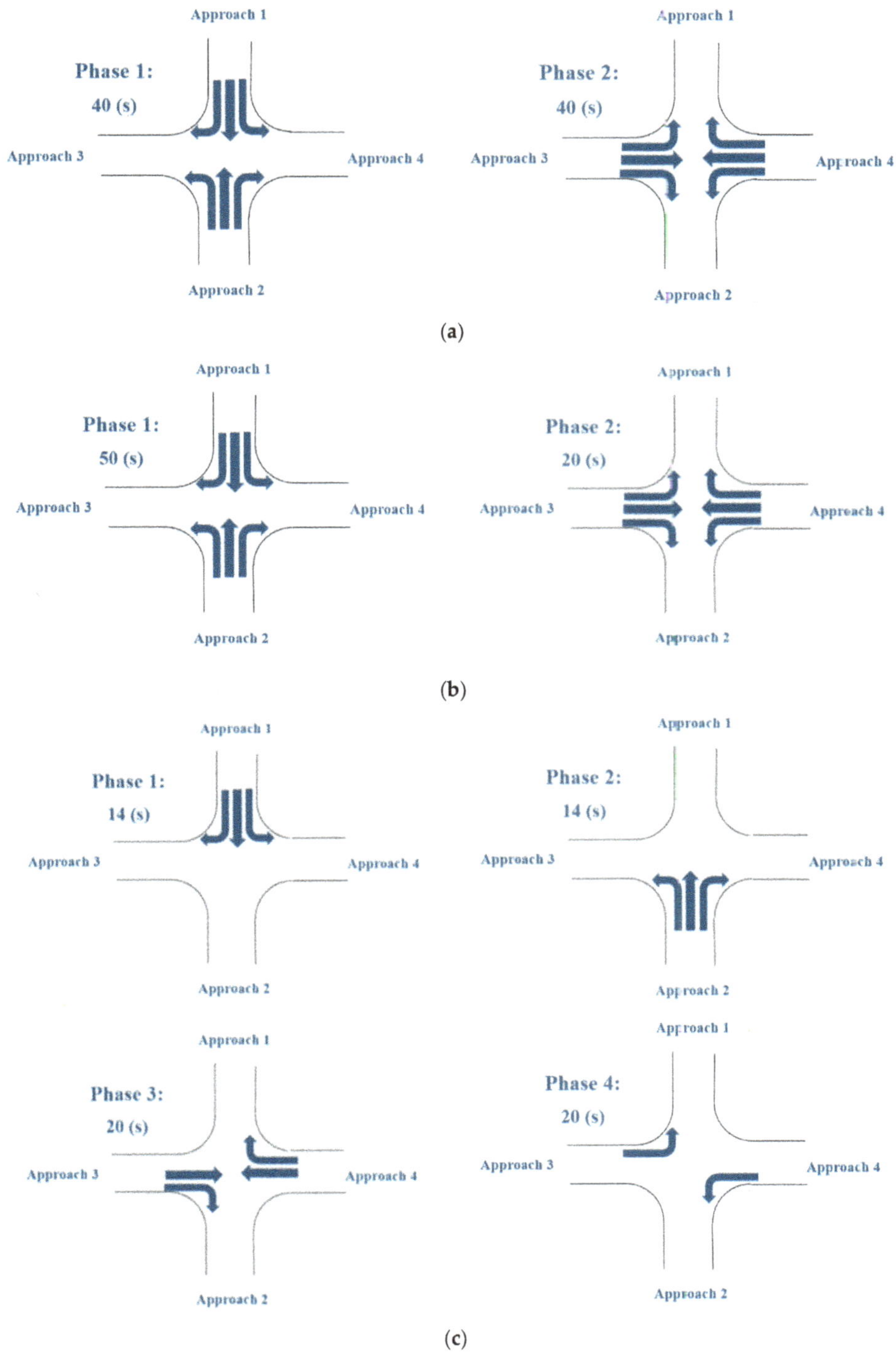

Figure 4. The phase timing plan diagram for: (**a**) Case Study 1; (**b**) Case Study 2; (**c**) Case Study 3. Note: 1—Approaches 1 to 4, for each case study, are the same as the approaches illustrated in Figure 2b,d,f. 2—The total yellow times in a cycle length for case studies 1, 2, and 3 are 10, 10, and 12 s, respectively.

3.2. *Research Methodology*

The case studies in this research were analyzed using two methodologies: a heuristic approach, based on the HCM [18], and a microsimulation approach using AIMSUN Next [22]. Each methodology is described in the following sections. It is important to note that signalized intersections were assumed to operate as TWSC intersections following storm events due to power outages. The only storm-related impact considered in this study is the electricity cutoff that occurs after a storm and persists for a certain period.

3.2.1. The Heuristic Approach

This heuristic approach is analytical, and the control delay time is considered the key parameter for assessing the efficiency and resilience of each type of traffic control in pre- and post-storm situations. Control delay measures the time vehicles are slowed or stopped at intersections due to traffic control devices [18]. To calculate the total control delay time for the intersection, the first step is to calculate the control delay time per lane, followed by the control delay time per approach, and finally the control delay time for the entire intersection. The formulas for calculating the control delay time in each lane for roundabouts (Chapter 22 of the HCM [18]), signalized intersections (Chapter 19 of the HCM [18]), and TWSC intersections (Chapter 20 of the HCM [18]) are presented as Equations (1)–(3) in Table 2. Furthermore, the variables, along with a schematic sketch for each traffic control mode, are also introduced in Table 2.

Table 2. Introducing the control delay time equations and variables for each traffic control mode.

Traffic Control Type/Scheme [1]	Control Delay Time Equation		Description of Variables
Roundabouts 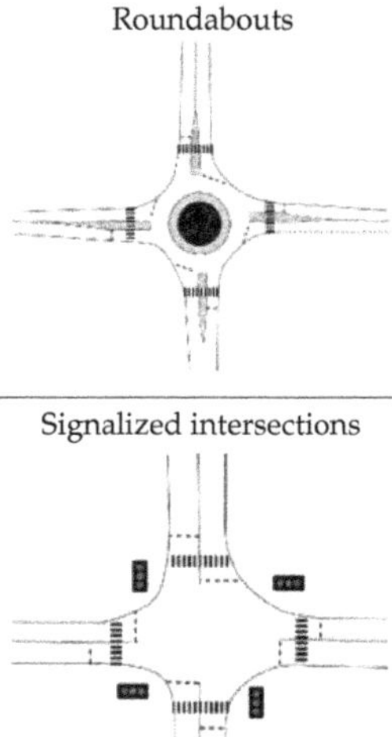	$d_l = \frac{3600}{K} + 900 \cdot t \cdot \left[(y-1) + \sqrt{(y-1)^2 + \frac{\left(\frac{3600}{K}\right) \cdot y}{450 \cdot t}} \right] + 5 \cdot min[y, 1]$	(1)	d_l: Control delay per lane (s/veh), K: Capacity of the subject lane (veh/h), t: Analysis time period (h), y: Volume-to-capacity ratio of the subject lane
Signalized intersections 	$d_l = f_p \cdot \frac{0.5 \cdot C_l \cdot \left(1 - \frac{g_e}{C_l}\right)^2}{1 - \left[min(1, y) \cdot \frac{g_e}{C_l}\right]} + 900 \cdot t \cdot \left[(y_a - 1) + \sqrt{(y_a - 1)^2 + \frac{8 \cdot f_{id} \cdot f_u \cdot y_a}{K_a \cdot t}} \right]$	(2)	d_l: Control delay per lane (s/veh), f_p: Progression adjustment factor, C_l: Cycle length (s), g_e: Effective green time (s), y: Volume-to-capacity ratio of the subject lane, t: Analysis time period (h), y_a: Average volume-to-capacity ratio, f_{id}: Incremental delay factor, f_u: Upstream filtering/metering adjustment factor, K_a: Average capacity (veh/h) [2]

Table 2. *Cont.*

Traffic Control Type/Scheme [1]	Control Delay Time Equation		Description of Variables
TWSC intersections 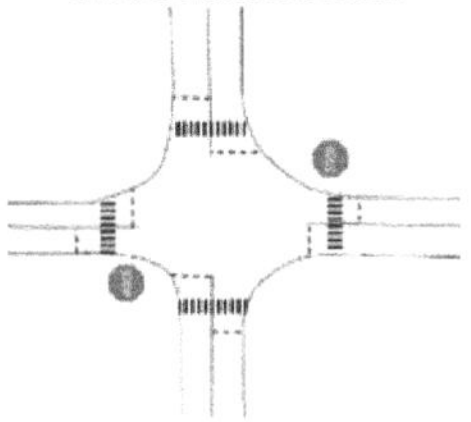	$d_l = \frac{3600}{K_x} + 900 \cdot t \cdot \left[\left(\frac{r_x}{K_x} - 1 \right) + \sqrt{\left(\frac{r_x}{K_x} - 1 \right)^2 + \frac{\left(\frac{3600}{K_x} \right) \cdot \left(\frac{r_x}{K_x} \right)}{450 \cdot t}} \right] + 5$	(3)	d_l: Control delay per lane (s/veh), K_x: Capacity of movement x (veh/h), t: Analysis time period (h), r_x: Flow rate of movement x (veh/h)

[1] The general schematic sketches of the traffic control modes are presented in this table. Each examined case study in this research might differ from this general form in terms of lane numbers and other geometric characteristics.
[2] The average capacity equals the capacity if no initial queue exists in any lane group at the intersection. Otherwise, it should be computed using the procedure outlined in Section 4, Chapter 19 of the HCM [18].

The control delay per approach is calculated as a weighted average of the control delay time for each lane in that approach, with the volume or flow rate of each lane serving as the weight. Similarly, the total control delay time for the intersection is calculated as a weighted average of the control delay time for each approach, with the volume or flow rate of each approach serving as the weight.

Another important parameter in the control delay analysis for each traffic control type, required both for delay time calculations and calibrating the microsimulations, is the capacity (see Table 3, Equations (4)–(6)).

Table 3. Capacity equations and variables for each traffic control mode.

Traffic Control Type	Capacity Equation		Description of Variables
Roundabouts	$K_e = A \cdot B \cdot e^{C \cdot r_c}$	(4)	K_e: Lane capacity (veh/h), A: Adjusting factor related to the entry and circulating lanes [1], B: Heavy vehicles adjustment factor, C: Adjusting factor related to the entry and circulating lanes [1], r_c: Conflicting flow rate (veh/h)
Signalized intersections	$K = L_n \cdot \frac{g_e}{C_l} \cdot r_s$	(5)	K: Lane group capacity (veh/h), L_n: Number of lanes in a lane group, g_e: Effective green time (s), C_l: Cycle length (s), r_s: Adjusted saturation flow rate (veh/h/ln)
TWSC intersections	$K_{p,x} = r_{c,x} \cdot \frac{e^{-r_{c,x} \cdot w_{c,x}/3600}}{1 - e^{-r_{c,x} \cdot w_{f,x}/3600}}$	(6)	$K_{p,x}$: Potential capacity of movement x (veh/h), $r_{c,x}$: Conflicting flow rate of movement x (veh/h), $w_{c,x}$: Critical headway for minor movement x (s), $w_{f,x}$: Follow-up headway for minor movement x (s)

[1] A and C are adjusting factors that are related to the number of entry and circulating lanes. For a one-lane entry conflicting with a single-lane roundabout: A = 1380 and $C = (-1.02 \times 10^{-3})$, and for a two-lane entry conflicting with a two-lane roundabout: A = 1420 and $C = (-0.85 \times 10^{-3})$ for the right entry lane, and A = 1350 and $C = (-0.92 \times 10^{-3})$ for the left entry lane.

The adjusted saturation flow rate (r_s), expressed in vehicles per hour per lane (veh/h/ln), is calculated by multiplying the base saturation flow rate by the adjustment factors. The base saturation flow rate (r_{bs}) was set to 1750 passenger cars per hour per lane (pc/h/ln) for Case Study 1, and 1900 (pc/h/ln) for Case Studies 2 and 3, following the default values in Chapter 19 of the HCM [18], which are suggested based on city population. The adjustment factors (f_i) correct the base saturation flow rate by accounting for the geometry and traffic conditions of the specific lane, such as lane width, heavy vehicles, grades, area type, turning movements, and parking activity adjacent to the lane group. Equation (7) presents the adjusted saturation flow rate, where m is the number of adjustment factors applied to the base saturation flow rate, and all other parameters were introduced earlier. In this study, adjustment factors applied to the base saturation flow rates include heavy vehicles, right-turn movements, and left-turn movements for Case Study 1, with area type added for Case Studies 2 and 3.

$$r_s = r_{bs} \cdot \prod_{i=1}^{m} f_i \tag{7}$$

The efficiency of each case study under pre-storm scenario was evaluated by calculating total control delay for alternative control strategies, and the results were compared to determine whether the existing configuration represents the most efficient option. Resilience was assessed by comparing total control delay under pre- and post-storm scenarios using the same approach, assuming that after a storm, signalized intersections operate as TWSC during power outages. This comparative analysis identifies the most efficient and resilient control strategies for each case study relative to their current configurations.

The level of service for a lane group, approach, or entire intersection was determined from the calculated or observed control delay time [18]. LOS is categorized from A to F, with A representing the best level of service and F the worst. For motorized vehicles, LOS thresholds depend primarily on the volume-to-capacity ratio and control delay. When the volume-to-capacity exceeds 1 (indicating oversaturation), LOS is rated as F regardless of control delay. Otherwise, LOS is assigned according to control delay thresholds, as summarized in Table 4.

Table 4. Level of service threshold for motorized vehicles in undersaturated conditions based on [18].

Metric	Traffic Control Type	Level of Service					
		A	B	C	D	E	F
Control delay time (s/veh)	Roundabout	0–10	>10–15	>15–25	>25–35	>35–50	>50
	Signalized intersection	0–10	>10–20	>20–35	>35–55	>55–80	>80
	TWSC intersection	0–10	>10–15	>15–25	>25–35	>35–50	>50

Note: the ">" symbol in the table means "more than". For example, ">10–15" means the control delay time of more than 10 seconds per vehicle and up to 15 seconds per vehicle.

As mentioned earlier in the Introduction Section [20] and in previous paragraphs, the heuristic approach in this study uses control delay time and LOS as indicators of intersection functionality. The values under normal conditions represent the efficiency of the examined intersections. Intersection resilience is defined by comparing these metrics before and after storm events. Finally, a comparison of these results, along with LOS assessments for the examined traffic control types, indicates whether each case study is efficient and resilient under pre- and post-storm scenarios or whether the traffic control type should be reconsidered.

3.2.2. Microsimulation

This section focuses on microsimulation using AIMSUN Next [22]. AIMSUN Next, a widely used transport simulation tool, was selected for this research due to its ability to simulate isolated intersections and road segments [63–65], its flexibility in representing diverse intersection layouts, traffic control types, and configurations [63,64], and the availability of a licensed version for this study. In this analysis, only human-driven vehicles were considered in the traffic flow.

Data collected during peak hours were used to develop the Origin-Destination (O-D) matrices. For the studied intersections in Poland, observed traffic flows served as the basis for constructing these matrices. The next step involved replicating the actual geometry of each intersection in AIMSUN Next [22]. This process included defining road sections, specifying the number of lanes per approach, road type, lane width, and speed limit, and then connecting road sections with nodes. Centroids, representing trip start and end points for each approach, were assigned to populate the O-D matrix cells for each intersection.

Depending on the traffic control strategy under investigation, the linking node was defined as a roundabout, a signalized intersection, or a TWSC intersection, with the corresponding input parameters for each mode. For example, signalized intersections required a detailed signal timing plan, including predefined signal groups and phases, turning movements, cycle length, control plan duration, and yellow intervals. For TWSC intersections, movement priorities were specified, while for roundabouts, parameters such as central island diameter, circulating lane width and number, and entry priorities were defined. Detectors were placed at selected points along road sections to record measures such as speed, count, density, flow, and headway.

A dynamic microsimulation scenario was implemented using the stochastic route choice (SRC) method, in which vehicles are probabilistically assigned to routes based on a defined distribution. This allows vehicle trajectories to be observed and their performance statistics analyzed [22]. Each case study model, representing real-world conditions, was simulated for one-hour periods, and the resulting flows were compared against observed data to assess the accuracy of AIMSUN Next in reproducing real traffic conditions. To validate the model, Geoffrey E. Havers' statistic (GEH index) was used [63]. A model is considered acceptable if 85% or more of the flows have a GEH value below 5. Nearly all models showed GEH results below this threshold. Additionally, the percentage difference between simulated and observed data was examined, averaging around 10% across all case studies. Therefore, calibration of model parameters was required in all scenarios.

A sensitivity analysis of model parameters was conducted for each model to assess the impact of changing each parameter on the simulation results. The sensitivity analysis was performed to identify which parameters significantly affected the simulation outcomes and should be considered for calibration. These parameters are: 1—Speed limit acceptance, reflecting drivers' compliance with speed limits, where values above 1 indicate exceeding the limit and values below 1 indicate driving under it. 2—Reaction time (s), which is the duration a driver needs to respond to changes in the speed of the vehicle ahead. 3—Reaction time at stop (s), which is the time a stationary vehicle requires to respond to the acceleration of the vehicle ahead. 4—Reaction time at traffic light (s), which is the time the first vehicle waiting at a traffic light takes to respond when the light turns green. 5—Maximum acceleration (m/s^2) is the vehicle's highest possible acceleration, applicable in all conditions, as defined in the Gipps car-following model. 6—Gap (s), allows setting a custom headway between vehicles, overriding the car-following model's calculated distance, with the effective gap being the lesser of the specified value and the default distance [22].

The most sensitive parameters were calibrated by iteratively adjusting their values to align the simulated capacity data with the observed or calculated capacities, ensuring a more accurate representation of traffic behavior. Calibration was performed for each type of traffic control in every case study. For roundabouts, the conflicting flow ranged from free flow to 2000 vehicles per hour (based on the entry mechanisms described in [66]); for signalized intersections, the effective green time ranged from 10 to 60 s; and for TWSC intersections, the conflicting flow also ranged from free flow to 2000 (veh/h). The default and calibrated model parameter values for each case study are presented in Table 5.

Table 5. The default and calibrated values of the model parameters.

Case Study	Type of Traffic Control	Model Parameter	Default Value	Calibrated Value
Case study 1	Roundabout	Speed limit acceptance	1.10	1.00
		Gap (s)	0.00	1.55
		Reaction time (s)	0.80	1.05
	Signalized intersection	Speed limit acceptance	1.10	1.30
		Maximum acceleration (m/s^2)	3.00	3.20
		Reaction time (s)	0.80	0.90
		Reaction time at stop (s)	1.20	1.35
		Reaction time at traffic light (s)	1.60	1.20
	TWSC intersection	Speed limit acceptance	1.10	1.20
		Maximum acceleration (m/s^2)	3.00	2.60
		Gap (s)	0.00	1.93
		Reaction time (s)	0.80	1.45
		Reaction time at stop (s)	1.30	1.85
Case study 2	Roundabout	Speed limit acceptance	1.10	1.00
		Gap (s)	0.00	1.45
		Reaction time (s)	0.80	1.05
	Signalized intersection	Speed limit acceptance	1.10	1.30
		Reaction time (s)	0.80	0.90
		Reaction time at stop (s)	1.20	1.45
		Reaction time at traffic light (s)	1.60	1.20
	TWSC intersection	Speed limit acceptance	1.10	1.20
		Maximum acceleration (m/s^2)	3.00	2.60
		Gap (s)	0.00	1.95
		Reaction time (s)	0.80	1.25
		Reaction time at stop (s)	1.20	1.75
Case study 3	Roundabout	Speed limit acceptance	1.10	1.00
		Gap (s)	0.00	1.65 (1.55) [1]
		Reaction time (s)	0.80	0.95
	Signalized intersection	Speed limit acceptance	1.10	1.30
		Maximum acceleration (m/s^2)	3.00	3.20
		Reaction time (s)	0.80	0.85 (0.80) [2]
		Reaction time at stop (s)	1.20	1.30 (1.25) [2]
		Reaction time at traffic light (s)	1.60	1.20
	TWSC intersection (post-storm) [3]	Speed limit acceptance	1.10	1.20
		Maximum acceleration (m/s^2)	3.00	2.60
		Gap (s)	0.00	1.90 (1.95) [2]
		Reaction time (s)	0.80	1.25 (1.30) [2]
		Reaction time at stop (s)	1.20	1.80
	TWSC intersection (lane adjustment) [3]	Speed limit acceptance	1.10	1.20
		Maximum acceleration (m/s^2)	3.00	2.60
		Gap (s)	0.00	1.50 (1.90) [4]
		Reaction time (s)	0.80	1.10 (120) [4]
		Reaction time at stop (s)	1.20	1.80 (1.85) [4]

[1] Value outside parentheses corresponds to the right entry lane; value in parentheses corresponds to the left entry lane. [2] Values outside parentheses correspond to the exclusive right-turn lane; values in parentheses correspond to the shared left-turn and through lane. [3] These terms are explained later in the text. [4] Values outside parentheses correspond to the shared right-turn and through lane; values in parentheses correspond to the exclusive left-turn lane.

It is worth noting that, when Case Study 3 was conceptualized as a TWSC intersection, the movement groups were adjusted according to [18] to align with TWSC right-of-way rules. In the signalized intersection under pre- and post-storm conditions, the major streets had two lanes: one shared left-turn and through lane and one exclusive right-turn lane. In the TWSC intersection, these were modified to one exclusive left-turn lane and one shared through and right-turn lane. Based on these adjustments, sensitivity analyses and calibration were necessary for both TWSC intersections in Case Study 3, and the calibrated results are presented in Table 5 as TWSC intersection (post-storm) and TWSC intersection (lane adjustment).

To confirm the validity of the calibration process, the GEH index was used [63]. In all cases, more than 90% of the flows had a GEH value below 5, and the models were therefore accepted based on this. Furthermore, scattergram analysis and *t*-test were performed to ensure the reliability of the calibration. Figure 5 illustrates the scattergram for the current configuration of Case Study 2, approach 1 (see Figure 2d) as an example. Tables 6–8 present the corresponding statistics for comparing simulated versus observed or calculated capacities across all case studies. Due to right-of-way priorities at the TWSC intersection with lane adjustment (explained earlier in the previous paragraph), the calculated capacities for the shared right-turn and through lane were equal across all conflicting flow rates, and the simulated capacity values were nearly identical. Therefore, a *t*-test was deemed unnecessary for this lane group and is not reported in Table 8.

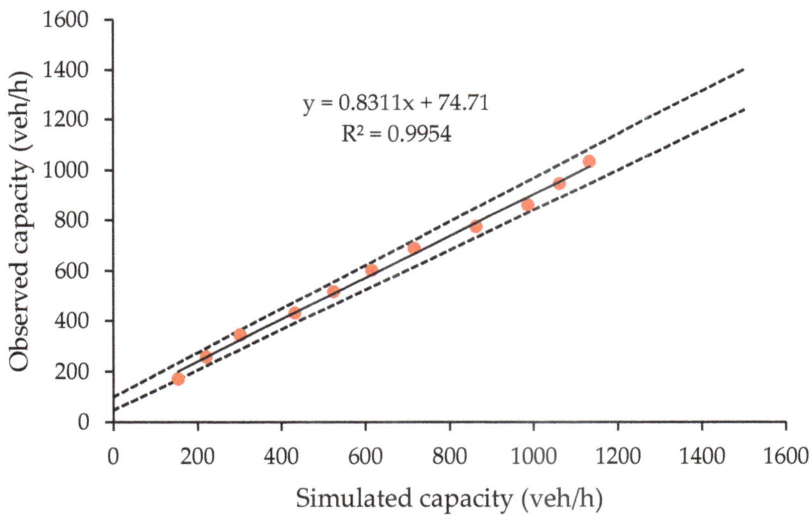

Figure 5. Scattergram for Case Study 2 in its current configuration.

Table 6. The statistics of the calibration process in Case Study 1.

Type of Traffic Control	Type of Data	Mean	s.e.	R^2	t-Statistics	t-Critical	$p(\alpha)$-Value
Roundabout	Observed/Calculated Simulated	553.000 605.727	108.304 119.010	0.964	0.328	2.086	0.747
Signalized intersection	Observed/Calculated Simulated	484.545 557.818	69.227 74.952	0.994	0.718	2.086	0.481
TWSC intersection	Observed/Calculated Simulated	1087.636 1138.273	103.055 85.983	0.964	0.377	2.086	0.710

Note: Mean indicates the mean values of the observed and simulated samples, while s.e. stands for the standard error for the mean; R^2 indicates the percentage of variance in the dependent variable that is explained by the independent variable in a regression model; t-statistic is derived from the *t*-test with N degrees of freedom at the significance level $\alpha = 0.05$; $p(\alpha)$-value is the probability under the null hypothesis of equal means at the significance level $\alpha = 0.05$.

Table 7. The statistics of the calibration process in Case Study 2.

Type of Traffic Control	Type of Data	Mean	s.e.	R^2	t-Statistics	t-Critical	p(α)-Value
Roundabout	Observed/Calculated Simulated	601.727 631.636	117.752 1301.221	0.984	0.170	2.086	0.867
Signalized intersection	Observed/Calculated Simulated	604.273 640.000	86.291 102.491	0.995	0.267	2.086	0.792
TWSC intersection	Observed/Calculated Simulated	1131.727 1141.364	101.174 66.883	0.976	0.080	2.086	0.937

Note: Mean indicates the mean values of the observed and simulated samples, while s.e. stands for the standard error for the mean; R^2 indicates the percentage of variance in the dependent variable that is explained by the independent variable in a regression model; t-statistic is derived from the *t*-test with N degrees of freedom at the significance level $\alpha = 0.05$; p(α)-value is the probability under the null hypothesis of equal means at the significance level $\alpha = 0.05$.

Table 8. The statistics of the calibration process in Case Study 3.

Type of Traffic Control	Type of Data	Mean	s.e.	R^2	t-Statistics	t-Critical	p(α)-Value
Roundabout (RL entry)	Observed/Calculated Simulated	677.545 690.545	111.814 128.862	0.998	0.076	2.086	0.940
Roundabout (LL entry)	Observed/Calculated Simulated	614.636 686.000	109.373 126.234	0.987	0.427	2.086	0.674
Signalized intersection (RT movement)	Observed/Calculated Simulated	1208.455 1071.636	172.645 159.919	0.997	−0.581	2.086	0.567
Signalized intersection (Shared LT & T movements)	Observed/Calculated Simulated	1208.455 1103.364	172.645 170.171	0.998	−0.433	2.086	0.669
TWSC intersection (RT movement)	Observed/Calculated Simulated	1120.273 1147.636	102.223 90.936	0.933	0.200	2.086	0.843
TWSC intersection (Shared LT & T movements)	Observed/Calculated Simulated	1120.273 1126.364	102.223 93.099	0.910	0.044	2.086	0.965
TWSC intersection (LT movement)- Adjusted	Observed/Calculated Simulated	786.272 839.364	132.780 131.335	0.957	0.284	2.086	0.779

Note: Mean indicates the mean values of the observed and simulated samples, while s.e. stands for the standard error for the mean; R^2 indicates the percentage of variance in the dependent variable that is explained by the independent variable in a regression model; t-statistic is derived from the *t*-test with N degrees of freedom at the significance level $\alpha = 0.05$; p(α)-value is the probability under the null hypothesis of equal means at the significance level $\alpha = 0.05$; RL stands for right lane, LL stands for left lane, RT stands for right-turn. LT stands for left-turn, and T stands for through.

After calibration, the models were run using mean peak flow rates as inputs. The results and discussion are presented in Sections 4 and 5. Delay time, travel time, speed, and average approach delay were used as metrics to evaluate the efficiency and resilience of the case studies under pre- and post-storm scenarios.

- Delay time (s/km) is calculated as the difference between travel time under free-flow conditions and actual travel time [22];
- Travel time (s/km) is the average time a vehicle takes to cover one kilometer within the network, calculated as the mean of all individual travel times (exit time minus entrance time) [22];
- Speed (km/h) is the average speed of all vehicles upon exiting the system, computed from each vehicle's mean travel speed [22];
- Average approach delay (s/veh) is computed by summing the delays of each link weighted by their flow and dividing by the total traffic flow [22].

Finally, the level of service for the average approach delay of each case study was determined according to the HCM [18] classification (see Table 4).

Similar to the heuristic approach, delay time, travel time, speed, average approach delay, and LOS are considered indicators of intersection functionality in the microsimulations. Under normal conditions, these metrics reflect the efficiency of the intersections, whereas their comparison before and after storm events represents intersection resilience.

4. Results

The results section presents findings from two complementary methods used to evaluate intersection efficiency and resilience: a heuristic analytical approach and microsimulations. For the heuristic approach, the average control delay time and the corresponding level of service are reported for all traffic control types and case studies. Microsimulation results include delay time, travel time, speed, average approach delay, and level of service based on the HCM [18] classification. While each method differs in calculation procedures and assumptions, both approaches yield comparable trends and LOS classifications. Presenting the results of both methods separately demonstrates their consistency and convergence, supporting the robustness and reliability of the proposed framework.

4.1. The Heuristic Approach Results

The analysis for Case Studies 1 to 3 considers both pre- and post-storm situations. The pre-storm scenario represents the normal situation before any storm, while the post-storm scenario depicts the situation after a storm, resulting in an electricity outage. Each intersection is assumed to perform as a roundabout, a signalized, or a TWSC intersection, and the baseline of this analysis is Figure 2b,d,f. As noted earlier in the text and in accordance with [18], adjustments were made to the movement groups in Case Study 3 to ensure consistency with right-of-way regulations at the TWSC intersection. In the signalized intersection, the major streets consisted of two lanes: one shared lane designated for left-turn and through movements, and one exclusive right-turn lane. In the TWSC intersection, these were reconfigured into one exclusive left-turn lane and one shared through and right-turn lane. The average control delay time is used as a metric to assess the performance efficiency and resilience of roundabouts, signalized, and TWSC intersections before and after a storm, respectively.

Total control delay is typically used to compare delays across different traffic control types. Control delay, which measures the time vehicles are slowed or stopped at intersections due to traffic control devices, is a crucial metric for determining an intersection's level of service [18]. In this study, total control delay was analyzed to evaluate the efficiency and resilience of roundabouts, signalized, and two-way stop-controlled intersections before and after a disaster event. The corresponding level of service in each case was also determined to provide a clearer understanding of efficiency and resilience. Figure 6 presents the average control delay for each intersection under the three traffic control types.

Figure 6 illustrates the average control delay time for the entire intersection across the case studies under three traffic control types. The y-axis denotes the average delay in seconds per vehicle (s/veh), while the x-axis distinguishes the case studies and their associated control strategies in both pre- and post-storm conditions. Blue columns correspond to roundabout operations, red columns to signalized intersections, and green columns to TWSC intersections.

Case Study 1 evaluates a roundabout in its current configuration. In both pre- and post-storm conditions, and under the study's assumption that storm impacts are limited to electricity outages, the average control delay remains constant at 9.09 (s/veh), corresponding to LOS A. When the same site is modeled as a signalized intersection, the outcomes diverge: under pre-storm conditions, signal control results in 23.95 s per vehicle (LOS C), whereas in the post-storm scenario, the assumed power outage renders the signals inoperative, forcing TWSC operation. This shift produces a substantial increase in delay, reaching 76.54 (s/veh), corresponding to LOS F. When directly modeled as a TWSC, the intersection performs identically in pre- and post-storm conditions, consistent with the study's assumptions, with a delay of 76.54 s per vehicle (LOS F). These results demonstrate the resilience of the roundabout configuration compared to the vulnerability of signalized control under power outages.

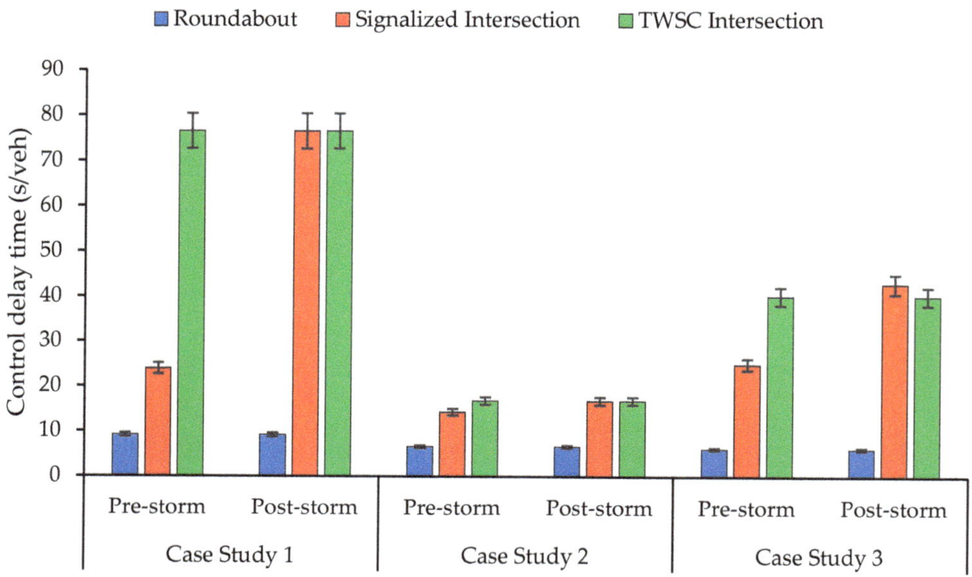

Figure 6. The average control delay time (s/veh) of the entire intersection in pre- and post-storm conditions for case studies 1 to 3.

Case Study 2 examines a signalized intersection in its existing layout. In the pre-storm scenario, the average control delay is 14.23 s per vehicle (LOS B). Following the electricity outage in post-storm conditions, the intersection defaults to TWSC operation, yielding 16.75 (s/veh), corresponding to LOS C. The marginal difference between pre- and post-storm delays indicates that performance is largely preserved despite the operational shift. When the intersection is modeled as a single-lane roundabout, the delay decreases notably to 6.49 s per vehicle (LOS A), with no change between scenarios due to the study's assumptions. Finally, when modeled directly as a TWSC, performance remains the same in pre- and post-storm conditions, with a control delay of 16.75 s per vehicle (LOS C). Overall, this case illustrates that certain signalized intersections may maintain acceptable service levels even when converted to TWSC operation during electricity outages.

Case Study 3 considers another signalized intersection. Under pre-storm conditions, the average control delay time is 24.90 (s/veh), corresponding to LOS C. In the post-storm scenario, the assumed outage disables the traffic signals, and the intersection operates as a TWSC with delays increasing to 42.82 s per vehicle (LOS E), marking a substantial deterioration in performance. When reconfigured as a two-lane roundabout, the average delay decreases sharply to 6.05 s per vehicle (LOS A), remaining unchanged between scenarios under the study's assumptions. Finally, when directly modeled as a TWSC with adjusted movement groups (as described earlier), the intersection again shows identical results in both scenarios, reporting 40.04 (s/veh), corresponding to LOS E. These findings highlight that, for intersections with higher pre-storm delays, the transition to TWSC under outage conditions can produce significant reductions in service quality.

4.2. The Microsimulation Results

The models were reproduced in AIMSUN Next [22], as explained in Section 3.2.2. After completing the calibration process and verifying its reliability, the models are simulated using AIMSUN Next [22], and the results are presented in this section. The delay time, travel time, speed, and average approach delay are considered as metrics to evaluate the efficiency and resilience of different traffic control types in each case study under normal and post-storm conditions. Additionally, the level of service in each situation was determined based on the HCM classification [18], to provide a clearer understanding of efficiency and resilience in the examined case studies. Figures 7–10 illustrate the aforementioned values for each case study. In Figures 7–10, consistent with Figure 6, the blue columns present the results for roundabouts, the red columns show the values for signalized intersections, and the green columns depict the results of the simulations for the TWSC intersections under pre-and post-storm conditions for each examined case. Similarly, the x-axis represents the case studies and their associated control types in pre- and post-storm conditions, while the y-axis represents the delay time (s/km), travel time (s/km), speed (km/h), and average approach delay (s/veh), respectively.

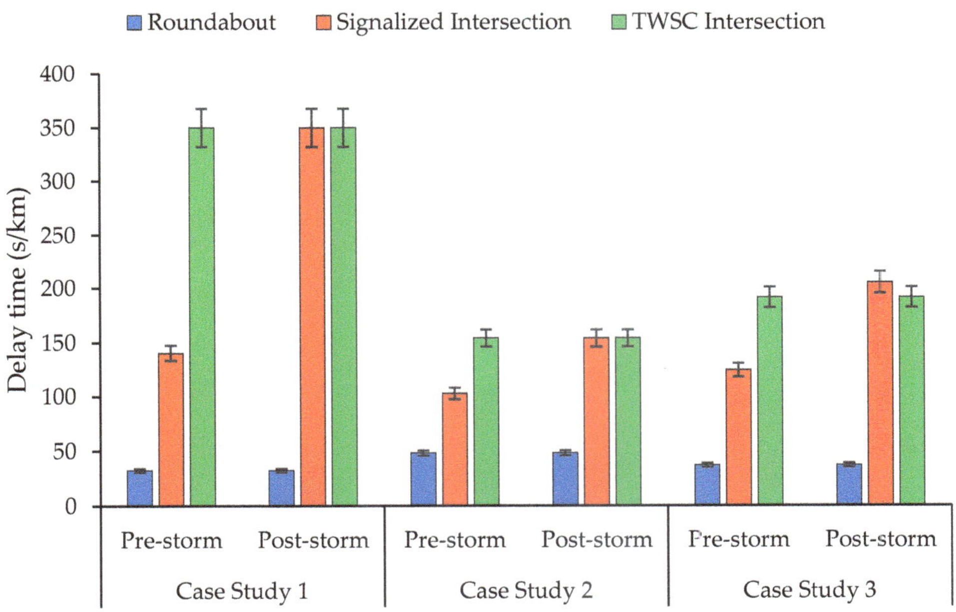

Figure 7. The comparison of delay time (s/km) in pre- and post-storm situations for case studies 1 to 3.

Figure 7 presents the simulated delay time (s/km) for all case studies under three types of traffic control and in both pre- and post-storm scenarios.

In Case Study 1, the current roundabout configuration shows no change in performance between pre- and post-storm conditions, consistent with the study's assumptions that storm effects are limited to electricity outages. The average delay remains 32.41 s per kilometer. When modeled as a signalized intersection, delay increases markedly, reaching 140.47 (s/km) under pre-storm conditions and 350.00 (s/km) under post-storm conditions. The intersection is then modeled as a TWSC, which operates identically across both scenarios, with a delay time of 350.00 (s/km).

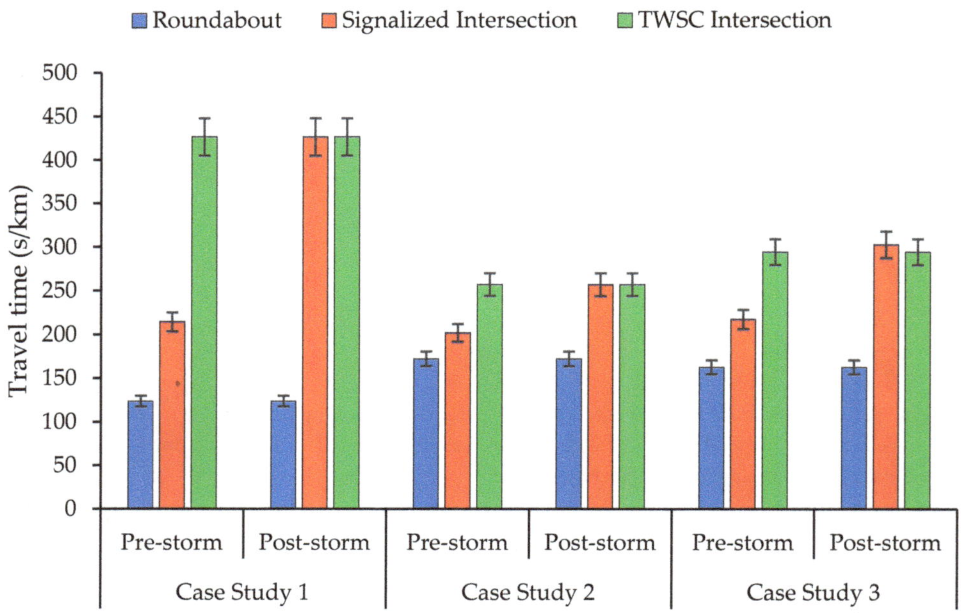

Figure 8. The comparison of travel time (s/km) in pre- and post-storm situations for case studies 1 to 3.

Case Study 2, a signalized intersection in its current form, shows average control delay times of 102.64 (s/km) pre-storm and 153.91 (s/km) post-storm. When reconfigured as a single-lane roundabout, delay time decreases significantly to 48.07 (s/km) in both scenarios. Under TWSC control, the intersection again performs identically across conditions, reporting 153.91 (s/km), which is consistent with the post-storm signalized case.

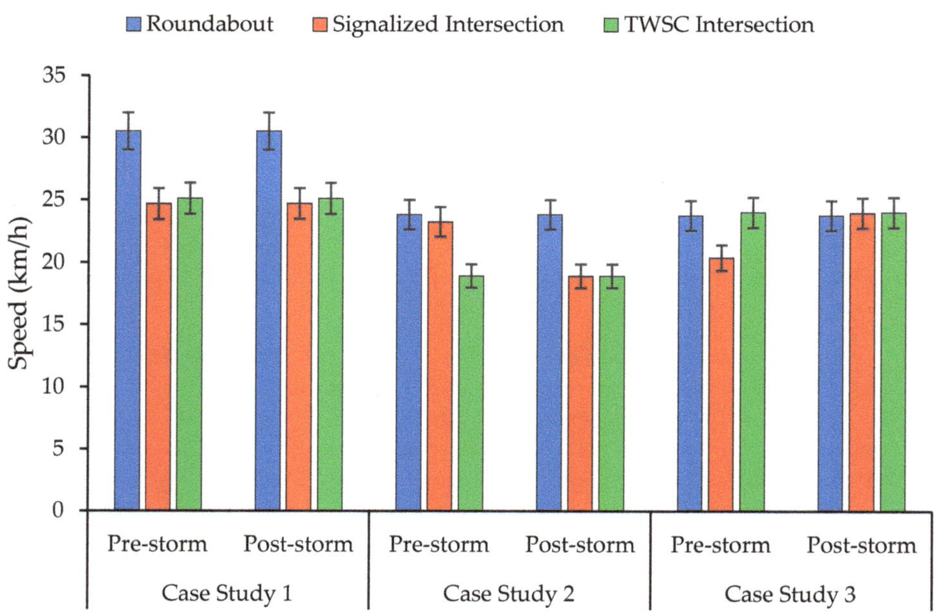

Figure 9. The comparison of speed (km/h) in pre- and post-storm situations for case studies 1 to 3.

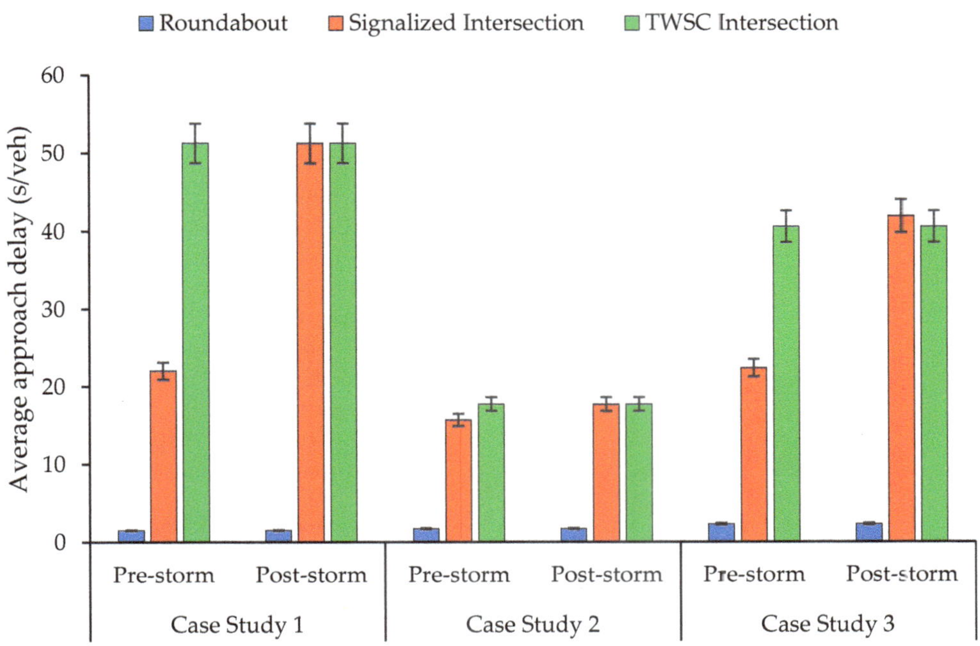

Figure 10. The comparison of average approach delay (s/veh) in pre- and post-storm situations for case studies 1 to 3.

Case Study 3, also a signalized intersection, reports 124.36 (s/km) under pre-storm conditions. In the post-storm scenario, the assumed power outage converts operations to TWSC, where delay nearly doubles to 205.19 (s/km). When modeled as a two-lane roundabout, delay decreases substantially to 36.74 (s/km), with no difference between scenarios. After adjusting movement groups to reflect TWSC right-of-way regulations, the modeled TWSC delay equals 191.12 (s/km), which is more than five times the roundabout value but slightly lower than the post-storm condition of the signalized model.

Across all case studies, roundabouts consistently exhibit the lowest delay, regardless of the number of circulating lanes or scenarios. Signalized intersections rank second in pre-storm conditions but deteriorate under the assumed power outage, often matching or exceeding TWSC delays.

Figure 8 shows the simulated travel time (s/km). The bar chart follows the same structure as Figure 7. The results mirror the delay time patterns, though the values differ. For example, in Case Study 1, roundabout configuration produces the shortest travel time at 123.51 (s/km), compared to 214.39 (s/km) under the signalized pre-storm scenario and 426.44 (s/km) under both the signalized post-storm and TWSC scenarios. Similar trends are observed across all case studies.

Simulated speeds (km/h) are depicted in Figure 9. In most cases, roundabout configurations yield higher average speeds than the other control types. However, in Case Study 3 under post-storm conditions, the two-lane roundabout exhibits slightly lower speeds compared to the signalized and TWSC models, suggesting that speed outcomes may not align directly with delay or travel time metrics and warrant further investigation.

Finally, Figure 10 presents the average approach delay (s/veh). The trends are consistent with those observed for delay time and travel time. To better assess the operational performance and resilience of these case studies, these results are also reported here based on the LOS classifications suggested by the HCM [18].

In Case Study 1, the current roundabout configuration maintains an approach delay of 1.46 (s/veh) in both pre- and post-storm conditions, corresponding to LOS A. When modeled as a signalized intersection, this value increases to 22.03 (s/veh) under pre-storm conditions (LOS C) and 51.27 (s/veh) under post-storm conditions (LOS F). Similarly, when the intersection operates as a TWSC, the intersection yields 51.27 (s/veh) in both scenarios (LOS F), matching the post-storm signalized case.

Considering Case Study 2, which is a signalized intersection, the average approach delay equals 15.69 (s/veh) under pre-storm conditions and 17.73 (s/veh) in the post-storm scenario, corresponding to LOS B and LOS C, respectively. The metric's value when the intersection is modeled as a TWSC intersection remains unchanged compared to the post-storm signalized model. On the other hand, when the intersection is modeled as a single-lane roundabout, the average approach delay decreases to 1.70 (s/veh) under both scenarios, corresponding to LOS A. Although the control delay time (from the heuristic approach) and the average approach delay (from microsimulations) are not directly comparable, Table 9 presents both sets of results for Case Study 2, shown in Figures 6 and 10, along with LOS, as an example that clearly illustrates the findings of this research.

Table 9. The results obtained from the heuristic approach and microsimulations for Case Study 2.

Type of Traffic Control	The Heuristic Approach Control Delay Time (s/veh)/LOS	Microsimulations Average Approach Delay (s/veh)/LOS
Roundabout	6.49/LOS A	1.70/LOS A
Signalized intersection	14.42/LOS B	15.69/LOS B
TWSC intersection	16.75/LOS C	17.73/LOS C

Case Study 3 is another signalized intersection that under normal conditions reports an average approach delay of 22.42 (s/veh), corresponding to LOS C. However, this metric nearly doubled in post-storm conditions to 42.01 (s/veh), when the intersection operates as a TWSC due to power outages. This delay value corresponds to LOS E. The performance and resilience of a TWSC model under normal and post-storm conditions are 40.64 (s/veh), representing LOS E. When the intersection was reconfigured to a double-lane roundabout, the average approach delay substantially decreased to 2.27 (s/veh) for both scenarios, corresponding to LOS A.

5. Discussion

The operational efficiency of road infrastructure has long been a crucial topic due to the strong dependency of economic and life activities on reliable mobility [1,67]. Although the concept of road resilience is more recent, its importance is not inferior to operational performance and has become a timely research focus [1,2,6,16,68,69]. However, measuring the resilience of road infrastructure is not straightforward; it may depend on the objectives and priorities of decision-makers as well as on the type of disruption to which the road network is exposed.

This study intentionally integrates two complementary methods—a heuristic analytical approach and microsimulations—to evaluate intersection performance under consistent conditions. Presenting both allows for a direct comparison between independent approaches, providing a built-in validation of the results and strengthening the credibility of the conclusions (see Table 9).

In this study, the efficiency and resilience of road intersections were assessed using a heuristic analytical framework and microsimulations. Pre-storm values of performance metrics are used to assess efficiency (operational performance), whereas comparisons of pre- and post-storm values are employed to evaluate resilience.

Although both the heuristic analytical method and AIMSUN Next [22] microsimulations were applied; it is important to acknowledge that their results are not strictly comparable due to methodological differences and variations in performance measure definitions. The heuristic framework reports control delay based on the HCM [18], which aggregates all forms of delay experienced by vehicles at the intersection, including deceleration, stopped delay, and acceleration [18]. In contrast, microsimulation platforms such as AIMSUN typically compute approach delay based on vehicle trajectories, capturing localized delays at stop lines but treating mid-block or re-acceleration time differently [63,70]. These conceptual distinctions explain the numerical discrepancies observed, particularly for roundabouts, where the heuristic approach estimates an average control delay of approximately 9.00 (s/veh) (LOS A), while AIMSUN reports an average approach delay of about 1.50 (s/veh) (LOS A), as observed in Case Study 1.

Despite these numerical differences, this comparison serves as an internal validation of the proposed framework, as the two approaches demonstrate strong convergence when results are interpreted in terms of LOS (see Table 9) and overall delay trends. Across all case studies and traffic control types, both methods consistently identified the relative performance of roundabouts, signalized intersections, and TWSCs under pre- and post-storm scenarios. This suggests that while absolute delay values may differ in some cases, the approaches provide complementary insights: the heuristic framework offers standardized and widely recognized benchmarks, while microsimulation captures more granular and dynamic vehicle interactions [71–73]. Accordingly, LOS classification emerges as a robust common ground for integrating analytical and simulation-based findings. At the same time, these differences underscore the need for caution when directly comparing delay magnitudes across methodologies, as highlighted in prior comparative studies [74–76].

The combined results yield several important insights regarding the operational performance and resilience of different intersection control types under normal conditions and storm-related electricity outages. Roundabouts consistently exhibited stable performance across pre- and post-storm conditions, underscoring their independence from electrical systems and highlighting their suitability as resilient traffic control solutions during power outages. Furthermore, in all case studies and scenarios, both average control delay (heuristic approach results) and average approach delay (microsimulation results) were lowest for roundabouts, confirming them as the most efficient and resilient option in this study. These findings are consistent with previous studies [6,77–80], which have also shown that roundabouts are less vulnerable to system disruptions compared to signalized intersections.

Signalized intersections, by contrast, displayed clear vulnerabilities. In pre-storm scenarios, signals typically achieved moderate levels of service (LOS B–C, which are considered acceptable for urban road networks). However, under post-storm conditions, the assumed power outage forced a transition to TWSC operation, which substantially degraded performance—particularly in Case Studies 1 and 3, where delays nearly doubled, indicating reduced resilience. These findings align with earlier research emphasizing the operational risks of power outages at signalized intersections [14,81,82]. Microsimulation results reinforced this vulnerability, showing sharp increases in both delay and travel time when signals became inoperative.

TWSC intersections, whether directly modeled or as a fallback for failed signals, consistently produced higher delays than roundabouts and, in many cases, even higher than signalized intersections under normal conditions. This finding reinforces both HCM evidence [18], and simulation-based studies [14,83,84], which demonstrate that TWSCs perform poorly under heavy or unbalanced traffic due to their reliance on gap acceptance.

Notably, Case Study 2 revealed that under moderate traffic, TWSC operation was only marginally less efficient and resilient than signals, suggesting that context-specific factors strongly influence efficiency and resilience assessments. Similar context-sensitive findings have also been reported in case studies in Italy [14].

Periodic reassessment of intersections is often warranted to maintain efficiency, safety, and resilience when evolving traffic demand and driver behavior exceed the capacity of existing control schemes. One real-world example involved converting a TWSC intersection into a turbo roundabout in 2015 [85], a transformation that eliminated congestion, improved capacity, and significantly enhanced safety. Empirical studies enforced this pattern: for instance, retrofitting signalized and stop-controlled intersections to roundabouts in Oxford [86], as well as converting signalized intersections into mini-roundabouts in Michigan [87]. These findings support the argument that traffic control should be periodically reassessed and redesigned in response to shifts in traffic patterns to maintain both efficiency and resilience.

Another noteworthy observation arises from the speed results. While roundabouts generally supported higher average speeds, Case Study 3 showed an exception, where the two-lane roundabout in the post-storm scenario recorded slightly lower average speeds than signalized and TWSC configurations. This suggests that roundabout benefits are not universal and may vary depending on geometry, traffic composition, and flow patterns. Further research should therefore investigate the interplay of speed, safety, and resilience under disrupted conditions.

Finally, the dual-method approach of combining analytical calculations with AIMSUN microsimulations strengthens the analysis. The heuristic framework provided benchmark values and LOS classifications, while microsimulations captured more nuanced patterns in delay, travel time, and speed. The consistency of findings across both approaches increases confidence in the results, although the simplifying assumption that storm impacts are limited to electricity outages should be acknowledged. Other storm-related disruptions, such as flooding, debris, or traffic demand surges, should be explored in future research to generate more comprehensive resilience assessments. Importantly, the convergence between analytical and microsimulation results, together with agreement with published findings, demonstrates that the proposed framework builds upon and extends current knowledge rather than standing in isolation.

6. Conclusions

The main objective of this study was to evaluate the efficiency and resilience of road intersections under pre-storm and post-storm conditions, focusing on roundabouts, signalized intersections, and TWSC intersections. Two complementary approaches were applied: a heuristic analytical method based on the control delay suggested by the HCM [18], and microsimulations using AIMSUN Next 20.0.5 [22]. Three case studies from Poland (a single-lane roundabout and two signalized intersections in their current configurations) were analyzed using data collected between September and November 2023. Each of these intersections was investigated under all three traffic control types. For example, Case Study 1, originally a single-lane roundabout, was also modeled as a signalized intersection and a TWSC intersection for comparison purposes.

The results of both approaches consistently indicated that roundabouts outperformed signalized and TWSC intersections in terms of efficiency and resilience. Their independence from external power supply makes them particularly advantageous in post-storm scenarios involving electricity outages. Signalized intersections generally provided acceptable service under normal conditions but showed substantial performance degradation when signals became inoperative, reverting to TWSC operation. TWSC intersections, whether modeled directly or as a fallback, performed the worst overall, particularly under heavy or unbalanced traffic.

While heuristic and microsimulation methods are not directly comparable due to differences in delay estimation, their convergence in level of service categories confirms LOS as a reliable common basis for integrating results. This convergence reinforces confidence in the consistency and validity of the findings.

From a practical perspective, the results identify roundabouts as the most storm-resilient traffic control strategy, especially in regions prone to electricity disruptions. For signalized intersections, contingency planning should be prioritized, including backup power systems, temporary traffic management strategies, or long-term conversion to more resilient designs.

The findings highlight that intersection efficiency and resilience are site-specific and should be periodically reassessed to adapt to changing conditions. Climatic variations, technological advances in vehicles (e.g., connected and automated vehicles), and evolving traffic demands may render previously optimal control strategies less effective over time. The proposed framework addresses these challenges by integrating efficiency and resilience within a unified approach that relies on a standardized analytical metric. This metric is simple to compute, requires minimal data collection, and is not restricted by climate, environment, or geometry, enabling its application to any intersection worldwide. The framework thus provides a practical, adaptable, and universal tool for maintaining intersection performance in dynamic urban contexts.

The main contribution of this study lies in integrating analytical and simulation-based methods to evaluate intersection resilience under storm conditions. However, it is limited by its core assumption that storm impacts are restricted to electricity outages. Future research should address additional storm-related disruptions (e.g., flooding, debris, or fluctuating demand) and incorporate safety performance metrics. Moreover, due to the negligible presence of pedestrians and cyclists in the analyzed intersections, these modes were excluded from the current analysis. Further studies should examine intersections with substantial pedestrian and cyclist activity to assess their influence on intersection resilience.

In conclusion, while signalized intersections can function efficiently under normal conditions, they remain highly vulnerable to power outages. Roundabouts, by contrast, consistently demonstrated superior operational performance and resilience. These findings underscore the importance of the periodic reassessment of intersection control strategies to ensure long-term efficiency and resilience. For stakeholders, quantitative metrics such as control delay or approach delay provide precise performance measures, while LOS classification offers a practical benchmark for decision-making, particularly when exact delay values are not required.

Author Contributions: Conceptualization, N.Z., E.M. and A.G.; methodology, N.Z., E.M. and A.G.; software, N.Z. and M.L.T.; validation, N.Z. and M.L.T.; formal analysis, N.Z. and E.M.; investigation, N.Z., E.M. and A.G.; resources, N.Z., E.M. and A.G.; data curation, M.L.T., N.Z., E.M. and A.G.; writing—original draft preparation, N.Z.; writing—review and editing M.L.T., N.Z., E.M. and A.G.; visualization, M.L.T., N.Z., E.M. and A.G.; supervision, M.L.T., E.M. and A.G.; project administration, E.M. and N.Z.; funding acquisition N.Z., E.M. and A.G. All authors have read and agreed to the published version of the manuscript.

Funding: This research received no external funding.

Data Availability Statement: The data presented in this study are available on request from the corresponding author due to licensing restrictions of the AIMSUN Next 20.0.5 software used to produce the simulation data.

Acknowledgments: This research has been partially supported by the European Union—Next Generation EU—National Sustainable Mobility Center, Italian Ministry of University and Research, Spoke 9.

Conflicts of Interest: The authors declare no conflicts of interest.

Abbreviations

The following abbreviations are used in this manuscript:

GEH	Geoffrey E. Havers
HCM	Highway Capacity Manual
HDV	Human-Driven Vehicle
IoV	Internet of Vehicles
LOS	Level of Service
LSTM	Long Short-Term Memory
O-D	Origin-Destination
SDGs	Sustainable Development Goals
SRC	Stochastic Route Choice
TWSC	Two-Way Stop-Controlled

References

1. Kasmalkar, I.G.; Serafin, K.A.; Miao, Y.; Bick, I.A.; Ortolano, L.; Ouyang, D.; Suckale, J. When floods hit the road: Resilience to flood-related traffic disruption in the San Francisco Bay Area and beyond. *Sci. Adv.* **2020**, *6*, eaba2423. [CrossRef]
2. Xu, W.; Zhang, Y.; Proverbs, D.; Zhong, Z. Enhancing the resilience of road networks to flooding. *Int. J. Build. Pathol. Adapt.* **2024**, *42*, 213–236. [CrossRef]
3. Zare, N.; Talebbeydokhti, N. Policies and governance impact maps of floods on metropolitan Shiraz (the first step toward resilience modeling of the city). *Int. J. Disaster Risk Reduct.* **2018**, *28*, 298–317. [CrossRef]
4. World Bank Group. Urban Development. Available online: https://www.worldbank.org/en/topic/urbandevelopment/overview#1 (accessed on 15 July 2025).
5. CRED. *2023 Disasters in Numbers: A Significant Year of Disaster Impact*; Emergency Events Database: Brussels, Belgium, 2024; p. 8.
6. Zare, N. Harmonizing Efficiency and Resilience in Road Intersection Performance: A Novel Heuristic for Mobility Infrastructure Assessment. Ph.D. Thesis, University of Palermo, Palermo, Italy, 2025; p. 183.
7. Li, Z.; Yan, W.; Wang, L. Measuring mobility resilience with network-based simulations of flow dynamics under extreme events. *Transp. Res. Part D Transp. Environ.* **2024**, *135*, 104362. [CrossRef]
8. Carvallo, J.P.; Casey, J. Weather-Related Power Outages Rising. Available online: https://www.climatecentral.org/climate-matters/weather-related-power-outages-rising (accessed on 15 September 2024).
9. Stankovski, A.; Gjorgiev, B.; Locher, L.; Sansavini, G. Power blackouts in Europe: Analyses, key insights, and recommendations from empirical evidence. *Joule* **2023**, *7*, 2468–2484. [CrossRef]
10. Batica, J. Methodology for Flood Resilience Assessment in Urban Environments and Mitigation Strategy Development. Ph.D. Thesis, Université Nice Sophia Antipolis, Nice, France, 2015.
11. Rezvani, S.M.; de Almeida, N.M.; Falcão, M.J. Climate adaptation measures for enhancing urban resilience. *Buildings* **2023**, *13*, 2163. [CrossRef]
12. Lv, Y.; Sarker, M.N.I. Integrative approaches to urban resilience: Evaluating the efficacy of resilience strategies in mitigating climate change vulnerabilities. *Heliyon* **2024**, *10*, e28191. [CrossRef]
13. United Nations Statistics Division. *SDG Indicators: Metadata Repository*; United Nations: Nairobi, Kenya, 2021.
14. Zare, N.; Macioszek, E.; Granà, A.; Giuffrè, T. Blending efficiency and resilience in the performance assessment of urban intersections: A novel heuristic informed by literature review. *Sustainability* **2024**, *16*, 2450. [CrossRef]
15. Therrien, M.-C.; Normandin, J.-M.; Paterson, S.; Pelling, M. Mapping and weaving for urban resilience implementation: A tale of two cities. *Cities* **2021**, *108*, 102931. [CrossRef]
16. Espinet, X.; Schweikert, A.; van den Heever, N.; Chinowsky, P. Planning resilient roads for the future environment and climate change: Quantifying the vulnerability of the primary transport infrastructure system in Mexico. *Transp. Policy* **2016**, *50*, 78–86. [CrossRef]
17. Bengigi, E. Efficiency and Resilience Trade-Offs for Roadway Intersection Design in the US Virgin Islands. Master's Thesis, Naval Postgraduate School, Monterey, CA, USA, 2020.

18. National Academies of Sciences, Engineering, and Medicine. *Highway Capacity Manual: A Guide for Multimodal Mobility Analysis*, 6th ed.; The National Academies Press: Washington, DC, USA, 2016. [CrossRef]
19. Sharifi, A. Resilient urban forms: A review of literature on streets and street networks. *Build. Environ.* **2019**, *147*, 171–187. [CrossRef]
20. Boeing, G.; Ha, J. Resilient by design: Simulating street network disruptions across every urban area in the world. *Transp. Res. Part A Policy Pract.* **2024**, *182*, 104016. [CrossRef]
21. Sun, W.; Bocchini, P.; Davison, B.D. Resilience metrics and measurement methods for transportation infrastructure: The state of the art. *Sustain. Resilient Infrastruct.* **2020**, *5*, 168–199. [CrossRef]
22. Aimsun Next. *Version20 Dynamic Simulator User Manual*; TSS-Transport Simulation Systems: Barcelona, Spain, 2020.
23. Berrang-Ford, L.; Pearce, T.; Ford, J.D. Systematic review approaches for climate change adaptation research. *Reg. Environ. Change* **2015**, *15*, 755–769. [CrossRef]
24. Klemt-Albert, K.; Hartung, R.; Bahlau, S. Enhancing resilience of traffic networks with a focus on impacts of neuralgic points like urban tunnels. In *Resilience Engineering for Urban Tunnels*; ASCE: Reston, VA, USA, 2018, pp. 55–70. [CrossRef]
25. Zhao, Z.; Tang, L.; Yang, X.; Zhang, H.; Li, G.; Li, Q. Identifying critical urban intersections from a fine-grained spatio-temporal perspective. *Travel Behav. Soc.* **2024**, *34*, 100649. [CrossRef]
26. McDaniels, T.L.; Chang, S.E.; Hawkins, D.; Chew, G.; Longstaff, H. Towards disaster-resilient cities: An approach for setting priorities in infrastructure mitigation efforts. *Environ. Syst. Dec.* **2015**, *35*, 252–263. [CrossRef]
27. Pratelli, A.; Leandri, P.; Aiello, R.; Souleyrette, R.R. Intersection redesign for network resilience and safety. *Promet-Traffic Transp.* **2021**, *33*, 297–308. [CrossRef]
28. Yang, P.P.-J. Energy resilient urban form: A design perspective. *Energy Procedia* **2015**, *75*, 2922–2927. [CrossRef]
29. Sharafeldin, M.; Farid, A.; Ksaibati, K. Investigating the impact of roadway characteristics on intersection crash severity. *Eng* **2022**, *3*, 412–423. [CrossRef]
30. Hettiarachchi, S.; Wasko, C.; Sharma, A. Rethinking urban storm water management through resilience–The case for using green infrastructure in our warming world. *Cities* **2022**, *128*, 103789. [CrossRef]
31. Mattinzioli, T.; Butt, A.A.; Harvey, J. Literature review on pavements and electric vehicle interaction: A research roadmap. *Transp. Res. Part D Transp. Environ.* **2023**, *122*, 103886. [CrossRef]
32. Mondal, M.A.; Rehena, Z. Priority-based adaptive traffic signal control system for smart cities. *SN Comput. Sci.* **2022**, *3*, 417. [CrossRef]
33. Lv, Y.; Zheng, N. Monash Research Center of Automated and Resilient Road Systems [Its Research Lab]. *IEEE Intell. Transp. Syst. Mag.* **2023**, *15*, 166–175. [CrossRef]
34. Ghaffarpasand, O.; Burke, M.; Osei, L.K.; Ursell, H.; Chapman, S.; Pope, F.D. Vehicle telematics for safer, cleaner and more sustainable urban transport: A review. *Sustainability* **2022**, *14*, 16386. [CrossRef]
35. Mageto, J.; Twinomurinzi, H.; Luke, R.; Mhlongo, S.; Bwalya, K.; Bvuma, S. Building resilience into smart mobility for urban cities: An emerging economy perspective. *Int. J. Prod. Res.* **2024**, *62*, 5556–5573. [CrossRef]
36. Tang, J.; Heinimann, H.R. A resilience-oriented approach for quantitatively assessing recurrent spatial-temporal congestion on urban roads. *PLoS ONE* **2018**, *13*, e0190616. [CrossRef] [PubMed]
37. Chen, G.; wan Zhang, J. Intelligent transportation systems: Machine learning approaches for urban mobility in smart cities. *Sustain. Cities Soc.* **2024**, *107*, 105369. [CrossRef]
38. Magalhaes, R.P.; Lettich, F.; Macedo, J.A.; Nardini, F.M.; Perego, R.; Renso, C.; Trani, R. Speed prediction in large and dynamic traffic sensor networks. *Inf. Syst.* **2021**, *98*, 101444. [CrossRef]
39. Abduljabbar, R.; Dia, H.; Liyanage, S. Machine Learning Traffic Flow Prediction Models for Smart and Sustainable Traffic Management. *Infrastructures* **2025**, *10*, 155. [CrossRef]
40. Topilin, I.; Jiang, J.; Feofilova, A.; Beskopylny, N. Traffic Flow Prediction via a Hybrid CPO-CNN-LSTM-Attention Architecture. *Smart Cities* **2025**, *8*, 148. [CrossRef]
41. Toba, A.-L.; Kulkarni, S.; Khallouli, W.; Pennington, T. Long-Term Traffic Prediction Using Deep Learning Long Short-Term Memory. *Smart Cities* **2025**, *8*, 126. [CrossRef]
42. Mishra, P.; Singh, G. Internet of Vehicles for Sustainable Smart Cities: Opportunities, Issues, and Challenges. *Smart Cities* **2025**, *8*, 93. [CrossRef]
43. Rejeb, A.; Rejeb, K.; Zaher, H.F.; Simske, S. Blockchain and Smart Cities: Co-Word Analysis and BERTopic Modeling. *Smart Cities* **2025**, *8*, 111. [CrossRef]
44. Zhang, L.; Xu, M.; Wang, S. Mitigating vulnerability of a multimodal public transit system for sustainable megacities: A real-time operational control method. *Sustain. Cities Soc.* **2024**, *101*, 105142. [CrossRef]
45. Wassmer, J.; Merz, B.; Marwan, N. Resilience of transportation infrastructure networks to road failures. *Chaos* **2024**, *34*. [CrossRef] [PubMed]

46. Azolin, L.G.; da Silva, A.N.R.; Pinto, N. Incorporating public transport in a methodology for assessing resilience in urban mobility. *Transp. Res. Part D Transp. Environ.* **2020**, *85*, 102386. [CrossRef]
47. Tang, J.; Lin, H.; Fan, X.; Yu, X.; Lu, Q. A topology-based evaluation of resilience on urban road networks against epidemic spread: Implications for COVID-19 responses. *Front. Public Health* **2022**, *10*, 1023176. [CrossRef]
48. Lara, D.V.R.; Pfaffenbichler, P.; da Silva, A.N.R. Modeling the resilience of urban mobility when exposed to the COVID-19 pandemic: A qualitative system dynamics approach. *Sustain. Cities Soc.* **2023**, *91*, 104411. [CrossRef] [PubMed]
49. Liu, Y.; Bu, S.; Zhang, S.; Xu, C. Research on the Socio-Spatial Resilience Evaluation and Evolution of the Central Area of Beijing in Transitional China. *Sustainability* **2024**, *16*, 7098. [CrossRef]
50. Shafiei-Dastjerdi, M.; Lak, A. Towards resilient place emphasizing urban form: An assessment framework in urban design. *Sustain. Cities Soc.* **2023**, *96*, 104646. [CrossRef]
51. Garcia, L.; Hafezi, M.; Lima, L.; Millett, C.; Thompson, J.; Wang, R.; Akaraci, S.; Goel, R.; Reis, R.; Nice, K.A. Future-proofing cities against negative city mobility and public health impacts of impending natural hazards: A system dynamics modelling study. *Lancet Planet. Heath* **2025**, *9*, e207–e218. [CrossRef]
52. Vichiensan, V.; Wasuntarasook, V.; Komkong, T.; Takano, T.; Wongsa, S.; Nakamura, S. Assessing the impact of pluvial flooding adaptation measures on urban transport in Bangkok. *Asian Transp. Stud.* **2025**, *11*, 100167. [CrossRef]
53. Fan, X.; Zhang, X.; Wang, X.; Yu, X. A deep reinforcement learning model for resilient road network recovery under earthquake or flooding hazards. *J. Infrastruct. Preserv. Resil.* **2023**, *4*, 8. [CrossRef]
54. Cheek, W.; Chmutina, K. Measuring resilience in the assumed city. *Int. J. Disaster Risk Sci.* **2022**, *13*, 317–329. [CrossRef]
55. Tarek, S.; Ouf, A.S.E.-D. Biophilic smart cities: The role of nature and technology in enhancing urban resilience. *J. Eng. Appl. Sci.* **2021**, *68*, 40. [CrossRef]
56. Arimah, B. Infrastructure as a Catalyst for the Prosperity of African Cities. *Procedia Eng.* **2017**, *198*, 245–266. [CrossRef]
57. Cordero, D.; Rodriguez, G. Merger of network graph indicators to estimate resilience in Latin American cities. *IEEE Access* **2022**, *10*, 81071–81093. [CrossRef]
58. Tariverdi, M.; Nunez-Del-Prado, M.; Leonova, N.; Rentschler, J. Measuring accessibility to public services and infrastructure criticality for disasters risk management. *Sci. Rep.* **2023**, *13*, 1569. [CrossRef]
59. Santos, T.; Silva, M.A.; Fernandes, V.A.; Marsden, G. Resilience and vulnerability of public transportation fare systems: The case of the city of Rio De Janeiro, Brazil. *Sustainability* **2020**, *12*, 647. [CrossRef]
60. Karagkouni, A.; Dimitriou, D. The Overton Window in Smart City Governance: The Methodology and Results for Mediterranean Cities. *Smart Cities* **2025**, *8*, 98. [CrossRef]
61. Bąk, R.; Gaca, S.; Ostrowski, K.; Tracz, M.; Woźniak, K. *Wytyczne Projektowania Skrzyżowań Drogowych, Część II: Ronda*; WR-D-31-3; Ministry of Infrastructure: Warsaw, Poland, 2022.
62. Bąk, R.; Gaca, S.; Kieć, M.; Ostrowski, K. *Guidelines for the Design of Road Intersections. Part 2: Ordinary and Channelized Intersections [Wytyczne Projektowania Skrzyżowań Drogowych. Część 2: Skrzyżowania Zwykłe i Skanalizowane]*; Ministry of Infrastructure: Warsaw, Poland; Department of Public Roads: Warsaw, Poland, 2022.
63. Barceló, J. *Fundamentals of Traffic Simulation*; Springer: New York, NY, USA, 2010; Volume 145.
64. Acuto, F.; Coelho, M.C.; Fernandes, P.; Giuffre, T.; Macioszek, E.; Grana, A. Assessing the environmental performances of urban roundabouts using the VSP methodology and AIMSUN. *Energies* **2022**, *15*, 1371. [CrossRef]
65. Tumminello, M.L.; Macioszek, E.; Granà, A.; Giuffrè, T. Evaluating traffic-calming-based urban road design solutions featuring cooperative driving technologies in energy efficiency transition for smart cities. *Energies* **2023**, *16*, 7325. [CrossRef]
66. Tumminello, M.L.; Macioszek, E.; Granà, A.; Giuffrè, T. Simulation-based analysis of "what-if" scenarios with connected and automated vehicles navigating roundabouts. *Sensors* **2022**, *22*, 6670. [CrossRef]
67. Berdica, K. An introduction to road vulnerability: What has been done, is done and should be done. *Transp. Policy* **2002**, *9*, 117–127. [CrossRef]
68. Jamal, T.B.; Hasan, S.; Abdul-Aziz, O.I.; Mozumder, P.; Meyur, R. Strengthening Infrastructure Resilience to Hurricanes by Modeling Transportation and Electric Power Network Interdependencies. *Nat. Hazard. Rev.* **2025**, *26*, 04025024. [CrossRef]
69. Saei, S.; Tajik, N. Scenario-Based Optimization of Network Resilience: Integrating Vulnerability Assessments and Traffic Flow. *arXiv* **2025**, arXiv:2503.23251. [CrossRef]
70. Alexiadis, V.; Jeannotte, K.; Chandra, A.; Skabardonis, A.; Dowling, R. *Traffic Analysis Toolbox Volume I, Traffic Analysis Tools Primer*; United States Federal Highway Administration Office of Research: McLean, VA, USA, 2004.
71. Dowling, R.; Skabardonis, A.; Alexiadis, V. *Traffic Analysis Toolbox, Volume III: Guidelines for Applying Traffic Microsimulation Modeling Software*; Department of Transportation Federal Highway Administration: McLean, VA, USA, 2004.
72. Hollander, Y.; Liu, R. The principles of calibrating traffic microsimulation models. *Transportation* **2008**, *35*, 347–362. [CrossRef]
73. Ahn, K.; Rakha, H.; Trani, A.; Van Aerde, M. Estimating vehicle fuel consumption and emissions based on instantaneous speed and acceleration levels. *J. Transp. Eng.* **2002**, *128*, 182–190. [CrossRef]

74. Dion, F.; Rakha, H.; Kang, Y.-S. Comparison of delay estimates at under-saturated and over-saturated pre-timed signalized intersections. *Transp. Res. Part B Methodol.* **2004**, *38*, 99–122. [CrossRef]
75. Kim, S.-O.; Benekohal, R.F. Comparison of control delays from CORSIM and the Highway Capacity Manual for oversaturated signalized intersections. *J. Transp. Eng.* **2005**, *131*, 917–923. [CrossRef]
76. Ghanim, M.S.; Shaaban, K.; Allawi, S. Operational performance of signalized intersections: HCM and microsimulation comparison. In Proceedings of the 2022 Intermountain Engineering, Technology and Computing (IETC). Orem, UT, USA, 13–14 May 2022. [CrossRef]
77. Taglieri, D.; Liu, H.; Gayah, V.V. Network-wide implementation of roundabouts versus signalized intersections on urban streets: Analytical and simulation comparison. *Transp. Res. Rec.* **2024**, *2678*, 719–735. [CrossRef]
78. Demir, H.G.; Demir, Y.K. A Comparison of traffic flow performance of roundabouts and signalized intersections: A case study in Nigde. *Open Transp. J.* **2020**, *14*, 120–132. [CrossRef]
79. Taglieri, D. Urban Traffic Network Operations: Roundabouts vs. Signalized Intersections. Master's Thesis, Penn State University, University Park, PA, USA, 2022.
80. Macioszek, E. Roundabout entry capacity calculation—A case study based on roundabouts in Tokyo, Japan, and Tokyo surroundings. *Sustainability* **2020**, *12*, 1533. [CrossRef]
81. Ansariyar, A.; Jeihani, M. Real-time traffic control and safety measure analysis using LiDAR sensor during traffic signal failures. In Proceedings of the International Conference on Transportation and Development 2024, Atlanta, GA, USA, 15–18 June 2024; pp. 798–810.
82. Chandler, B.E.; Myers, M.; Atkinson, J.E.; Bryer, T.; Retting, R.; Smithline, J.; Trim, J.; Wojtkiewicz, P.; Thomas, G.B.; Venglar, S.P. *Signalized Intersections Informational Guide*; Department of Transportation Federal Highway Administration: Washington, DC, USA, 2013.
83. Yang, Z.; Zhang, Y.; Grembek, O. Combining traffic efficiency and traffic safety in countermeasure selection to improve pedestrian safety at two-way stop controlled intersections. *Transp. Res. Part A Policy Pract.* **2016**, *91*, 286–301. [CrossRef]
84. Kaczorek, M.; Jacyna, M. Fuzzy logic as a decision-making support tool in planning transport development. *Arch. Transp.* **2022**, *61*, 51–70. [CrossRef]
85. Macioszek, E.; Kurek, A. A case study analysis of roundabouts entry capacity localised on one of the main road in Sosnowiec city (Poland). *Zesz. Naukowe. Transp./Politech. Śląska* **2019**, *105*, 139–156. [CrossRef]
86. Uddin, W.; Headrick, J.; Sullivan, J.S. Performance Evaluation of Roundabouts for Traffic Flow Improvements and Crash Reductions at a Highway Interchange in Oxford, MS. In Proceedings of the Transportation Research Board 91st Annual Meeting, Washington, DC, USA, 22–26 January 2012.
87. Qawasmeh, B.; Kwigizile, V.; Oh, J.-S. Performance and safety effectiveness evaluation of mini-roundabouts in Michigan. *J. Eng. Appl. Sci.* **2023**, *70*, 36. [CrossRef]

smart cities

MDPI

Article

Modeling Informal Driver Interaction and Priority Behavior in Smart-City Traffic Systems

Alica Kalašová, Peter Fabian *, Ľubomír Černický and Kristián Čulík

Department of Road and Urban Transport, University of Žilina, Univerzitná 1, 01026 Žilina, Slovakia; alica.kalasova@uniza.sk (A.K.); lubomir.cernicky@uniza.sk (Ľ.Č.); kristian.culik@uniza.sk (K.Č.)

* Correspondence: peter.fabian@stud.uniza.sk

Highlights

What are the main findings?

- Traditional methodologies systematically overestimate waiting times at uncontrolled intersections because they ignore the phenomenon of psychological priority among drivers.
- Regression analysis and supplementary data demonstrated that incorporating psychological priority yields results that correspond much more closely to real traffic conditions.

What is the implication of the main finding?

- The consideration of psychological priority increases the accuracy of traffic modeling, enabling a more realistic assessment of intersection capacity and more effective transport planning in the urban environment.
- The integration of this approach into intelligent transport systems of Smart Cities will enable more accurate simulations, more flexible traffic management, and contribute to improving the quality of life of residents.

Abstract

Accurate traffic modeling is essential for effective urban mobility planning within Smart Cities. Conventional capacity assessment methods assume rule-based driver behavior and therefore neglect psychological priority, an informal interaction in which drivers negotiate right-of-way contrary to traffic regulations. This study investigates how the absence of this behavioral factor affects the accuracy of delay and capacity evaluation at unsignalized intersections. A 12 h field observation was conducted at an intersection in Prešov, Slovakia, and 28 driver interactions were analyzed using linear regression modeling. The derived model ($R^2 = 0.83$, $p < 0.05$) demonstrates that incorporating psychological priority significantly improves the agreement between calculated and observed waiting times. Unrealistic results occurring under oversaturated conditions in standard methodologies were eliminated. The findings confirm that behavioral variability has a measurable impact on traffic performance and should be reflected in analytical and simulation models. Integrating these behavioral parameters into Smart City traffic modeling contributes to more realistic and human-centered decision-making in intersection design and capacity management, supporting the development of safer and more efficient urban mobility systems.

Keywords: psychological priority; traffic modeling; smart cities; unsignalized intersections

1. Introduction

The concept of Smart Cities represents a comprehensive and systemic approach to urban development, encompassing culture, infrastructure, environment, energy, and social services [1,2]. At its core, a Smart City can be understood as a combination of technological, organizational, and social innovations designed to improve the quality of life for urban residents through sustainable development principles [3].

Among the core components of Smart Cities is smart mobility, which focuses on optimizing the movement of people and goods while enhancing traffic safety, reducing environmental burdens, promoting alternative transport modes, and ensuring efficient use of infrastructure [4–6].

Within this context, intersections play a crucial role as microsystems of the Smart City, where technological, safety, and environmental factors converge [7].

Properly functioning intersections have a direct influence on road safety, fluency, and urban livability [8–10].

However, as noted by Nama and Pardo [3], Smart Cities cannot rely solely on technological innovations; understanding human behavior and psychological factors is equally essential. This forms the foundation for the present study, which addresses how driver behavior, specifically psychological priority, affects the accuracy of traffic modeling within the framework of smart mobility.

Psychological priority is defined as a situation in which a driver at an intersection asserts priority that is not granted by traffic regulations.

In contrast to legal priority, which arises from formal traffic regulations and right-of-way rules, psychological priority reflects spontaneous driver interaction based on perception, intuition, and social signaling. It can be observed through behavioral indicators such as eye contact, short head movements, flashing headlights, vehicle positioning, or micro-accelerations used to communicate intent. These actions represent implicit negotiation between drivers and serve as the operational basis for identifying psychological priority in real traffic observation [11].

It represents informal behavior based on expectations, experience, or assumed rules among road users. This phenomenon occurs most frequently at small uncontrolled intersections, where visual cues, narrow roadways, or traffic intensity lead to dynamic interactions between drivers [12–14]. Driver behavior is shaped by a combination of traffic regulations, visual stimuli, and implicit social signals [12].

Psychological priority arises from the following factors:

- expectations of the behavior of other participants, where the driver assumes that others will yield;
- experience and cognitive shortcuts, where the driver relies on previous situations or intuition;
- social and visual perception, such as eye contact, vehicle movements, or the speed of another car signaling available space for entry.

Psychological priority is a double-edged phenomenon:

- Positive: it contributes to greater efficiency and fluency of traffic when participants communicate through implicit signals [12].
- Negative: it represents a safety risk because it is unpredictable and may conflict with legislation [3,4].

Psychological priority differs from related concepts such as gap acceptance, social yielding, and cooperative driving. Gap acceptance focuses on measurable time or distance gaps that drivers consider safe to enter traffic, while social yielding refers to intentional courtesy or compliance with social norms. Cooperative driving involves deliberate coordination among drivers, whereas psychological priority represents spontaneous and informal decision-making, where drivers assert or yield right-of-way based on mutual perception and momentary expectation rather than explicit rules [11,15].

Recent studies emphasize that cognitive and psychological mechanisms, including perception, attention, and implicit negotiation between drivers, have a key role in sustainable and age-friendly transport systems [15]. These insights strengthen the behavioral interpretation of psychological priority in traffic modeling. Additional research highlights that nonverbal communication, eye contact, and cooperative behavior strongly influence how drivers negotiate right-of-way in real traffic environments [11].

The significance of psychological priority is most evident at small uncontrolled intersections, which are typical of historic city centers [12–14]. This phenomenon has been examined in the context of traffic behavior theories that emphasize the importance of informal traffic rules. According to Kaparias et al. and Pettitt, Christie, and Tyler, implicit social agreements are formed among drivers in urban environments, which may temporarily substitute or even outweigh official traffic regulations [16,17]. While this phenomenon has the potential to improve traffic fluency, it simultaneously introduces the risk of greater unpredictability.

Risser points out that drivers base their decisions not only on formal rules but also on cognitive shortcuts and subjective risk assessment [18]. From the perspective of traffic psychology, the expectation of other participants' behavior also plays an important role. According to Fuller, this may lead to a safety paradox: when all participants assume the same course of action, traffic becomes more fluent, yet when expectations diverge, the risk of collisions increases [19].

In the design of intersections and traffic solutions, it is necessary to take psychological priority into account. Ignoring these aspects may result in underestimating factors that significantly influence safety. The integration of insights from traffic psychology into the planning of smart mobility leads to more efficient and safer solutions.

The concept of Smart Cities emphasizes the harmonization of technological innovations with social and behavioral factors [20,21].

Although the informal negotiation of right-of-way has been addressed in previous research, the present study introduces the concept of psychological priority as a formalized and measurable behavioral factor that can be incorporated into quantitative traffic modeling. This approach extends existing knowledge by linking behavioral interaction with mathematical representation, allowing the phenomenon to be expressed statistically and applied within capacity and delay analyses.

However, most existing traffic models still rely on idealized, rule-based assumptions about driver behavior and therefore ignore psychological and behavioral aspects that occur in real traffic operations. This represents a significant research gap that limits the accuracy of current traffic modeling and motivates the present study.

Intersections, as microsystems of the Smart City, demonstrate that beyond traffic engineering parameters, it is also necessary to consider psychological patterns of behavior [3,22,23]. Traffic models are a key tool for transport planning and management, yet most of them assume strictly rational driver behavior. Real-world operation shows that drivers often apply psychological priority, which contributes to greater fluency and reduced delays at intersections. The limitation of current models is that they fail to account for this phenomenon, which may result in overestimating waiting times and underestimating capacity.

The present study is therefore grounded in behavioral traffic psychology and human factors research, emphasizing that cognitive, perceptual, and social processes fundamentally influence driver decision-making in real traffic conditions [11,15].

Research hypotheses of the article:

- H1: Traffic models that do not consider psychological priority systematically overestimate vehicle waiting times.
- H2: The consideration of psychological priority leads to improved intersection capacity and a reduction in overall delay time.

The aim of the article is to highlight this methodological shortcoming and to discuss its implications for the accuracy of traffic modeling, which constitutes a fundamental element in the development of smart mobility within Smart Cities.

2. Materials and Methods

This chapter describes the procedure of data collection, the selection of sites, and the method of processing calculations that made it possible to compare the standard approach with the adjusted computation that considers psychological priority.

2.1. Foundations of the Analysis

The analysis was based on a comparison of the results of intersection capacity assessments conducted using standard methodologies, which assume strictly rule-based driver behavior, with calculations adjusted by incorporating the factor of psychological priority. The objective was to demonstrate the difference between the idealized model approach and the actual behavior of road users [24–26].

Standard calculations were carried out in accordance with applicable technical conditions and methodological procedures used in road transport. Such an approach to evaluating intersections represents a global standard [27–29]. The adjusted calculations were derived from observations of real traffic operation and accounted for situations in which drivers allowed other vehicles to pass even when not formally required by traffic regulations.

2.2. Survey Location

The subject of the analysis was a selected uncontrolled intersection in the city of Prešov, located in a traffic-intensive area with frequent conflicts between major and minor traffic flows. Such an environment is ideal for observing the application of psychological priority by drivers, as the gap between the theoretical model and actual behavior is most pronounced in this context.

Because the analysis focused on a single intersection, the research design inherently minimized the influence of external factors such as geometry, signal control, or differences in local driver behavior. Data collection was limited to stable daytime and weather conditions to ensure consistency. Situations influenced by pedestrians, temporary obstructions, or atypical traffic events were excluded, allowing the study to isolate the behavioral effect of psychological priority as accurately as possible.

The observation lasted for a total of 12 h during regular weekday traffic. In total, 28 instances of driver interaction were classified as psychological priority based on clearly observable behavioral indicators such as eye contact, head movement, or vehicle positioning, suggesting informal negotiation. All evaluations were performed by the same observer to maintain consistency in classification and interpretation. Each event was time-stamped during observation to maintain chronological order and consistency of data recording. Outlier cases with unrealistic or ambiguous driver behavior that could not be reliably classified were excluded from further analysis.

The probability p represents the share of driver interactions in which priority was informally asserted or yielded contrary to traffic regulations. It was calculated according to the expression $p = k/n$, where n denotes the total number of observed driver–driver interactions and k the subset identified as psychological priority. Classification was based on clearly observable behavioral indicators, including eye contact, head movement, vehicle positioning, or short accelerations indicating implicit negotiation. During the 12 h observation, a total of 28 eligible interactions were recorded, of which 22 were classified as psychological priority, resulting in $p = 22/28 = 0.79$. This value represents a full-day aggregate under comparable daytime and weather conditions, with all events influenced by pedestrians, temporary obstructions, or atypical traffic situations excluded to ensure data consistency and validity.

Figure 1 presents a detailed view of the analyzed intersection. For the purposes of calculations and analyses, the traffic situation was further schematically illustrated in Figure 2, where the levels of subordination and the traffic signage are indicated.

Figure 1. Location of the analyzed uncontrolled intersection [30].

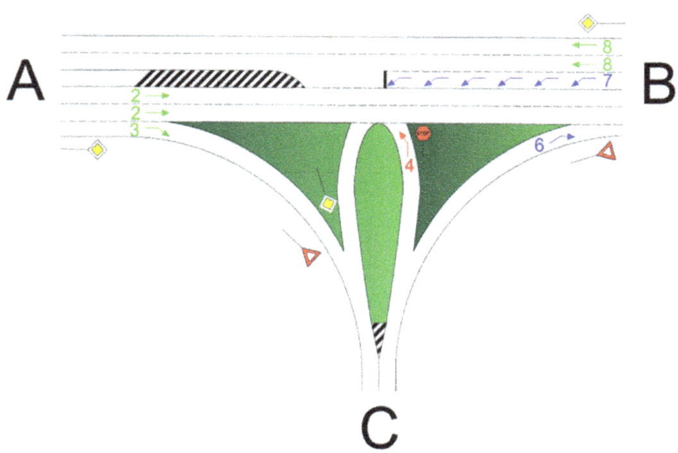

Figure 2. Schematic representation of the intersection geometry and traffic flow hierarchy [31].

Particular attention was devoted to two traffic flows in which psychological priority was most frequently observed. These flows were evaluated independently, and the results were subsequently compared with the standard calculation based on rule-compliant driver behavior.

2.3. Data Collection

An overview of the observed situations is provided in Table 1. Each row represents a single vehicle, with records including the vehicle type (PC—passenger car, HGV—heavy goods vehicle, B—bus), the time of passing through the intersection, the waiting time, and the manner of merging. In the final column, the abbreviation N (normal) is used when the vehicle merged in accordance with traffic regulations, and P (permitted) is used for cases in which the vehicle was allowed entry by another road user.

Table 1. Observed driver interactions classified by vehicle type.

Vehicle Type	Time of Arrival [hh:mm:ss]	Delay Time [hh:mm:ss]	Merging Behavior
B	6:07:53	0:00:18	N
B	6:12:22	0:00:36	P
B	6:16:15	0:00:40	P
B	6:20:17	0:00:46	P
B	6:24:15	0:00:16	P
HGV	6:26:54	0:00:14	P
HGV	6:29:52	0:00:38	P
PC	6:30:15	0:00:16	P
B	6:32:28	0:00:38	P
PC	6:33:15	0:00:09	P
PC	6:34:10	0:00:11	N
PC	6:40:33	0:00:20	P
PC	6:42:30	0:00:17	N
B	6:44:37	0:00:02	N
PC	6:46:22	0:00:53	P
B	7:00:27	0:00:13	P
B	7:14:22	0:01:00	P
PC	7:29:45	0:01:00	P
HGV	7:32:55	0:00:20	P
HGV	7:34:35	0:01:10	P
PC	7:52:30	0:01:07	P
B	7:52:35	0:00:31	P
B	7:58:33	0:01:30	P
PC	7:59:10	0:01:35	P
B	7:59:15	0:00:16	P
PC	8:22:32	0:01:27	P
PC	8:23:10	0:01:33	N
PC	8:23:15	0:01:08	N

Table 1 clearly shows that the occurrence of psychological priority is not an isolated phenomenon but recurs in a significant proportion of the observed situations. This finding highlights the gap between the theoretical model and reality, which must be considered in traffic calculations.

The choice of transport mode significantly affects the way vehicles merge as it can be observed that buses make up a considerable share of vehicles joining, thanks to psychological priority. Therefore, it can be said that these findings have the potential to influence mode choice [32].

Based on the collected data, calculations and comparisons were subsequently carried out, as presented in the Results chapter. In addition, regression analysis and supplementary data from the Google Maps application were employed to verify the relationships between parameters and to compare real traffic operation with the model-based approach.

3. Results

The research results are divided into several sections according to the assessment methods applied. The objective is to compare calculations based on the technical standard TP 102, which represents a standardized calculation process, regression analysis derived from the measured data, and estimates obtained from the Google Maps application [24]. Such a multi-level approach makes it possible not only to evaluate the accuracy of individual methods but also to demonstrate how the original calculation can be adjusted to better reflect actual driver behavior and thereby the real conditions at the analyzed intersection [24,25,33].

3.1. Results According to the Standard Methodology

This approach represents the official standard applied in traffic capacity assessments of intersections. Although it was developed for Slovak conditions, its logic is universal and corresponds to procedures used in many other countries. A common feature of these methodologies is the assumption of ideal driver behavior without considering psychological factors. This means that drivers strictly follow traffic regulations and do not allow other vehicles to pass beyond their formal right of way [34,35].

The methodology represents the standard approach to traffic capacity assessment of uncontrolled intersections. It is based on calculating the capacity of individual traffic flows and the average vehicle waiting times derived from it. The calculation assumes that drivers strictly adhere to traffic regulations and that no psychological yielding of other vehicles occurs [24].

An important limitation of the methodology is that the calculation of average waiting time can only be applied if the degree, of saturation $y = \frac{q}{Q} \leq 1$. When the traffic flow intensity exceeds capacity, an overloaded condition occurs, denoted as level F. In such a case, methodology no longer recommends calculating the average waiting time, as the results would yield unrealistically high values on the order of thousands or even millions of seconds. These results do not represent reality but are a consequence of applying formulas outside their valid range. Nevertheless, in this study, they are included in Table 2 below, which follows the sample calculation of a traffic flow with an acceptable outcome.

$$t_w = \frac{3600}{q} \times \frac{y}{1-y} = \frac{3600}{522} \times \frac{0.832}{1-0.832} = 34.1544\ [\text{s}] \tag{1}$$

Table 2. Results of standardized methodology.

Traffic Flow	Quality Level	Waiting Time [s]
7	F	N/A
6	D	34.1
4	F	N/A
7+8	F	N/A
4+6	F	N/A

Note: Values marked as N/A indicate cases where the calculation exceeded the valid range of the applied formula under oversaturated conditions.

The presented results clearly indicate that the methodology has significant limitations. For traffic flows with a degree of saturation $y \leq 1$, the calculation provides specific outcomes, such as 34.1 s for traffic flow 7. However, for flows with values of $y > 1$, the methodology produces unrealistically high results on the order of thousands or even millions of seconds, which do not correspond to actual traffic operation and arise from applying formulas beyond their valid range.

These findings confirm that the standard calculation without considering psychological priority systematically overestimates waiting times and does not provide a realistic picture of intersection performance. Therefore, the next section presents the results of regression analysis, which makes it possible to correct these distortions and bring the calculations closer to real values.

3.2. Results of the Regression Analysis

As the previous calculations have shown, the standardized methodology leads to unrealistic results at higher values of the degree of saturation. For this reason, regression analysis was applied in the next step, as it allows a better representation of the relationship between the observed waiting times and the values calculated.

The analysis was conducted for two observed traffic flows (4 and 7), with a separate regression model constructed for each. The outcome was a linear equation of the form.

$$t_{real} = a * t_{standard} + b \quad (2)$$

where

- t_{real}—waiting time based on observation [s],
- $t_{standard}$—waiting time according to standardized methodology [s],
- a,b—regression coefficients determined from the measured data.

The regression model was estimated using the ordinary least squares (OLS) method based on 28 observed cases (n = 28). The coefficient of determination reached R^2 = 0.83, indicating a strong relationship between the observed and calculated values. All regression coefficients were statistically significant at the 0.05 level, confirming the reliability of the model fit.

Although the applied regression model is intentionally limited to a simple linear form with one explanatory variable, this structure was selected to demonstrate the direct proportionality between the observed and calculated waiting times. The residual analysis confirmed that the deviations were randomly distributed around zero, with no apparent trend, supporting the adequacy of the linear specification. Despite its simplicity, the model achieved a coefficient of determination R^2 = 0.83 and statistically significant parameters (p < 0.05), indicating a strong and consistent fit between the measured and predicted values.

The resulting linear regression can be expressed as $t_{real} = 0.87\ t_{standard} + 3.54$, where the slope coefficient (0.87) reflects the proportional adjustment between theoretical and observed delay, and the constant term (3.54 s) represents the average systematic deviation. The model achieved a good fit quality, with residuals showing low variance and no systematic bias. The overall uncertainty of the fitted parameters remained within acceptable limits for a sample of this size (n = 28), confirming the stability of the regression outcome.

Sample calculation (traffic flow 7)

$$t_{real} = 0.87 \times 34.1 + 3.54 = 33.22 \text{ s}$$

The regression correction reduced the average waiting time from 34.1 s, obtained using the standardized methodology, to approximately 33.2 s according to the adjusted regression model. This value corresponds more closely to the actually observed waiting times recorded during the field survey.

The results presented in Table 3 show that regression analysis provides values closer to reality than the standardized methodology, although in some cases, deviations from the actual observed times still occur.

Table 3. Calculated waiting time.

	Interval	t_w Measured [s]	t_w Calculated [s]
Morning rush hour	7:15–7:30	33	67.93
	7:30–7:45	56	75.50
	7:45–8:00	82	67.84
	8:00–8:15	60	48.97
Afternoon rush hour	15:00–15:15	62	48.86
	15:15–15:30	59	52.00
	15:30–15:45	29	37.93
	15:45–16:00	39	20.97

The results of the regression analysis can be better illustrated through a graphical representation in Figure 3, that compares the actual observed waiting times with the values calculated using regression analysis. Such visualization makes it possible to clearly track the progression throughout the day and to reveal the differences between the individual methods.

Figure 3. Comparison of waiting times in traffic flow 7 between the observed time and the time calculated by regression.

The graph clearly shows that during the morning and afternoon rush hour periods, the observed values (orange curve) are significantly higher than the average in other time intervals. The regression analysis (green curve) smooths these fluctuations and provides values that, in most intervals, are closer to real traffic operation than the standardized methodology. Deviations appear mainly in intervals with very high or very low waiting times, but the overall trend demonstrates that regression analysis can effectively eliminate extreme overestimations.

These results confirm that the regression model provides a more realistic representation of the traffic situation at the intersection and serves as a suitable tool to complement traditional calculation methods.

3.3. Results According to Google Maps

As a supplementary source of information, data from the Google Maps application was used, providing travel time estimates based on a combination of real-time and historical traffic flow data. This tool is widely accessible and frequently used by both drivers and transport planners, and therefore represents a suitable independent validation of the obtained results [36,37].

For the selected intersection, times were obtained during the morning and afternoon rush hour periods, when traffic load is highest and the differences between the individual methodological approaches are most evident. These values were subsequently compared with the actually observed waiting times and with the results of the regression analysis. Addressing rush hours is a standard in traffic evaluation, as the conditions during this time are the least favorable [37,38].

The results showed that the deviations between the Google Maps data and the actual measured times were within only a few seconds. In some intervals, the values were almost identical, confirming the high accuracy of the application's estimates. A graphical representation of the results is provided in Figure 4, which illustrates the relationships between the Google Maps values and the observed data for the individual traffic flows.

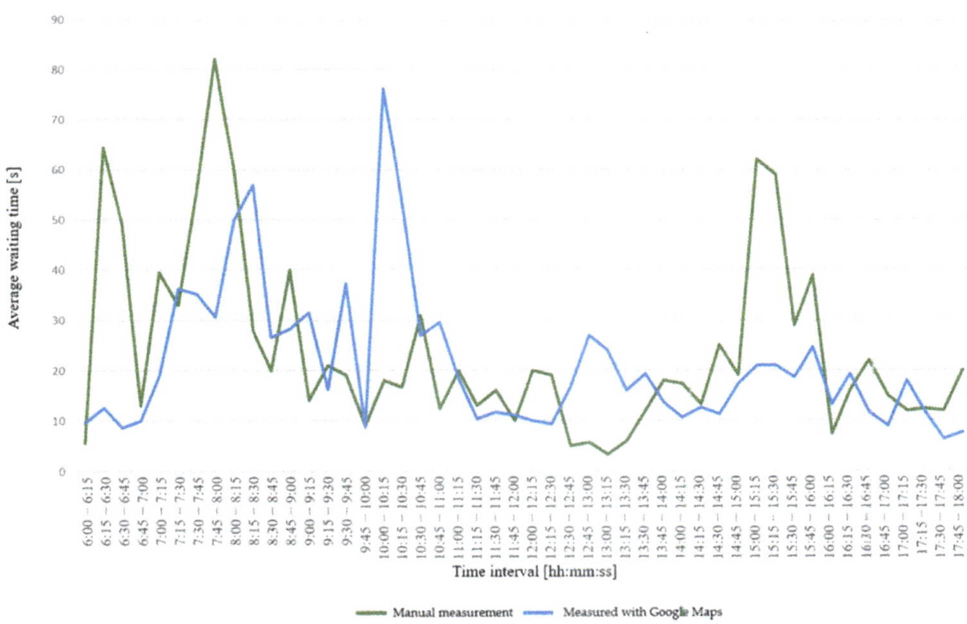

Figure 4. Comparison of waiting times in traffic flow 7 between the observed time and the time measured by Google Maps.

The graph shows that the progression of both curves is similar. The Google Maps times (blue line) in most intervals are close to the values measured directly in the field (green line). Deviations appear mainly at extreme values during the morning rush hour, where Google Maps estimates slightly lower delays than those actually observed.

The differences are caused by the fact that Google Maps processes a large amount of data collected from a wide range of smartphones. As a result, the data are not entirely accurate, since this method of data collection shows certain limitations [23].

The data obtained through Google Maps were compared with the results of the regression analysis, which was carried out based on the actual measured data. Figure 5 illustrates the progression of waiting times in traffic flow 7, displaying the values calculated by regression analysis together with the confidence interval, as well as the estimates provided by Google Maps. This approach makes it possible to verify the extent to which an independent external data source corresponds to the values derived from the statistical model.

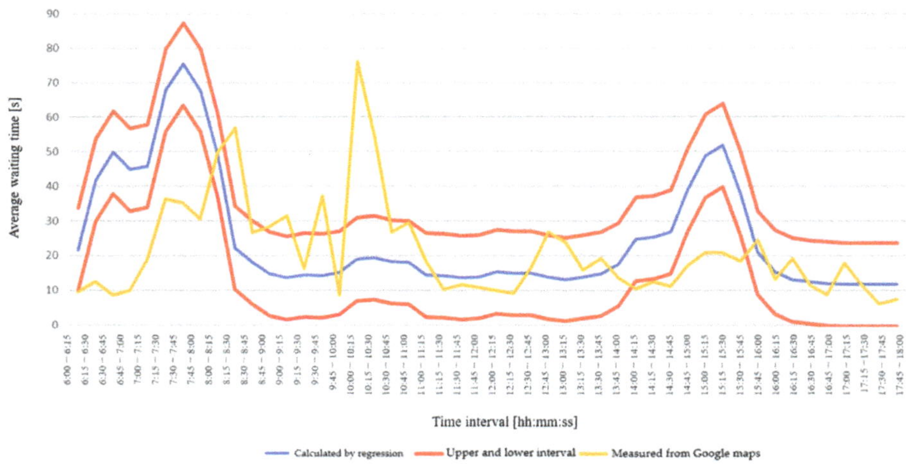

Figure 5. Comparison of waiting times in traffic flow 7 between the calculated time and the time measured by Google Maps.

The graph shows that the progression of waiting times calculated by regression analysis (blue line) follows the trend of real traffic conditions and displays higher values during rush periods. The confidence interval of the model (red lines) defines the range within which the real values are expected to occur. The Google Maps data (yellow line) in most cases fall within this interval, confirming their consistency with the regression model.

The largest deviations can be observed during the morning rush hour, when Google Maps estimated lower waiting times than those indicated by the regression analysis. In the afternoon rush hour, by contrast, the values approached the upper boundary of the interval. Overall, however, the differences amounted to only a few seconds, confirming that the data from Google Maps are suitable for validating the accuracy of regression-adjusted results. Nevertheless, it is important to note that Google Maps data represent aggregated travel-time estimates influenced by multiple intersections, speed limits, and route-level conditions rather than direct measurements of intersection delay. Therefore, this method should be interpreted only as indicative validation, useful primarily for trend comparison rather than precise calibration. Despite these limitations, the consistency between the application's estimates and the regression-based results confirms the robustness of the proposed behavioral model.

The results revealed significant differences between the individual approaches to assessing waiting times at the intersection. The standardized methodology calculations systematically overestimated average waiting times and, at higher levels of saturation, even produced unrealistically high values beyond the range of real conditions. The reason lies in the methodology's assumption that drivers behave strictly according to traffic regulations and do not apply psychological priority, which does not hold true in practice.

The regression analysis, carried out on the basis of real measurements, provided results that corresponded much more closely with the observed traffic operation. This approach was able to eliminate extreme overestimations and bring the calculations closer to actual values.

The comparison with Google Maps data confirmed the validity of the regression model. The values obtained from this independent application in most cases fell within the confidence interval of the regression analysis, indicating a high level of consistency between the two approaches.

Based on these findings, it can be concluded that while the standardized methodology does not reflect actual driver behavior and leads to overestimation of waiting times, the combination of regression analysis and validation using Google Maps provides a realistic representation of intersection operation. This confirms the need to extend traditional calculation methods with elements that take into account the psychological factors of driver behavior.

4. Discussion

The results presented in the previous chapter clearly demonstrated that the methodology for assessing uncontrolled intersections systematically overestimates waiting times and, in the case of oversaturated flows, even produces unrealistically high values. The reason lies in the assumption of ideal driver behavior, which is not fulfilled in practice, as road users frequently apply so-called psychological priority. In contrast, the regression analysis based on real data provided results that corresponded substantially better with the observed traffic operation, and their validity was further supported by the data obtained from Google Maps.

These findings are of fundamental importance for transport planning and traffic management in the context of Smart Cities. When methodologies that do not reflect actual driver behavior are applied, this may lead to an overestimation of problems, misjudgment of intersection capacity, and consequently inappropriate investment decisions [39,40].

Regression analysis made it possible to quantify these deviations and to develop a correction model. By comparing calculated and actually observed values, it was shown that regression can eliminate extreme overestimations and bring the results closer to reality. The deviations between the real values and the regression model were within a few seconds, which represents an acceptable level of accuracy in traffic modeling.

The Google Maps data supported this conclusion. Their progression in most cases remained within the confidence interval defined by the regression analysis, indicating consistency between the two approaches. Minor differences appeared mainly during rush hours, which can be explained by the different methods of data processing. Google Maps relies on historical and real-time data from GPS devices, whereas regression analysis is based directly on observations of the specific intersection.

It follows that standardized methodology overestimates waiting times because it does not reflect the psychological behavior of drivers, whereas regression and Google Maps indirectly capture this behavior. The results, therefore, confirm the need to extend traditional methodologies with factors that better represent reality.

4.1. Adjustment of the Calculation

The principle of the adjustment lies in incorporating the probability of a vehicle being allowed to pass in conflict situations. This probability was determined based on the survey at the selected intersection and reflects the percentage of cases in which drivers applied psychological priority. Instead of evaluating all situations as strictly rule-based, the adjusted calculation considers a portion of vehicles as permitted to proceed, which reduces the resulting average waiting time and increases the capacity of the intersection.

Mathematically, the adjustment can be expressed as follows:

$$t_{w,adjusted} = (1 - p) \times t_{w,standard} + pt_{w,real} \quad (3)$$

where

- $t_{w,adjusted}$: average waiting time after the adjustment;
- $t_{w,standard}$: average waiting time according to the standardized methodology;
- $t_{w,real}$: observed or regression-derived average value;
- p: probability of a vehicle being allowed to proceed.

By applying this approach, the waiting time values became significantly closer to real conditions, as confirmed by the results of the regression analysis and the comparison with Google Maps data. The adjusted calculation thus eliminates extreme deviations in oversaturated flows and provides a more realistic representation of the traffic situation.

4.2. Sample Adjusted Calculation (Traffic Flow 7)

From the calculation for traffic flow 7 in the given interval, the average waiting time is, approximately,

$$t_{w,TP102} \approx 34.1 \text{ s}$$

From the behavioral survey: during the morning and afternoon rush periods, 79 percent of mergers occurred through yielding by other drivers, while 21 percent followed the standard rule-based procedure.

$$p = 0.79$$

After incorporating the probability of psychological priority into the adjusted calculation, the resulting waiting time is obtained.

$$t_{w,adjusted} = (1 - 0.79) \times 34.1 + 0.79 \times 29.1 = 30.15 \text{ s}$$

The result shows that the original value of 34.1 s according to the methodology was adjusted to 30.15 s, thereby approaching the real value of 29.1 s. The difference compared to actual driver behavior was thus significantly reduced, and the outcome is methodologically more accurate. This approach confirms that considering psychological priority makes it possible to eliminate the systematic overestimation of waiting times and to provide a more realistic representation of intersection operation.

Based on the regression analysis and the incorporation of the psychological priority factor, new average waiting time values were calculated. These results are presented in Tables 4 and 5, which compare four approaches: the original values, the measured times, the values after regression adjustment, and the estimates provided by Google Maps.

Table 4. Summary of results for traffic flow 4.

Traffic Flow 4	Manually Measured [s]	Calculated with Regression [s]	Measured with Google Maps [s]	Standardized Methodology [s]
Morning rush hour	51	46	43	N/A
Afternoon rush hour	39	44	47	N/A

Note: Values marked as N/A indicate cases where the calculation exceeded the valid range of the applied formula under oversaturated conditions.

Table 5. Summary of results for traffic flow 7.

Traffic Flow 7	Manually Measured [s]	Calculated with Regression [s]	Measured with Google Maps [s]	Standardized Methodology [s]
Morning rush hour	57	65	39	N/A
Afternoon rush hour	46	40	29	N/A

Note: Values marked as N/A indicate cases where the calculation exceeded the valid range of the applied formula under oversaturated conditions.

The tables show that the standardized methodology significantly overestimates waiting times. In the case of traffic flow 4, the values reached the order of hundreds of millions of seconds, which is a consequence of applying the methodology beyond its valid range. These results are unrealistic and do not reflect actual traffic conditions. In the case of traffic flow 7, the standardized methodology values were lower but still considerably overestimated.

In contrast, the values obtained from the regression analysis were close to the actual measured times. For flow 4, the morning rush hour was 46 s compared to the observed 51 s, and for flow 7, it was 65 s compared to 57 s. In the afternoon rush hour, the regression model even matched the observations very closely (traffic flow 4: 44 s compared to 39 s, traffic flow 7: 40 s compared to 46 s). These results confirm that incorporating psychological priority significantly reduces the discrepancies between calculations and reality.

The supplementary comparison with Google Maps data showed that the application's estimates in most cases fell within the confidence interval of the regression model. For example, in traffic flow 4 during the morning rush hour, the Google Maps time was 43 s, which is very close to the observed 51 s and the regression-adjusted 46 s, similarly, in traffic flow 7.

Overall, it can be concluded that the adjusted calculation provides a more realistic representation of intersection performance than the original methodology. The consideration of psychological priority eliminated unrealistic overestimations and brought the results closer to both real traffic operation and independent external data. This approach, therefore, represents a practical improvement of traditional methodologies and may serve as a suitable basis for planning traffic measures in the context of Smart Cities.

This finding is not limited to Slovak conditions, as similar methodological frameworks across Europe and beyond are based on comparable assumptions of strictly rule-based driver behavior. Therefore, the adjustment proposed in this study can be considered relevant for a wider range of urban contexts and traffic environments.

Although the present research does not directly employ sensor networks, simulation environments, or digital urban infrastructure, its findings are highly relevant for Smart City development. The quantified behavioral factor of psychological priority can be integrated into microscopic and macroscopic traffic models, improving the accuracy of digital twins and intelligent transport systems that rely on real-time or predictive data. By incorporating behavioral variability into analytical and simulation-based tools, the results contribute to a more human-centered and realistic approach to Smart City traffic management.

5. Conclusions

The aim of this study was to highlight the limitations of traditional methodologies in the assessment of uncontrolled intersections and to propose an adjustment of the calculation that reflects actual driver behavior through the phenomenon of psychological priority. Based on regression analysis and the incorporation of this factor, new average waiting time values were calculated, which were significantly closer to the actual measured data, and their validity was further confirmed through comparison with Google Maps data. The adjusted calculation thus eliminates extreme deviations and provides a more realistic representation of intersection performance.

The results confirmed both hypotheses of the study:

- H1: The standardized methodology systematically overestimates average waiting times at intersections,
- H2: The incorporation of psychological priority through regression analysis provides results that are significantly closer to real conditions.

These findings have practical implications for transport planning in the context of Smart Cities. If aspects of driver psychological behavior are not considered in the assessment of intersection capacity, there is a risk of misinterpreting the traffic situation and consequently making suboptimal investment decisions [41,42]. The adjusted methodology, therefore, provides a suitable basis for more accurate modeling of traffic conditions and may contribute to more efficient traffic management in cities.

The phenomenon of psychological priority reflects informal driver interactions that occur spontaneously in real traffic situations, where drivers negotiate right-of-way contrary to formal regulations. While such behavior is not defined or permitted by traffic law, it represents a real and measurable aspect of road user dynamics. The present study does not endorse or encourage such actions but seeks to understand them as part of human behavior that influences intersection performance. Recognizing these patterns enables planners and engineers to design control strategies and infrastructure that account for actual driver responses, reduces the potential for unsafe situations, and ensures that behavioral factors are addressed without promoting unsafe or extra-legal practices in Smart City environments.

The study was carried out at a specific urban intersection, which introduces certain contextual limitations. The conclusions are based on the local setting, yet the quantifiable difference between the traditional and adjusted calculations indicates that a similar effect is highly likely to occur at other intersections as well. Psychological priority was analyzed during rush periods, which confirmed its importance, particularly under conditions of increased traffic intensity.

The study was conducted at a single intersection, which represents a contextual limitation. However, the observed gap between standardized calculations and actual driver behavior is systematic in nature and has been documented in multiple international studies. For this reason, the results can be regarded as indicative of a broader methodological shortcoming rather than a purely local phenomenon.

One of the limitations of the study is that it did not focus on pedestrians. In such cases, psychological priority may be influenced by additional factors such as reduced mobility, age, or other individual characteristics, which could further shape interactions at intersections and affect overall traffic dynamics [43]. Future research could therefore also focus on this topic, with particular attention to the role of psychological priority in pedestrian behavior and its impact on intersection performance.

Future research should focus on extending the analysis to multiple intersections, diverse traffic conditions, and different cultural contexts. A promising direction is also the use of more advanced regression models, nonlinear approaches, and the integration of data from sensors or traffic applications. Such integration would enable continuous validation of calculations and their calibration according to the actual behavior of road users.

The study achieved its main objective of demonstrating that incorporating driver behavior, specifically the phenomenon of psychological priority, significantly improves the accuracy of waiting time estimation and capacity assessment at unsignalized intersections. The results confirmed that conventional calculation methods systematically overestimate delay when behavioral factors are ignored, whereas the proposed regression-based correction aligns more closely with observed traffic conditions.

Nevertheless, several limitations should be acknowledged. The research was conducted on a single intersection, without replication under different geometric or traffic conditions, and did not include a full sensitivity analysis. The model uncertainty was assessed only through residual variance and confidence intervals derived from regression results, which provide indicative rather than comprehensive bounds of reliability.

Future research should therefore extend this approach to multiple urban contexts, apply larger datasets, and incorporate simulation or sensor-based validation to further verify the behavioral correction proposed in this study. Such efforts would enhance the generalizability of the findings and strengthen their applicability for smart mobility planning in Smart Cities.

An important contribution of the study is the identification of a fundamental shortcoming of traditional traffic capacity methodologies, namely the absence of consideration of psychological priority. For the field of Smart Cities, this highlights the need to supplement traffic models with aspects of human behavior so that assessment results more accurately reflect reality. The integration of these elements into intelligent transport systems will enable more precise planning, more efficient traffic management, and contribute to improving the quality of life of urban residents.

For comparison, various methodologies across Europe were also identified, and it was found that the Slovak Republic is not an exception in disregarding psychological priority. In addition to the Slovak Republic [24], psychological priority was not considered in any of the methodologies of Germany [25], the Netherlands [44,45], or the United Kingdom [26].

Although the empirical data were collected in a single Slovak city, the behavioral mechanisms analyzed in this study are not geographically specific. Informal priority negotiation and driver interaction patterns are observable in urban environments worldwide, especially at unsignalized intersections where regulation and perception intersect. The methodological framework and regression-based adjustment proposed here can therefore be applied to diverse urban contexts, providing a transferable foundation for integrating behavioral variability into Smart City traffic modeling on an international scale.

Author Contributions: Conceptualization, P.F. and Ľ.Č.; methodology, Ľ.Č.; software, K.Č.; validation, A.K., and K.Č.; formal analysis, K.Č.; investigation, P.F.; resources, P.F.; data curation, P.F. and Ľ.Č.; writing—original draft preparation, P.F.; writing—review and editing, A.K.; visualization, Ľ.Č.; supervision, A.K.; project administration, K.Č.; funding acquisition, A.K. All authors have read and agreed to the published version of the manuscript.

Funding: This work was supported by the Slovak Research and Development Agency under the Contract no. SK-CN-23-0009.

Data Availability Statement: The raw data presented in this study are available on request from the corresponding author.

Acknowledgments: During the preparation of this manuscript, the authors used AI soft-ware such as Grammarly for the purposes of grammar and spelling correction. The authors have reviewed and edited the output and take full responsibility for the content of this publication.

Conflicts of Interest: The authors declare no conflicts of interest.

References

1. Rachmawati, R.; Haryono, E.; Rohmah, A.A. Developing Smart City in the New Capital of Indonesia. In Proceedings of the 2021 IEEE International Smart Cities Conference (ISC2), Manchester, UK, 7–10 September 2021; pp. 1–7.
2. Kim, J. Smart City Trends: A Focus on 5 Countries and 15 Companies. *Cities* **2022**, *123*, 103551. [CrossRef]
3. Nam, T.; Pardo, T.A. Conceptualizing Smart City with Dimensions of Technology, People, and Institutions. In *Proceedings of the 12th Annual International Digital Government Research Conference: Digital Government Innovation in Challenging Times, College Park, MD, USA, 12–15 June 2011*; Association for Computing Machinery: New York, NY, USA, 2011; pp. 282–291.
4. Enabling Technologies for Urban Smart Mobility: Recent Trends, Opportunities and Challenges. Available online: https://www.mdpi.com/1424-8220/21/6/2143 (accessed on 22 September 2025).
5. Mutavdžija, M.; Kovačić, M.; Buntak, K. Moving towards Sustainable Mobility: A Comparative Analysis of Smart Urban Mobility in Croatian Cities. *Sustainability* **2024**, *16*, 2004. [CrossRef]
6. Brčić, D.; Slavulj, M.; Šojat, D.; Jurak, J. The Role of Smart Mobility in Smart Cities. In Proceedings of the 5th International Conference on Road and Rail Infrastructure, Zadar, Croatia, 17–19 March 2018.
7. Assaying Traffic Settings with Connected and Automated Mobility Channeled into Road Intersection Design. Available online: https://www.mdpi.com/2624-6511/8/3/86 (accessed on 22 September 2025).

8. Fernandes, P.; Coelho, M.C. Can Turbo-Roundabouts and Restricted Crossing U-Turn Be Effective Solutions for Urban Three-Leg Intersections? *Sustain. Cities Soc.* **2023**, *96*, 104672. [CrossRef]
9. Lungu, M.A. Smart Urban Mobility: The Role of AI in Alleviating Traffic Congestion. *Proc. Int. Conf. Bus. Excell.* **2024**, *18*, 1441–1452. [CrossRef]
10. Enhancing Urban Intersection Efficiency: Utilizing Visible Light Communication and Learning-Driven Control for Improved Traffic Signal Performance. Available online: https://www.mdpi.com/2624-8921/6/2/31 (accessed on 22 September 2025).
11. Tinella, L.; Bosco, A.; Koppel, S.; Lopez, A.; Spano, G.; Ricciardi, E.; Traficante, S.; Napoletano, R.; Grattagliano, I.; Caffò, A.O. Sociodemographic and Psychological Factors Affecting Motor Vehicle Crashes (MVCs): A Classification Analysis Based on the Contextual-Mediated Model of Traffic-Accident Involvement. *Curr. Psychol.* **2024**, *43*, 25683–25703. [CrossRef]
12. Gap Acceptance Behaviour of Drivers at Uncontrolled T-Intersections under Mixed Traffic Conditions. Available online: https://www.researchgate.net/publication/320297740_Gap_Acceptance_Behaviour_of_Drivers_at_Uncontrolled_T-Intersections_under_Mixed_Traffic_Conditions (accessed on 22 September 2025).
13. Lemonnier, S.; Désiré, L.; Brémond, R.; Baccino, T. Drivers' Visual Attention: A Field Study at Intersections. *Transp. Res. Part F Traffic Psychol. Behav.* **2020**, *69*, 206–221. [CrossRef]
14. Costa, M.; Bichicchi, A.; Nese, M.; Lantieri, C.; Vignali, V.; Simone, A. T-Junction Priority Scheme and Road User's Yielding Behavior. *Transp. Res. Part F Traffic Psychol. Behav.* **2019**, *60*, 770–782. [CrossRef]
15. Tinella, L.; Bosco, A.; Traficante, S.; Napoletano, R.; Ricciardi, E.; Spano, G.; Lopez, A.; Sanesi, G.; Bergantino, A.S.; Caffò, A.O. Fostering an Age-Friendly Sustainable Transport System: A Psychological Perspective. *Sustainability* **2023**, *15*, 13972. [CrossRef]
16. Behavioural Analysis of Interactions between Pedestrians and Vehicles in Street Designs with Elements of Shared Space. Available online: https://www.researchgate.net/publication/274405800_Behavioural_analysis_of_interactions_between_pedestrians_and_vehicles_in_street_designs_with_elements_of_shared_space (accessed on 22 September 2025).
17. Armitage, S.; Rodwell, D.; Lewis, I. Applying an Extended Theory of Planned Behaviour to Understand Influences on Safe Driving Intentions and Behaviours. *Transp. Res. Part F Traffic Psychol. Behav.* **2022**, *90*, 347–364. [CrossRef]
18. Risser, R. Behavior in Traffic Conflict Situations. *Accid. Anal. Prev.* **1985**, *17*, 179–197. [CrossRef] [PubMed]
19. Towards a General Theory of Driver Behavior. Available online: https://www.researchgate.net/publication/7951328_Towards_a_general_theory_of_driver_behavior (accessed on 22 September 2025).
20. Wirsbinna, A.; Grega, L.; Juenger, M. Assessing Factors Influencing Citizens' Behavioral Intention towards Smart City Living. *Smart Cities* **2023**, *6*, 3093–3111. [CrossRef]
21. Patel, Y.; Doshi, N. Social Implications of Smart Cities. *Procedia Comput. Sci.* **2019**, *155*, 692–697. [CrossRef]
22. Smart Cities at the Intersection of Public Governance Paradigms for Sustainability-Giuseppe Grossi, Olga Welinder, 2024. Available online: https://journals.sagepub.com/doi/10.1177/00420980241227807 (accessed on 22 September 2025).
23. Smart Intersections and Connected Autonomous Vehicles for Sustainable Smart Cities: A Brief Review. Available online: https://www.mdpi.com/2071-1050/17/7/3254 (accessed on 22 September 2025).
24. TP 102 Výpočet Kapacít Pozemných Komunikácií. Available online: https://www.slov-lex.sk/vestniky/rezortne-zbierky/59852c17-9371-4681-bb88-a5fd83d92207 (accessed on 22 September 2025).
25. Forschungsgesellschaft für Straßen- und Verkehrswesen e. V. (FGSV). *Handbuch für die Bemessung von Straßenverkehrsanlagen (HBS), Ausgabe 2015*; FGSV Verlag: Köln, Germany, 2015; Available online: https://www.fgsv-verlag.de/pub/media/pdf/299.r.26082015.pdf (accessed on 10 November 2025)ISBN 978-3-86446-103-3.
26. TfL. *Traffic Modelling Guidelines Version 4.0*; Transport for London: London, UK, September 2021. Available online: https://content.tfl.gov.uk/traffic-modelling-guidelines.pdf (accessed on 10 November 2025).
27. Uribe-Chavert, P.; Posadas-Yagüe, J.-L.; Poza-Lujan, J.-L. Evaluating Traffic Control Parameters: From Efficiency to Sustainable Development. *Smart Cities* **2025**, *8*, 57. [CrossRef]
28. Frontiers | Evaluating Urban Traffic Dynamics: A Study of Mobility in Vellore City. Available online: https://www.frontiersin.org/journals/sustainable-cities/articles/10.3389/frsc.2025.1631748/full (accessed on 22 September 2025).
29. Evaluating the Effect of Road Surface Potholes Using a Microscopic Traffic Model. Available online: https://www.mdpi.com/2076-3417/13/15/8677 (accessed on 22 September 2025).
30. GKÚ Ortofotomozaika. Available online: https://www.gku.sk/gku/aktuality/ortofotomozaika.html (accessed on 22 September 2025).
31. Autodesk. *AutoCAD*; Version 2025; Autodesk Inc.: San Rafael, CA, USA, 2025. Available online: https://www.autodesk.com/products/autocad/ (accessed on 10 November 2025).
32. Gnap, J.; Poliak, M.; Semanova, S. The Issue of a Transport Mode Choice from the Perspective of Enterprise Logistics. *Open Eng.* **2019**, *9*, 374–383. [CrossRef]
33. Google Maps. Available online: https://www.google.com/maps/ (accessed on 22 September 2025).
34. 56/2012 Z.z.-Zákon o Cestnej Doprave. Available online: https://www.slov-lex.sk/ezbierky/pravne-predpisy/SK/ZZ/2012/56/ (accessed on 22 September 2025).

35. S-EPI 30/2020 Z. z. Vyhláška o dopravnom značení | Aktuálne znenie. Available online: https://www.zakonypreludi.sk/zz/2020-30 (accessed on 22 September 2025).
36. Rito, J.E.; Lopez, N.S.; Biona, J.B.M. Modeling Traffic Flow, Energy Use, and Emissions Using Google Maps and Google Street View: The Case of EDSA, Philippines. *Sustainability* **2021**, *13*, 6682. [CrossRef]
37. Cleland, C.L.; Ferguson, S.; Kee, F.; Kelly, P.; Williams, A.J.; Nightingale, G.; Cope, A.; Foster, C.; Milton, K ; Kelly, M.P.; et al. Adaptation and Testing of a Microscale Audit Tool to Assess Liveability Using Google Street View: MAPS-Liveability. *J. Transp. Health* **2021**, *22*, 101226. [CrossRef]
38. Adeyemi, O.; Paul, R.; Delmelle, E.; DiMaggio, C.; Arif, A. Road Environment Characteristics and Fatal Crash Injury during the Rush and Non-Rush Hour Periods in the U.S: Model Testing and Cluster Analysis. *Spat. Spatio-Temporal Epidemiol.* **2023**, *44*, 100562. [CrossRef] [PubMed]
39. Missing Data Filling Technology for Smart City Internet of Things Based on I-NMKNN Algorithm-Guifang Shen. 2025. Available online: https://journals.sagepub.com/doi/10.1177/14727978251361865 (accessed on 22 September 2025).
40. Data Augmentation Strategies to Improve Text Classification: A Use Case in Smart Cities | Language Resources and Evaluation. Available online: https://link.springer.com/article/10.1007/s10579-023-09685-w (accessed on 22 September 2025).
41. Samikwa, E.; Schärer, J.; Braun, T.; Di Maio, A. Machine Learning-Based Energy Optimisation in Smart City Internet of Things. In *Proceedings of the Twenty-fourth International Symposium on Theory, Algorithmic Foundations, and Protocol Design for Mobile Networks and Mobile Computing, Washington, DC, USA, 23–26 October 2023*; Association for Computing Machinery: New York, NY, USA, 2023; pp. 364–369.
42. Nagy, A.M.; Simon, V. Improving Traffic Prediction Using Congestion Propagation Patterns in Smart Cities. *Adv. Eng. Inform.* **2021**, *50*, 101343. [CrossRef]
43. Konečný, V.; Zuzaniak, M.; Jonasíková, D.; Abramović, B. Persons with Reduced Mobility and Consequences Related to Their Demand for Public Passenger Transport in the Conditions of Slovak Republic. *Transp. Res. Interdiscip. Perspect.* **2025**, *32*, 101518. [CrossRef]
44. Kruispunten-Welke aanbevelingen zijn er voor verschillende typen kruispunten? Available online: https://swov.nl/nl/fact/kruispunten-welke-aanbevelingen-zijn-er-voor-verschillende-typen-kruispunten (accessed on 22 September 2025).
45. CROW Kennisbank. Available online: https://kennisbank.crow.nl/public/gastgebruiker/WOBU/Handboek_Verkeersveiligheid/Kruispunten/21160?utm_source=chatgpt.com (accessed on 22 September 2025).

MDPI AG
Grosspeteranlage 5
4052 Basel
Switzerland
Tel.: +41 61 683 77 34

Smart Cities Editorial Office
E-mail: smartcities@mdpi.com
www.mdpi.com/journal/smartcities

www.ingramcontent.com/pod-product-compliance
Lightning Source LLC
Chambersburg PA
CBHW040311270926
51850CB00007BC/1026
* 9 7 8 3 7 2 5 8 6 8 6 4 3 *